Lecture Notes in Management and Industrial Engineering

Series editor

Adolfo López-Paredes, Valladolid, Spain

This book series provides a means for the dissemination of current theoretical and applied research in the areas of Industrial Engineering & Engineering Management. The latest methodological and computational advances that both researchers and practitioners can widely apply to solve new and classical problems in industries and organizations constitute a growing source of publications written for and by our readership.

The aim of this book series is to facilitate the dissemination of current research in the following topics:

- Strategy and Enterpreneurship;
- Operations Research, Modelling and Simulation;
- Logistics, Production and Information Systems;
- Quality Management;
- Product Management;
- Sustainability and Ecoefficiency;
- Industrial Marketing and Consumer Behavior;
- Knowledge and Project Management;
- Risk Management;
- Service Systems;
- Healthcare Management;
- Human Factors and Ergonomics;
- Emergencies and Disaster Management; and
- Education.

More information about this series at http://www.springer.com/series/11786

Ángel Ortiz · Carlos Andrés Romano
Raul Poler · José-Pedro García-Sabater
Editors

Engineering Digital Transformation

Proceedings of the 11th International Conference on Industrial Engineering and Industrial Management

Springer

Editors
Ángel Ortiz
Universitat Politècnica de València
Valencia, Spain

Raul Poler
Universitat Politècnica de València
Valencia, Spain

Carlos Andrés Romano
Universitat Politècnica de València
Valencia, Spain

José-Pedro García-Sabater
Universitat Politècnica de València
Valencia, Spain

ISSN 2198-0772 ISSN 2198-0780 (electronic)
Lecture Notes in Management and Industrial Engineering
ISBN 978-3-030-07127-1 ISBN 978-3-319-96005-0 (eBook)
https://doi.org/10.1007/978-3-319-96005-0

Softcover re-print of the Hardcover 1st edition 2019

This Springer imprint is published by the registered company Springer Nature Switzerland AG
The registered company address is: Gewerbestrasse 11, 6330 Cham, Switzerland

Preface

We are pleased to preface this book where you could find the selected papers of CIO 2017, XXI Congreso de Ingeniería de Organización/11th International Conference on Industrial Engineering and Industrial Management.

This conference was promoted by ADINGOR (Asociación para el Desarrollo de la Ingeniería de Organización), and it was hosted at Universitat Politècnica de València (Spain) from 5th to 6th July 2017. The CIO 2017 Conference motto was: "Engineering Digital Transformation".

The selected papers cover all branches of industry, information on the most recent and relevant research, theories and practices in Industrial Engineering, Management, and Operations.

In this book, the reader can find papers providing links between researchers and practitioners from different branches to enhance an interdisciplinary perspective of industrial engineering and management.

The contributions have been arranged in four chapters:

- Management
- Production
- Logistics and Supply Chain
- Methods and Applications

We hope this book gives you the opportunity to enjoy high-quality scientific papers.

Valencia, Spain
December 2017

Ángel Ortiz
Carlos Andrés Romano

Organization

Program Chairs

Ángel Ortiz	Universitat Politècnica de València, Spain
Carlos Andrés Romano	Universitat Politècnica de València, Spain

Program Committee

Ivan Garcia-Miranda	Universidad Carlos III de Madrid
Carlos Mataix	UPM
Carlos Andrés Romano	Universidad Politécnica de Valencia, Spain
Carlos Monroy	UPM, Spain
Luis Onieva	Universidad de Sevilla, Spain
Beatriz Andres	Universitat Politècnica de València, Spain
Juan José Laviós Villahoz	Universidad de Burgos, Spain
Claus Sørensen	Aarhus University, Denmark
Gonzalo Grau Gadea	Universitat Politècnica de València, Spain
Ramon Companys	ETSEIB, DOE
David Hermida	Universidad de Oviedo, Spain
Eugenia Babiloni	Universidad Politécnica de Valencia, Spain
Izaskun Alvarez	EHU/UPV, Spain
javier carrasco	UPM
German Gemar	Universidad de Malaga, Spain
Jochen Deuse	TU Dortmund
Türkay Dereli	Gaziantep University, Turkey

Alejandro Escudero-Santana	Universidad de Sevilla
David de la Fuente	University of Oviedo, Spain
Jesús Antonio Carrillocastrillo	Universidad de Sevilla, Spain
Isabel García Gutiérrez	Universidad Carlos III de Madrid, Spain
Julien Maheut	Universidad Politécnica de Valencia, Spain
Manuel Cardós Carboneras	Universidad Politecnica de Valencia, Spain
Lorenzo Gonzalez	Instituto Andaluz de Tecnología (IAT), Spain
Jesus Garcia Arca	Universidad de Vigo, Spain
Adolfo Lopez-Paredes	University of Valladolid, Spain
José Manuel Galán	Universidad de Burgos
Josefa Mula	Universitat Politècnica de València, Spain
Frano Barbic	Universidad Politecnica de Madrid, Spain
Isabel Fernandez	UNIOVI, Spain
Rodolfo De Castro	Universitat de Girona, Spain
Francisco Campuzano-Bolarín	Universidad Politécnica de Cartagena, Spain
Javier Santos	Tecnun—University of Navarra, Spain
Patxi Ruiz-de-Arbulo-López	Universidad del País Vasco (UPV/EHU)
José Antonio Pascual	Universidad de Valladolid, Spain
Jesús María Pinar-Pérez	Colegio Universitario de Estudios Financieros, Spain
Sofia Estelles Miguel	Universitat Politecnica de Valencia, Spain
Jesús Morcillo	Universidad Carlos III de Madrid, Spain
Joaquin Cestero	Universitat Politècnica de València, Spain
Pedro Sanz Angulo	University of Valladolid, Spain
Jordi Olivella Nadal	Universitat Politécnica de Catalunya
Pablo Cortés	Universidad de Sevilla
Raúl Poler	Universitat Politècnica de València, Spain
Pedro Gomez-Gasquet	Universitat Politècnica de Valeència, Spain
Alberto Albahari	Universidad de Málaga—Escuela de Ingenierías Industriales, Spain
Segismundo Izquierdo	Universidad de Valladolid, Spain
Raul Rodriguez	Unknown Affiliation
Antonio Hidalgo	Universidad Politécnica de Madrid, Spain
Eloy Hontoria	Technical University of Cartagena
Jose P. Garcia-Sabater	Universitat Politècnica de València, Spain
Carlos Dema Pérez	Universitat Politècnica de València, Spain
Victor Anaya	UPV
Raquel Sanchis	UPV
Emilio Ramírez-Juidías	Universidad de Sevilla
Ruiz Molina Antonio	Universidad de Málaga, Spain
Pablo Aparicio Ruiz	Universidad de Sevilla, Spain
Diego Ros McDonnell	UPCT, Spain

Miguel Palacios	ESCP Europe, Spain
Ernesto Cilleruelo-Carrasco	UPV/EHU, Spain
Juan Ramon Trapero	University of Castilla-La Mancha, Spain
Arturo J. Fernández González	Escuela de Ingeniería Industrial (sede Campus), University of Vigo, Spain
Ivan Roa	Universitat Abat Oliba CEU, Spain
Guillermo Montero	Universidad de Sevilla, Spain
Alfonso Duran Heras	UC3M, Spain
Jose Guadix	University of Seville, Spain
Faustino Alarcón Valero	Universitat Politècnica de València, Spain
Ricardo Chalmeta	Universidad Jaume I
Isabela Devonish	CEFET/RJ, Brazil
Iñaki Bildosola	UPV/EHU, Spain
Enara Zarrabeitia	Universidad del País Vasco, Spain
David Peidro	CIGIP
Julio J. Garcia-Sabater	Universidad Politécnica de Valencia, Spain
José I. Santos	Universidad de Burgos, Spain
Bopaya Bidanda	University of Pittsburgh, United States
Irene Martín-Rubio	ETSIDI—Technical University of Madrid, Spain
Miguel Ángel Manzanedo Del Campo	Universidad de Burgos, Spain
Eduardo Vicens-Salort	Universitat Politècnica de València, Spain
Lourdes Sáiz Bárcena	Universidad de Burgos, Spain
Giovanni Mummolo	Politecnico di Bari
Paloma Maria Teresa Martinez Sanchez	Bosque University, Colombia
Eva Ponce-Cueto	MIT
Juan A. Marin-Garcia	Universitat Politècnica de València, Spain
Esther Alvarez	Facultad de Ingeniería, Spain
José Carlos Prado Prado	Escuela Ingeniería Industrial Vigo, Spain
Marta Peris-Ortiz	Universitat Politècnica de València, Spain
María Victoria de la Fuente	Universidad Politécnica de Cartagena, Spain
Rafael Pastor	UPC
Ester Guijarro	Universitat Politècnica de València, Spain
Ángel Hernández-García	Universidad Politécnica de Madrid, Spain
Albert Corominas	UPC, Spain
María-José Verdecho	Universitat Politècnica de València, Spain
Amaia Lusa	Universitat Politècnica de Catalunya
Carme Martínez Costa	Universitat Politècnica de Catalunya, Spain
Jesús Muñuzuri	University of Seville, Spain
Rashed Sahraeian	Shahed University, Iran
Ana Suárez Sánchez	University of Oviedo, Spain
Jose Luis Roca Gonzalez	Centro Universitario de La Defensa en La Academia General del Aire, Spain

José Antonio Comesaña Benavides	University of Vigo, Spain
Ricardo del Olmo	Universidad de Burgos, Spain
María Carmen Carnero	University of Castilla-La Mancha, Spain
Cristóbal Miralles Insa	Universitat Politecnica de Valencia, Spain
Jesus Portillo	ETSI—Universidad de Sevilla, Spain
Pilar I. Vidal-Carreras	Universitat Politècnica de València, Spain
Francisco Javier Iglesias Rodríguez	Universidad de Oviedo, Spain
David Poza	INSISOC—University of Valladolid, Spain
Llanos Cuenca	Universitat Politècnica de València, Spain
Ana M. Mejías	University of Vigo, Spain
Juanjo Alfaro	Universitat Politècnica de València, Spain
Jayant Rajgopal	University of Pittsburgh, United States
Yuval Cohen	Tel-Aviv Afeka College of Engineering, Israel
Angel Gento	Universidad de Valladolid
Enrique Parra	University of Alcala, Spain
Ana Moreno Romero	ETSII, Spain
Manuel Mateo	Universitat Politècnica Catalunya, Spain
Andrés Boza	Universitat Politècnica de València, Spain
Lourdes Canós Darós	Universitat Politècnica de València, Spain
Simone Caberte Naimer	UFSM
Pedro L. Gonzalez-R	School of Engineers, Spain
María del Mar Alemany Díaz	Universitat Politècnica de València, Spain
Imma Ribas	Organización de Empresas, UPC, Spain
Borja Ponte	Cardiff University, United Kingdom
Carmen Avilés-Palacios	Universidad Politécnica de Madrid, Spain

Contents

Part II Production

Part III Methods and Applications

Part I
Management

New Perspective on Building Strategy in Public Education Institution

William Dias, Carine Nunes, Paloma Martinez and Úrsula Maruyama

Abstract The standard responses to public sector management issues in developing countries have failed in some respect to attend its agenda. Proceeding with so many intangibles and uncertainties managers need to find a new approach using better strategic planning indicating to institutions how to reach goals and achieve its intended vision. IFTO is a federal institute of education in the heart of Brazil. Its mission is to provide educational, scientific and technological development in the state of Tocantins through personal training and professional qualification. This case study presents how IFTO transformed itself by using a new planning tool.

Keywords Public administration · Public education · University management
Strategy · IFTO

1 Introduction

Strategic planning is a means by which an organization defines its future, indicating how to reach goals and achieve its intended vision. Its practice contributes to institu-

W. Dias (✉) · C. Nunes
Instituto Federal do Tocantins—IFTO, Pró-Reitoria de Desenvolvimento
Inntitucional, Avenida Joaquim Teotônio Segurado, Quadra 202 sul,
ACSU-SE 20, Conjunto 01, Lote 08, Piso 1, Palmas, TO CEP. 77.020-450, Brazil
e-mail: william.dias@ifto.edu.br

C. Nunes
e-mail: carine.nunes@ifto.edu.br

P. Martinez
Facultad de Ingeniería, Universidad El Bosque, av. Carrera 9 nº 131 A – 02, Bogotá, Colômbia
e-mail: palomams@gmail.com

Ú. Maruyama
Centro Federal de Educação Tecnológica Celso Suckow da Fonseca
Cefet/RJ, av. Maracanã 229, Maracanã, Rio de Janeiro, RJ CEP. 20.271-110, Brazil
e-mail: ursula.maruyama@cefet-rj.br

Á. Ortiz et al. (eds.), *Engineering Digital Transformation*, Lecture Notes
in Management and Industrial Engineering,
https://doi.org/10.1007/978-3-319-96005-0_1

tional self-knowledge, and provides changes in the conduct of its practices bringing effectiveness and strategy (Silva et al. 2015).

Federal Institute of Education, Science and Technology of Tocantins (IFTO) is the result Federal Palmas Technical School (ETF) and Federal Agricultural School of Araguatins (Eafa) integration. It was created by Law 11.892/2008, which established Federal Vocational, Scientific and Technological Education Network.

Today, IFTO has twelve campuses, as well as sixteen distance education poles. IFTO lives up to the best in the history of professional education in Brazil, since it forms professionals who meet both the development goals of the country and the demands of society.

Institutional leaders frequently feel challenged in their search for effective strategies to encourage faculty involvement in public service activities. Faculty are often deeply concerned about the lack of clear and comparable definition of terms such as service, outreach, public engagement, community service, service learning, internships, *practica*, and so on (Holland 1999, 2016).

This work aims to demonstrate how simple attitudes from school managers can transform education institutions building collaboratively goals and action plans.

2 Public Administration Strategy Context

Research focused on the role of networks has increased significantly in the field of public administration during the past two decades. This new research agenda has allowed scholars to explore the potential of alternative forms of collaboration, while it has also created several conceptual and empirical challenges associated with adapting the network concept to the specific needs of the public administration research community (Lecy 2014).

Meyer (2005) states there are many problems which currently challenge universities, such as intense competition, reduced demand, increased operating costs, lack of resources, reduced state support, student dropouts, declining teaching quality, and reduced student payment capacity. In order to universities being able to face challenges posed by a changing and competitive environment, it is necessary to have agile management and a planning model more appropriate to the reality of universities that are characterized by complexity, paradoxes, ambiguities and its conflicts.

According to Bryson et al. (2014), a new public administration movement is emerging to move beyond traditional public administration and New Public Management. The new movement is a response to the challenges of a networked, multi-sector, no-one-wholly-in-charge world and to the shortcomings of previous public administration approaches. This scenery it is not detached from universities challenges.

Contributing to this discussion Brinkerhoff and Brinkerhoff (2015) pose their questions about what new perspectives can shed a different light on public sector management, to either complement or confront the orthodox solution set? What alternative frameworks for theory and practice can move the discourse beyond New

Public Management? What experience and lesson can help to shape new explanations and responses? In order to answer some of these questions, this paper intends to present a Brazilian case applied to public education strategy management.

3 Balanced Scorecard Applied to Public Education Institution

Balanced Scorecard (BSC) is a measurement and performance management methodology developed by Harvard Business School (HBS) professors Robert Kaplan and David Norton in 1992. The methods used in business, service, Information Technology (IT) and ERP (Enterprise Resource Planning) software as support solutions, linking it to service management and business performance assurance (Kaplan 2010). The steps of these methodologies include: business strategy definition, business management, service management and quality management; These steps are implemented through performance indicators.

Figure 1 shows the architecture of a comprehensive BSC six stage closed-loop management system that links strategic planning with operational execution. (1) Develop the strategy (2) Translate the strategy (3) Align the organization (4) Plan operations (5) Monitor and learn (6) Test and adapt the strategy.

According to Kaplan (2010) intangible assets seldom have value by themselves. Generally, they must be bundled with other intangible and tangible assets to create

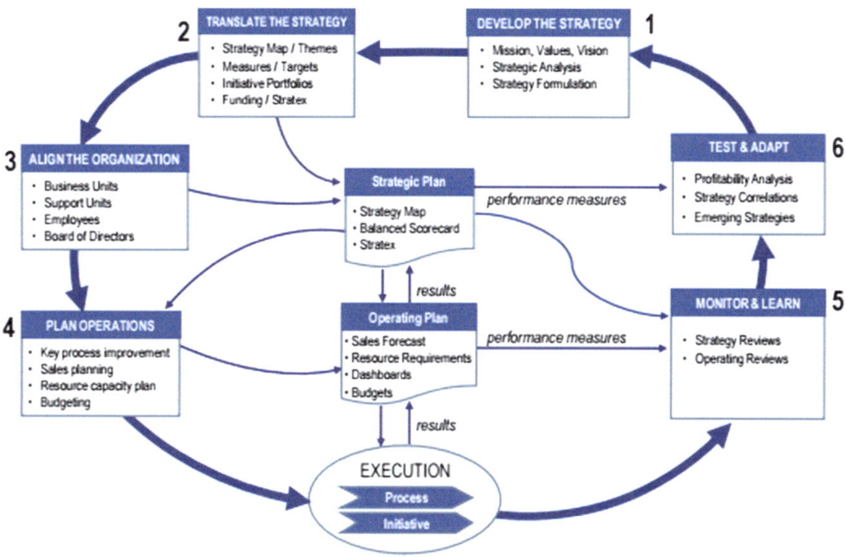

Fig. 1 Balanced scorecard management system for strategy execution

value. The value does not reside in any individual intangible asset. It arises from creating the entire set of assets along with a strategy that links them together. The value-creation process is multiplicative, not additive.

4 IFTO Case Study

This paper intends to present a case study in Technical Education Institution in the heart of Brazil, Tocantins. IFTO was designed to operate throughout the State, and is already synonymous with excellence in teaching throughout Tocantins, as it provides quality public education from basic to higher education.

The events were documented by another Technical Education Institution strategic director located in Rio by interviews and field research in 2016. The methodology was developed considering the following IFTO strategy drivers:

Mission: Provide educational, scientific and technological development in the state of Tocantins through personal training and professional qualification.
Vision: To be a reference in teaching, research and extension, with emphasis on technological innovation of products and services, providing sustainable regional development.
Guiding Principles: The Federal Institute of Tocantins, in its Statute, published on August 19, 2009, in its third article, states as guiding principles of the institution: (I) Commitment to social justice, equity, citizenship, ethics, preservation of the environment, transparency and democratic management; (II) Verticalization of teaching and its integration with research and extension; (III) Effectiveness in vocational training responses, dissemination of scientific and technological knowledge and support to local, social and cultural productive arrangements; (IV) Inclusion of people with special educational needs and specific disabilities; And (V) Public and free nature of education, under the responsibility of the Union.

Strategic Objectives:

1. Increase 'vacancies' offered by IFTO
2. Ensure conditions for students to stay and graduate successfully
3. Create conditions to meet physical and educational demands of the internal community with specific needs
4. Offer courses in distance mode (online courses)
5. Integrate teaching, research and extension actions as an educational and pedagogical act
6. Instituting practices and disseminating the culture of sustainability
7. Implement actions focusing on dissemination of associativism and entrepreneurship
8. Expand the relationship network with organizations in the productive sector
9. Expand the network of relationship with the community and social/cultural groups/arrangements

10. Expand the relationship network with public sector organizations
11. Promote continuous training and development of employees to achieve institutional results
12. Implement integrated system of institutional management
13. Increase fundraising
14. Improve the infrastructure of the campuses
15. Disseminate and implement actions to achieve the strategic objectives
16. Establish and strengthen collegiate bodies
17. Expand participation of extracurricular components in students' academic lives
18. Adapt IFTO vacancies in order to meet the legal demands and the rationalization of resources
19. To increase scientific and technological production and dissemination, as well as the realization and expansion of IFTO research and extension projects
20. Promote integration between the IFTO units, aiming at the exchange of experiences and knowledge in the different areas
21. Promote actions to disseminate basic human rights.

Based on the Balanced Scorecard, the Institution's macroprocesses were divided into Teaching (equivalent to the Customers perspective in the BSC), Research (equivalent to the Customers perspective in the BSC), Extension (equivalent to the Customers perspective in BSC), People Management And Growth in the BSC), Administration (Equivalent to the Financial perspective in the BSC) and Institutional Development (Equivalent to the Internal Processes perspective in the BSC). The prospective client gains a new subdivision given to the multifaceted clients of a public education institution, whose clients are students, graduates, the local community, social groups, businessmen, etc.

In order to meet the increase in the demand for higher education vacancies, IFTO invested in infrastructure and in the training of its employees, allowing the offer of more varied types of courses. It has also been invested for this purpose in online courses, which are able to provide cost reduction and increase capillarity, as it allows to reach a higher number of students per teacher through the use of technological resources, and also reduces the need for occupation of classrooms, as well as it reaches students who are in regions where there is no campus.

Considering the afore mentioned perspectives (macroprocesses), strategic objectives were elaborated for the period up to 5 years. Most of them are based on the Laws that direct the Institute, the main ones being: Basic Guidelines Law, Federal Institutes Creation Law and IFTO Statute. After that, indicators for these objectives were elaborated, including in these indicators established by Judgment of the TCU—Federal General Accounting Office, Brazilian institution provided for in the federal constitution to exercise the accounting, financial, budgetary, operational and patrimonial oversight of the country—and Term of Agreements and Goals between the MEC—Ministry of Education—and the IFTO. This strategic goals, indicators and goals are all included in the Institutional Development Plan—PDI 2015–2019, that is decomposed in the Annual Institutional Activity Plan—PAAI, which is decomposed in activities (Fig. 2).

Institutional Development Plan - PDI 2015-2019							
Perspectives	**Strategic goal**	**Indicator**	**Goal**				
			2015	2016	2017	2018	2019
Teaching, Research and Extension	05- Integrate teaching, research and extension actions as an educational and pedagogical act.	0503 - Number of activities (programs, projects, courses and events) of research, extension and programs to improve the quality of basic education maintained by IFTO.	15	17	19	20	22
[...]							

Annual Institutional Activity Plan – PAAI	
Year Reference	2017
Strategic goal	05
Indicator	0503
Goal	19
Description of action	Building IFTO's calendar of institutional events
Activity 1	Annual meeting for Building IFTO's calendar of institutional events
Start	01/03/2017
Duration	1 day
Cost	0
Accountable	List
Cooperator	List
Activity 2	[...]

Fig. 2 PAAI chart report sample

IFTO criteria for elaborating the other indicators, which are not foreseen in the legal agreements, were basically: Ability to measure the objective and possibility of measurement with available data. As the institution matures its systems and processes, indicators also tend to improve. Finally, the targets were defined for the duration of the IDP, also respecting the corresponding legislation, focusing on feasibility and feasibility.

Besides, the methodological procedure that was applied allowed to identify the critical work processes of the organizational units subject to vulnerabilities and gives a status of criticality that allows to know and classify the most important processes of the sectors.

5 Results

To prepare the Action Plan, a methodology was used that absorbed some concepts of some existing techniques. The first step was Problem Identification, using the concept advocated by the PDCA. For this, since the institute is very large, and a large-scale analysis such as SWOT, has already proved impracticable at another time, a web system was elaborated, in which, presenting the indicators with their respective goals, we made 4 Basic questions for the approximately 1200 IFTO employees (IFTO, 2016):

1. Do you visualize any problems that might prevent IFTO from reaching this indicator/goal?
2. What is your suggestion to remedy this problem (given that the institute is formed by a highly-qualified cadre of employees);
3. In what unit should your suggestion be applied?
4. Which sector in your opinion should implement your suggestion? With this we got several suggestions from the servers.

Thereafter, suggestions were directly to sector managers asking them to choose among three decisions: (1) Create an action based on suggestion; (2) Transfer to another sector, which would be the true responsible (in this step the managers were asked to look for the reason of the problem, a fact that usually uses techniques such as "5 Whys", "Ishikawa's Fishbone diagram" or "TOC's Current reality tree"); (3) ignore and file the request, which should be justified. The decision made by the manager returns as feedback to the server who gave the suggestion.

The second step, the preparation of the action plan itself. At that moment, the system already includes the actions elaborated based on the suggestions received by the managers of each sector. The action itself is equivalent to the "What" of the 5W2H tool. But at the action level, only "What" and "Where" are defined. The "Why", problem identified by the servers, "Who", responsible server, "When", start and duration, "How much", variable cost, are defined by activity and not by action. Activity is the equivalent of 5w2h's "How".

To define what to do, managers were encouraged to use techniques such as Brainstorming and techniques for productive meetings. In order to define the "When", it was decided to ask the beginning and duration based on a behavior identified by Theory of Constraints (TOC), when it addresses the critical chain project management (CCPM) known as "Student Syndrome", that is, instead of viewing the end date, and trying to get close to the end, the person responsible only sees the start date, and would be more likely to start the activity earlier.

Another point worth mention is that the system sends notifications to those responsible two days before their activity begins, trying to avoid delays in the beginning. Still, another added concept is that of Activity Collaborator.

The New Institutional Activities Annual Plan methodology (PAAI) is elaborated through its own electronic datasystem. The planning tool gathers the opinion of IFTO's servers. Thus, this mechanism arrives focusing on advising institution in the establishment of priorities and finding best investment of available resources. “In the beginning, it was difficult, but now it is much better with people participation in planning cycle so that we can identify problems and highlight solutions.”

6 Conclusion

In this system, employees portrays the problem it sees, indicates a possible solution and how it could be applied, and indicates which sector would be responsible for implementing that solution, in his opinion. It will be up to the manager to give feedback to this employee, whether positive or not. The System enables strategic alignment in the execution of almost all institutional activities, in a more structured and computerized way, what allow to monitor the status of activities in real time. Therefore, the integration between teaching, research and extension aimed at Local Productive Arrangements gains prominence in this institution, providing educational, scientific and technological development.

References

Brinkerhoff D, Brinkerhoff J (2015) Public sector management reform in developing countries: perspectives beyond NPM orthodoxy. Public Adm Dev 35:222–237 (Wiley Online Library)

Bryson J, Crosby B, Bloomberg L (2014) Public value governance: moving beyond traditional public administration and the new public management. Public Adm Rev 74(4):445–456

Holland B (1999) Factors and strategies that influence faculty involvement in public service. J Public Serv Outreach 4(1):37–43

Holland BA (2016) Factors and strategies that influence faculty involvement in public service. J Higher Educ Engagem 20(I). originally written by Hollan (1999) to Journal of Public Service and Outreach, 4(1), pp 37–43

IFTO (2016) PAAI. Plano Annual de Atividades Institucionais. Available at http://portal.ifto.edu.br/ifto/colegiados/consup/documentos-aprovados/plano-anual-de-atividades-institucionais/plano-anual-de-atividades-institucionais-paai-2016.pdf/view

Kaplan R (2010) Conceptual foundations of the balanced scorecard. Working paper. 10-074. Harvard Business School. Paper originally prepared for C. Chapman, A. Hopwood, and M. Shields (eds.), in (2009) Handbook of Management Accounting Research: vol 3 Elsevier

Lecy J et al (2014) Networks in public administration: current scholarship in review. Erschienen. Public Manag Rev 16(5):643–665

Meyer V (2005) University planning: rational, political or symbolic activity—a study of Brazilian universities. Alcance—UNIVALI 12(3):373–389

Silva F, Mario P (2015) O processo de formulação e monitoramento do planejamento estratégico de Tribunais de Contas sob a ótica da Nova Sociologia Institucional. Rev Adm Pública — Rio de Janeiro 49(6):1401–1427

Role Clarity and Satisfaction for Knowledge Workers

Francisco Jose Fernández-Ferreras, Ana Moreno-Romero and Gabriela Topa Cantisano

Abstract The objective of this work is to identify conditions for knowledge worker productivity and satisfaction. Two variables are proposed as key influencers: role clarity and knowledge content in the job position. To test the importance of these variables a questionnaire was prepared and 167 managers and knowledge professionals answered it. Strong and positive relationship between role clarity, knowledge content in the job position and satisfaction is confirmed.

Keywords Knowledge workers · Role clarity · Job content · Satisfaction
Networked organization · Productivity

1 Introduction

Information and Communication Technologies, ICT, are not producing the expected improvements in productivity into work processes, and there are studies that note that always-on models lead to a fall in productivity (Davenport 2011).

Productivity has two complementary approaches: entrepreneurial and human. The first one shows that low worker productivity is directly related with a decrease in business profitability (Miller 1987; Jain and Moreno 2015), and the other way round, if personal productivity rises, the same will happen with the organizational productivity (Goodman et al. 1994) and corporative mid-term and long-term financial results will also rise (Huselid 1995).

F. J. Fernández-Ferreras (✉) · A. Moreno-Romero
GIOS, Grupo de Investigación de Organizaciones Sostenibles, School of Industrial Engineering, Madrid Polytechnic University (UPM), C/José Gutiérrez Abascal, 2, 28006 Madrid, Spain
e-mail: franciscojose.fernandez@upm.es

A. Moreno-Romero
e-mail: ana.moreno.romero@upm.es

G. T. Cantisano
Departamento de Psicología Social y de las Organizaciones, UNED, Madrid, Spain

Á. Ortiz et al. (eds.), *Engineering Digital Transformation*, Lecture Notes in Management and Industrial Engineering,
https://doi.org/10.1007/978-3-319-96005-0_2

Managers in knowledge work contexts must ensure a stable workload by striking a balance between new projects, their operating schemes and individual assignments (Mintzberg 1989).

The analysis of work organization in knowledge-based positions should be made from three different realms: organization, team, and worker (Moreno et al. 2011). To understand team mechanisms is directly linked to the role clarity that has and indirect effect with executive performance (Hall 2008) and recent studies confirm the necessity to increase role clarity to improve well-being (Méndez et al. 2015) and work satisfaction (Orgambídez-Ramos et al. 2016).

2 Objectives

The main objective of this article is to confirm the high influence of two organizational variables, role clarity and knowledge content, for knowledge workers.

Once we have defined how to measure knowledge workers' productivity, the objective is to identify strong predictors in adhocratic organizational environments, both in job characteristics (role clarity) and in psychosocial variables (satisfaction).

This predictors should be the organizational constructs that could allow to evolve the organization structure and/or to develop human capital to align both elements and maximize productivity.

3 Methods

In order to study productivity in organizations that have a significant proportion of knowledge workers, a questionnaire was prepared and the following variables were included: knowledge content, productivity, role clarity and satisfaction.

Sample selection and data collection was done through a convenience non-probabilistical sampling. The study has been carried out with 167 professional Spanish workers. Some characteristics of the sample are:

- 68.8% male and 31.2% female,
- 49.0% Managers and 27.2% middle management,
- 90.0% with higher education,
- 94.8% with long-term contract,
- 97.0% work in private companies,
- 71.1% with less than 1.000 workers and
- 83.7% in the services sector.

Variables definition has been done as it follows:

- Knowledge content (decision-based job, knowledge is needed, autonomy, learning opportunities, non-repetitive, adaptive and able to design it): this variable was measured using five items from the CUBOT questionnaire (Moreno et al. 2011) referred to the worker's knowledge and with two additional items extracted from the U.S. Army Construction Engineering Research Laboratories' work components description (Thomas and Baron 1994). Some examples about the items are "I believe my job allows me to learn, evolve and adapt myself to the surrounding circumstances" and "My every day decisions are high complex". Valuation is done with a five point Lickert-type scale (1 = None; 5 = Much).

- Role clarity: this variable was measured using four items from the CUBOT-Mi productividad personal (GIOS) questionnaire. Some examples of the items are: "I know what era expected from me on my job" and "I feel that my personal characteristics are appropriate to do my work". Valuation is done with a five point Lickert-type scale (1 = None; 5 = Much).
- Satisfaction: this variable was measured with seven items from the Brief Index of Affective Job Satisfaction (BIAJS), that allows to obtain a global work satisfaction index (Thompson and Phua 2012). An example of the items is: "I feel close to the persons I work with". Also was measured with six item from ALQ questionnaire based on the Job Satisfaction Scale, JSS (Warr et al. 1979; MacIntyre et al. 1997). Some examples of the items are: "I really enjoy my work" and "I like my job more than most of the people". Valuation is done with a five point Lickert-type scale (1 = None; 5 = Much).

All these items are from a much bigger questionnaire used in a broader study involving 18 more variables. Knowledge content is one of the three variables in the Knowledge worker's block, role clarity is one of the three variables in the job characteristics block and satisfaction is one of the two variables in the psychosocial block. Productivity was the independent variable of the study.

4 Results

The statistical descriptions of the chosen variables are:

Variable	Minimum value	Average	Maximum value	Typical deviation
Knowledge content	1.00	3.59	5.00	0.74
Role clarity	1.00	3.85	5.00	0.72
Satisfaction	1.38	3.69	4.92	0.73

All averages are above 3.5 with a small typical deviation. The explanations of the results are:

- Knowledge content: The job gives knowledge workers autonomy to plan, control and regulate the environment, it allows them learn, develop and adapt to the surrounding circumstances, it shows a variety of experiences and activities, it makes them feel important in the organization as they are able to design and plan their job content. It also means that they have to make many high complex decisions every day.

- Role clarity: High knowledge of what they have to do on their jobs and high adaptation of their personal characteristics to their jobs.
- Satisfaction: Medium-high job satisfaction.

Cronbach's Alpha reliability of the scales is:

Scale	Cronbach's alpha	Number of elements
Knowledge content	0.858	7
Role clarity	0.759	4
Satisfaction	0.932	13

All values are higher than 0.6 so they are considered acceptable according to the academic criteria (Gliem and Gliem 2003). Satisfaction and knowledge content show very high values, role clarity also considering only four elements.

The influence amongst the variables is extremely powerful for knowledge workers as it is shown with the Pearson correlation:

		1	2	3
1	Knowledge content	1		
2	Role clarity	0.652**	1	
3	Satisfaction	0.597**	0.788**	1

**Significant correlation at level 0.01 (2 tails)

Strongest and positive relation is between Satisfaction and Role clarity ($r=0.788$; $p<0.01$). For knowledge workers in adhocratic organizations, more role clarity means more satisfaction, and also, if the knowledge worker is happier in his/her job, it will be easier to understand his/her role.

Next strongest positive relation is between Role clarity and Knowledge content ($r= 0.652$; $p<0.01$). The type of KW job needs role clarity, and role clarity is based on KW type of jobs.

Third one is between Satisfaction and Knowledge content ($r=0.597$; $p<0.01$). Having a decision-based job, where knowledge is needed, autonomy is possible, are learning opportunities, is non-repetitive, and adaptive and the KW is able to design it is equal to job satisfaction.

5 Conclusion

For knowledge workers in adhocratic organizations, role clarity importance has been confirmed and is the center of the three variables' triangle in order to clarify what knowledge workers, and particularly managers should do to foster productivity.

The more role clarity exists, the more satisfaction for knowledge workers. If the job is rich and meaningful, the clearer it would have the knowledge worker what to do and what the others do. And if the job has such characteristics, this type of worker will be more satisfied, happy and productive (Shoemaker 1999). Not only for one's job but also the ones' that work with you. In freedom to operate environments jobs' limits help to define neighboring roles.

Satisfaction feeds back knowledge workers boosting individual productivity having a significantly impact in corporate performance (Goodman et al. 1994).

References

Davenport T (2011) Rethinking knowledge work: a strategic approach. McKinsey Quarterly

Gliem RR, Gliem JA (2003) Calculating, interpreting, and reporting Cronbach's alpha reliability coefficient for Likert-type scales. In: Midwest research-to-practice conference in adult, continuing, and community education

Goodman P, Lerch J, Mukhopadhyay T (1994) Individual and organizational productivity: linkages and processes. In: Organizational linkages: understanding the productivity paradox. National Academies Press, pp 54–80

Hall M (2008) The effect of comprehensive performance measurement systems on role clarity, psychological empowerment and managerial performance. Acc Organ Soc 33(2):141–163

Huselid MA (1995) The impact of human resource management practices on turnover, productivity, and corporate financial performance. Acad Manag J 38(3):635–672

Jain AK, Moreno A (2015) Organizational learning, knowledge management practices and firm's performance: an empirical study of a heavy engineering firm in India. Learn Organ 22(1):14–39

MacIntyre PD, Thivierge KA, MacDonald JR (1997) The effects of audience interest, responsiveness, and evaluation on public speaking anxiety and related variables. Commun Res Rep 14(2):157–168

Méndez MD, Silva N, Palomo-Vélez G, da Costa S, Páez D (2015) Factores psicosociales del bienestar y la creatividad en una muestra de trabajadores de la región centro sur de chile. La felicidad de los chilenos: estudios sobre bienestar, pp. 215–236

Miller DM (1987) Note—analyzing total factor productivity with ROI as a criterion. Manage Sci 33(11):1501–1505

Mintzberg H (1989) Mintzberg on management: inside our strange world of organizations. The Free Press, New York. ISBN 0-02-921371-1

Moreno A, Mataix C, Mahou A (2011) Claves para la gestión de puestos de trabajo en las organizaciones en red, Revista de Dirección y Organización, Cepade

Orgambídez-Ramos A, Moura D, de Almeida H (2016) Estrés de rol y empowerment psicológico como antecedentes de la satisfacción laboral. Revista de Psicología 35(1):257–278

Shoemaker ME (1999) Leadership practices in sales managers associated with the self-efficacy, role clarity, and job satisfaction of individual industrial salespeople. J Pers Selling Sales Manag 1–19

Thomas BE, Baron JP (1994) Evaluating knowledge worker productivity: literature review (No. USACERL-IR-FF-94/27). Construction Engineering Research LAB (ARMY) Champaign IL

Thompson ER, Phua FT (2012) A brief index of affective job satisfaction. Group Org Manag 37(3):275–307
Warr P, Cook J, Wall T (1979) Scales for the measurement of some work attitudes and aspects of psychological well-being. J Occup Psychol 52(2):129–148

An Innovation Model for EPC/Turnkey Sector: The Case of Abengoa Solar New Technologies

Gaizka Garechana, Rosa María Río Belver, Ernesto Cilleruelo-Carrasco and Iñaki Bildosola

Abstract Abengoa Solar New Technologies (ASNT) is a relevant R&D actor in solar energy EPC/Turnkey sector, its innovation strategy being characterized by the aggressive patenting strategy they have deployed since year 2008. On that same year ASNT got its innovation management system certified under UNE 166002 standard, combining two divergent issues such as standardization and innovation. This paper presents the main features of the innovation model put into practice at ASNT, detailing its implications for knowledge management and R&D employee incentive system.

Keywords Innovation management · Management standards · Patent strategy

1 Introduction

EPC/Turnkey contracts in solar energy usually involve dealing in a technology-intensive area where patents play a key role both as a signalling instrument and a rival-blocking tool by protecting enabling technologies in solar power plants. Abengoa Solar New Technologies (ASNT) is the R&D&I subsidiary of Abengoa Solar (AS), a prominent player in the solar energy field that operates, maintains and has built some of the biggest and most advanced solar energy plants in the world.

We conducted a qualitative research in order to understand the innovation formula working at ASNT, and in this paper we describe an innovation model based on the standardization of innovation management and the use of patents as an innovation management tool. We consider this to be a particularly interesting case study from the point of view of the dichotomy of the innovation management standard UNE 166002, that brings together two apparently antagonistic managerial concepts like the standardization and homogenization of procedures, and the spontaneity and flexibility of innovation. The practical consequences of an incentive system based on patents are also discussed.

G. Garechana (✉) · R. M. Río Belver · E. Cilleruelo-Carrasco · I. Bildosola
University of the Basque Country UPV/EHU, Bilbao, Spain
e-mail: ggaretxana@gmail.com

Á. Ortiz et al. (eds.), *Engineering Digital Transformation*, Lecture Notes in Management and Industrial Engineering,
https://doi.org/10.1007/978-3-319-96005-0_3

2 Standardization of Innovation

The core of management standards lies in the assumption that the disciplined use of documented, standardized procedures helps to bring the processes that add value to the customer under control, thus improving the firm's ability to offer defect-free products and services (Anderson 1999).

The first experimental version of standard UNE 166002 was released in 2002, based on the recommendations stated in Oslo, Frascati and Bogota manuals. Innovation management guidelines do exist in the market, such as the BS 7000 in the UK but as far as the authors are concerned, this is the first fully certifiable innovation management standard backed by a national standardization agency (AENOR 2016) and a recognized body of auditors. The standard can fit well in organizations already working with management standards such as ISO 9001, and the systematization of information retrieval and spread phases in competitive intelligence leads to a better use of the information (Pineda 2015), but the system is prone to infoxicate users and IT tools become of the utmost importance to ease knowledge management and competitive intelligence processes (Mir-Mauri and Fa 2008). The presence of a strong innovation culture in the firm is deemed to be an important prerequisite because of the extra workload the standard involves for researchers, particularly regarding the maintenance of documentation and information systems (Mir-Mauri and Casadesús 2011). In addition to this, a pioneer paper recently (June 2016) published by Mir-Mauri et al. (2016) empirically proves the positive relationship between the certification under UNE 166002 and the company's innovative capability and business performance, adding that certified firms suffer in a lesser amount the effects of economic crisis on innovative capability.

On the negative side of management standards, the work overload usually associated with them is one of their most cited shortcomings: the redesigning of processes is particularly costly for complex organizations and the effort of codifying and recording almost everything takes attention away from the "real" work that must be done. The efforts of codification and recording can also improve the understanding of managers about how work is really conducted in the firm (Anderson 1999), but standardization irremediably leads to a reduction of ambiguity in tasks, hindering innovation. The whole mindset of promoting rational thinking supported by a set of predetermined data, tools and techniques leads to lower flexibility and less openness to change as a consequence of adhering oneself to repeated behavior (Pekovic and Galia 2009; Prajogo and Sohal 2004). Standardized systems look for the routinization of problem-solving procedures by the accumulation of data, and it seems unlikely that innovative solutions can come from such a way of working (Glynn 1996). The focus on documentation, standardization and conformity to rules and procedures results in a culture of attention to detail at the expense of organizational skills vital for product innovation such as learning from experimentation and product failure (Terziovski and Guerrero 2014; Naveh and Erez 2004).

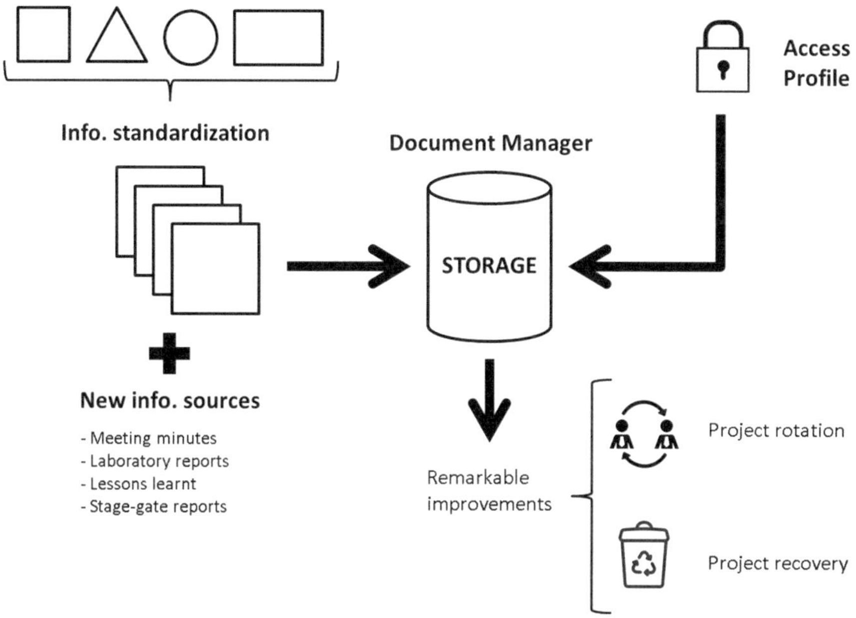

Fig. 1 Improvements in knowledge management from UNE 166002 implementation

3 Effects on Knowledge Management on a Project-Based Company

The main change that standardization brought to innovation management at ASNT has to do with knowledge management. The standard required a knowledge explicitation system to be put into practice, previously often reduced to the project final reports and which sometimes lacked detail or were quite heterogeneous.

The implementation of the procedures of UNE 166002 led to the elaboration of more detailed and regular project reports that make it possible to collect and preserve relevant information about how projects at ASNT were conducted and how the problems were solved. The same could be said about the technical/economical feasibility monitoring reports resulting from the stage-gate procedure, which were improved and incorporated to the knowledge management system. The habit of generating and keeping "lessons learnt" documents was also embraced by the company, as was the writing and filing of meeting minutes, especially when the meeting involved some kind of agreement with third parties, become almost obligatory under the pressure exerted by the standard. Knowledge generated in this process was preserved and protected similarly to any other intellectual property belonging to the firm. Other relevant documentation such as laboratory essay reports or project delay justification reports were generated by the systematic reporting habit instilled in many areas that

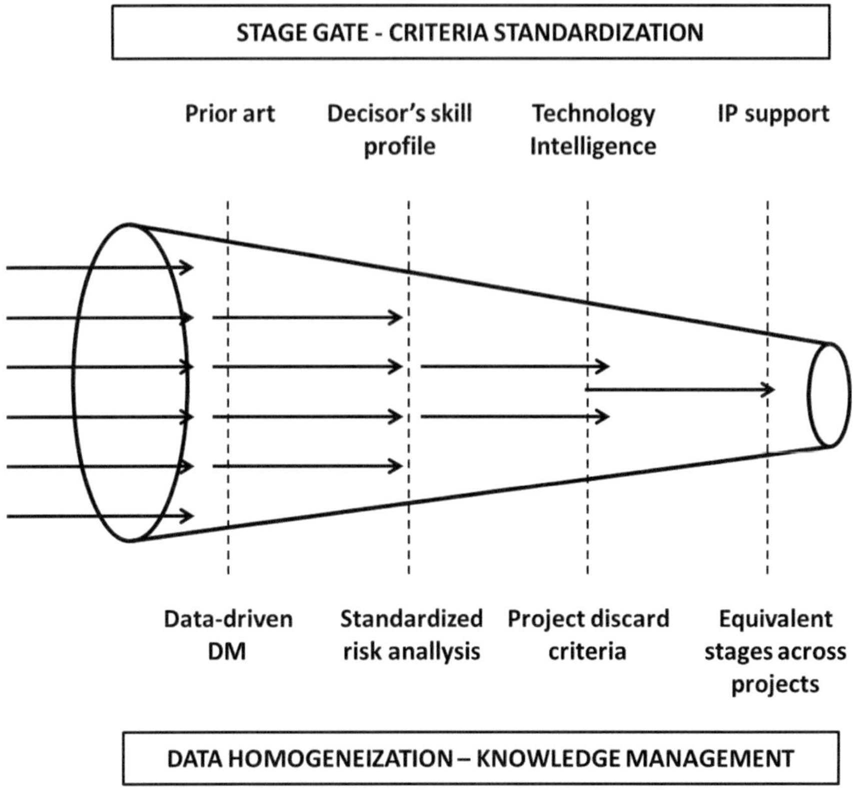

Fig. 2 Enhancements in stage-gate process management

had not previously or sporadically carried out such activities, or that did not keep or make such information available to the firm (Figs. 1 and 2).

EPC/Turnkey firms usually deal with certain amount of staff rotation coming from the reallocation of resources in different phases of the project. Complying with the exacting documentation requisites of standardization and making this information conveniently available through the use of a good knowledge management IT tool has significantly improved the traceability of the projects, allowing the quick updating of professionals that join a project in its latter stages. The knowledge transference between projects and between the business units of the parent company improves. Additionally, this rigor in information management has proven to be effective to increase the chance of "rescuing" promising projects that were killed off because of technical limitations whenever the evolution of technology eventually unlocks the situation. The very reasons why the project was killed are carefully documented in the system, making it easier to detect the opportunities to rearrange abandoned projects, and managers and researchers can resume the project at the exact point where it was left thanks to the volume and detail of the available information.

The standardized innovation system led to the implementation of data-driven, risk analysis procedures across all the stages of a project. A particularly difficult decision in R&D&I project management is that of abandoning a project that does not produce the expected results, since researchers are usually highly involved in their ideas. A standardized risk evaluation system based on objective indicators provides a foothold for decision making in this issue. Having too many research fronts open is a serious inefficiency in innovation management.

4 Patents as Management Tool

ASNT was incorporated in 2005, and it was after 2008, coinciding with the certification of ASNT's innovation management system, that an aggressive patenting strategy came about, filing 161 patent families in just six years.

The following analysis of the patent strategy at ASNT provides an insight into the perception of patent protection in a sector characterized by rapid technological evolution and a business model frequently based on Engineering, Procurement and Construction (EPC) or turnkey projects, which in our opinion have substantial distinctive features from a management of innovation point of view, as corroborated by Pellicer et al. (2012). The transversality of some developments, particularly those related with O&M business, makes the patents useful to establish an internal licensing system that gives valuable information to decision makers about the value of the innovations developed.

An internal patent office is in charge of evaluating the intangible asset value of patents, for this purpose, the office has established a scorecard (Fig. 3) that combines quantitative and qualitative information about the following variables:

- Legal robustness of the patent: The broadness of the claims is taken into account, as well as the positive evaluations from institutional patent offices.
- Technological analysis: The technical novelty of the patent is estimated, distinguishing between technologies that can be used in current projects or technologies that can be put into practice in the future.
- Market analysis: The patents are classified in those aimed at niche markets, those with a broad range of application and those which are not directly related with the core activities at Abengoa Group.
- Strategic analysis: The total amount of resources invested in the patent is considered, as well as the modularity of the technology.

As a result of this process an accounting value for the patent portfolio is determined, following consensus from the multidisciplinary team that evaluates the variables explained above.

Solar energy plants are often dependent on key components that are difficult to substitute or substantially modify should they be patented by third parties. The signaling of a strong patent portfolio in the solar energy sector is highly relevant for ASNT, since customers in this sector tend to associate patents to technology

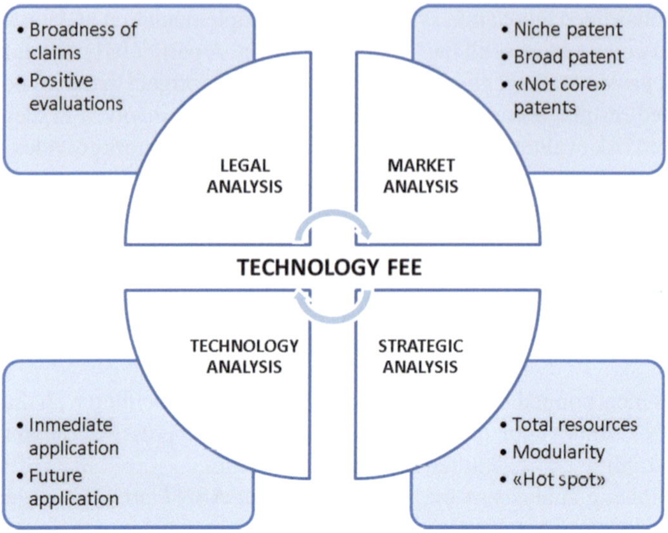

Fig. 3 Patent valuation scorecard

leadership. The following managerial changes were put into practice in order to avoid squandering money in an erratic patent frenzy:

– The researchers were trained in what can and cannot be protected by a patent, the meaning of inventive capacity and novelty, etc.
– An internal patent office was created. This office was in charge of conducting the aforementioned training and its staff work side by side with the researchers to create the patent. The people at the patent office have both a deep knowledge of the patent system and enough technical capability to understand the inventions developed by the researchers.
– A goal of a minimum number of filed patents per year was set for each researcher team. The minimum amount of patents to be required was defined depending on the total budget allocated for each activity, and the total amount of patents filed per year had an effect on the annual bonus of the researchers. Patent-filing based goals were also set for subsidiaries and business units.

This system has the additional advantage of setting the mindset of the researchers in the future commercialization of the solution being developed, and substantially improved the attention paid to prior art.

5 Conclusions and Implications for Management

The implementation of an innovation standard at ASNT does not manifest the innovation-blocking features that the state of art attributes to management standards.

On the other hand, the increase in bureaucratic work related to R&D&I is noticeable, and mainly takes the shape of heavy requests for reports and other documentary items aimed at producing a thorough record of R&D&I activities, this work overload is specially high when adapting the firm's previous procedures to the documentation requisites set by the standard. Part of this effort is leveraged when other bureaucratic tasks become considerably eased by the availability of complete and updated information concerning almost every aspect of R&D&I at the firm.

The stage-gate system for R&D&I project management improved its comparability across projects by the adoption of standard and objective indicators and methods, setting a solid cornerstone for the sensitive issue of project discarding and upgrading the allocation of resources to the most promising ideas. Strategic planning of R&D&I at ASNT was reinforced by the inclusion of several KPI and the feedback of UNE 166002 audit reports in strategic decision making. Knowledge management was greatly benefited from standardization, since it involved the strengthening of reporting on issues likely to generate valuable insights about project management and problem solving at ASNT. The thorough documentation requirements led to the expansion of an IT tool aimed at preserving and sharing the knowledge codified in these documents, which attained the status of any other intellectual property belonging to the firm. Additionally, the knowledge transfer between projects and business units was improved, as well as the updating of professionals that change from one project to another and the training of new staff, and top management decisions also received additional support from the availability of an expanded and detail-rich knowledge base. Other studies indicate that innovative practices benefit from the know-how derived from quality management (Perdomo-Ortiz et al. 2006), and our analysis at ASNT corroborates this fact.

Both certification under UNE 166002 and the strong patent portfolio developed by ASNT play a key role signaling the firm's commitment with R&D&I and a purposeful management of innovation efforts. Customers, potential business partners and particularly public institutions positively evaluate these features and associate them with technology leadership. The patent strategy at ASNT not only seeks for the obvious intellectual property protection, but also plays a key role in setting management incentives in R&D&I even at operative level. Patents are also used to put a technology transfer system between the business units of the parent company, Abengoa Group, into practice.

Our conclusion is that the innovation standardization is likely to lead to increased bureaucracy, however, no obstacles for innovation arising from the practices established by the standard have been detected. This study leads us to believe that the "sclerotization" of innovation attributed to management standards may come from an incorrect implementation of some of its aspects: data-driven management and decision making does not necessarily involve making robots out of the staff. The fact that this work analyses an R&D&I subsidiary may explain some of the features observed, since the particular features of this type of organization make it prone to adopt an enabling social structure (Adler 1999): Many tasks are non-routine, organizational structures tend to be flat, the work teams usually enjoy certain autonomy and the workforce is highly qualified.

Standardization of innovation also nurtured the development of a knowledge management system that put the vast organizational memory of ASNT at the service of the staff and a support point that reduced uncertainty in many decisions. These factors may explain most of the success from the implementation of a management standard in a highly dynamic process such as innovation management.

References

Adler P (1999) Building better bureaucracies. Acad Manag Executive 13(4):36–47

AENOR (2016) UNE 166002 R&D&I management systems. Retrieved from http://www.en.aenor.es/aenor/certificacion/innovacion/innovacion_sistemas_166002.asp#.Vy5DAoThDak

Anderson S (1999) Why firms seek ISO 9000 certification: regulatory compliance or competitive advantage? Prod Oper Manag 8(1):28–43

Glynn MA (1996) Innovative genius: a framework for relating individual and organizational intelligences to innovation. Acad Manag Rev 21(4):1081–1111

Mir-Mauri M, Casadesús M (2011) Standardised innovation management systems: a case study of the Spanish Standard UNE 166002:2006. Innovar 21(40):171–188

Mir-Mauri M, Fa M (2008) Une 166002: 2006: estandarizar y sistematizar la i + d + i la norma y la importancia de las tic en su implementación. DYNA-Ingeniería E Industria 83(6):325–331

Mir-Mauri M, Casadesús M, Petnji LH (2016) The impact of standardized innovation management systems on innovation capability and business performance: an empirical study. J Eng Technol Manag 41:26–44

Naveh E, Erez M (2004) Innovation and attention to detail in the quality improvement paradigm. Manag Sci 50(11):1576–1586

Pekovic S, Galia F (2009) From quality to innovation: evidence from two French employer surveys. Technovation 29(12):829–842

Pellicer E, Correa CL, Yepes V, Alarcón LF (2012) Organizational improvement through standardization of the innovation process in construction firms. Eng Manag J 24(2):40–53

Perdomo-Ortiz J, González-Benito J, Galende J (2006) Total quality management as a forerunner of business innovation capability. Technovation 26(10):1170–1185. https://doi.org/10.1016/j.technovation.2005.09.008

Pineda Á (2015) Sistematizar con éxito la innovación: UNE 166002 Integrated Circuits Málaga. AENOR: Revista de la normalización y la certificación. AENOR, Asociación Española de Normalización y Certificación

Prajogo DI, Sohal AS (2004) The multidimensionality of TQM practices in determining quality and innovation performance—an empirical examination. Technovation 24(6):443–453

Terziovski M, Guerrero JL (2014) ISO 9000 quality system certification and its impact on product and process innovation performance. Int J Prod Econ 158:197–207

How Does Working on University-Business Collaborative Projects Foster the Industrial Doctorates' Learning Process?

Jaione Ganzarain, Leire Markuerkiaga and Juan Ignacio Igartua

Abstract University and industry research collaboration hugely impacts on academic and industrial doctorates research career and curriculum development. The collaboration brings new opportunities of research areas of interest, real world problems and research challenges, and new skill and abilities in curriculum development. The collaboration also creates innovation and provides regional and national economic benefits. The paper presents the industrial Ph.D. training itinerary framework for creating and testing new business opportunities aligned with the research projects in the space university-business research collaboration. This itinerary provides a training journey based on different workshops along the process, testing the results obtained in their applied research with the real market. The framework encompasses the global learning itinerary with the driving criterion being ideation for industry and society, transforming knowledge into innovations Doctoral students follow the ideation for industry-society training itinerary in order to identify an initial business idea.

Keywords University-business collaborative projects · Industrial Ph.D. Triple helix model · Entrepreneurial university · Business model

1 Introduction

It is now generally accepted that innovation is the most important driver of long-term competitiveness, growth and employment (European Commission 2001). For

J. Ganzarain · L. Markuerkiaga (✉) · J. I. Igartua (✉)
Dpto. de Mecánica y producción Industrial, Escuela Politécnica Superior de Mondragon Unibertsitatea, C/Loramendi 4, 20500 Arrasate-Mondragón, Spain
e-mail: lmarkuerkiaga@mondragon.edu

J. I. Igartua
e-mail: jigartua@mondragon.edu

J. Ganzarain
e-mail: jganzarain@mondragon.edu

Á. Ortiz et al. (eds.), *Engineering Digital Transformation*, Lecture Notes in Management and Industrial Engineering,
https://doi.org/10.1007/978-3-319-96005-0_4

improving the innovation system, the collaboration between the three institutional spheres of industry, academia and government is critical. To achieve this, a model known as the Triple Helix model, which states that academia can play an enhanced role in innovation in increasingly knowledge-based societies (Etzkowitz and Leydesdorff 2000), is used. Furthermore, academic institutions such as universities are catalysts for the enhancement of new industry opportunities is local industry. This is true especially as regional and national governments view the high technology and knowledge-based sectors as a crucial source of direct and indirect employment opportunities for the future (Klofsten and Jones-Evans 2000).

Any university that embraces its role within the triple helix model and adopts the third mission is referred to as an 'Entrepreneurial University' (Philpott et al. 2011). Indeed, the Bayh-Doyle Act of 1980 spurred growth in academic entrepreneurship activities in universities across the United States (Osiri et al. 2013). In response to such government pressure and industry solicitations, universities assumed the additional role of promoting economic development; in pursuit of this goal, intellectual property from universities was exploited to create high technology companies (Etzkowitz 2003). Previous research into academic entrepreneurship activities tended to equate such an activity with the creation and development of a new organization, commonly known as spin-off firm (Klofsten and Jones-Evans 2000). However, the economic and social development was not simply about the outcome of launching spin-offs (Klofsten and Jones-Evans 2000; Grimaldi et al. 2011); it was only one type of academic entrepreneurship activities.

In addition, in recent years a rapid increase in technology based economic development initiatives occurred (Grimaldi et al. 2011), focused mainly on stimulating entrepreneurship in universities through academic entrepreneurship activities. From this baseline, many authors developed their own classifications, such as Louis et al. (1989), Kim (2008) and Philpott et al. (2011). They showed that there are many different academic entrepreneurship activities the fact that spin-off firm formation was only one type within these activities was corroborated; since, other mechanisms such as licensing, patenting, collaborative research, contract research and consulting, as well as ad hoc advice and networking with practitioners (Wright et al. 2008; Grimaldi et al. 2011) are as important as spin-off firm formation.

Therefore, this article shows an experimental methodology for working on one of these academic entrepreneurship activities; the collaborative research. In fact, a pilot project has been developed at the Engineering Faculty of Mondragon University, which promotes an experimental methodology based on tangible business-industry needs. The structure of the article that describes the pilot experience is as follows: in the introduction a reference to the University Business Collaboration approach has been made. Then it continues with a theoretical explanation about University-Business Collaborative Research and the Industrial Ph.D. Lastly, the conclusions obtained.

2 Industrial Relevance Industrial Ph.D. Research

2.1 *University-Business Collaborative Research*

According to Meyer-Krahmer and Schmoch (1998), the Collaborative Research is one of the most important interaction types between academia and Industry. It is also to be noted that "service like research" tends to be ranked higher in the fields which are the more applied ones. In this sense, the academic researchers perceive advantages of laboratories for interacting with industry, mainly in obtaining additional funding and exchanging knowledge, while the disadvantages reside in the short term orientation of industry's agendas.

In addition to this, Lee (2000) found that the most important reasons for collaborating were to test application of a theory and to gain insights in one's own academic research. Lee also studied the perceptions academics have of their gains from collaborating with industry. He showed that a large majority say that from research collaboration with industry they are gaining valuable insights into their research agendas and projects. And they are also agree that they find an opportunity to field test the practical application on their own research and theory. On the other hand, only a small size of faculty members believe that collaborative research with firms offer a significant window for business opportunity for them.

For this, there must be defined collaboration between the industrial sector with its role in managing and as an important actor in generating new employment related with the research, and the university which is responsible for create new research solutions in order to solve industrial future needs.

2.2 *Industrial Ph.D.*

Important expectations fall up upon doctoral programs and industrial doctorates. The growing demands for highly qualified personnel in knowledge based industries create a continuous increase in the number of Ph.D. students. The key role of Industrial Ph.D. involved in defining and creating new research solutions for the industry is crucial. Dialogue with the industry and working together University-Business very closely is a key element in defining new research applied solutions. Due to the strategic nature of industry, researchers may interpret needs in the markets and seek answers to provide innovative solutions to the market, as industry can define collaborating working spaces.

Increasing this University business collaboration often originate more aligned Industrial Ph.D. programs with industry future demands. There is an increase in university business collaboration, which can comprise a range of activities, from direct academic research commercialization such as university spin-off firms and licensing of university held patents, to technical consultancy by universities, with academics conducting contract research commissioned by firms to solve specific technical problems independently, so that a whole group of companies/members can benefit from the research outcome (Lee and Miozzo 2015). These activities may be associated with different outputs. With the emergence of these different University business collaborations, arises the emergence of "entrepreneurial academics" concepts; academics who do not necessarily want to set up their own venture, but who follow alternative paths to pursue their research interests (Meyer 2003).

This is in line with literature on incentives behind different collaborative activities. For projects initiated by firms, the objectives of such research are generally to seek a solution to a technical problem arising within a firm's R&D, manufacturing or other operations, to develop design significations or prototypes for new or improved products or processes, or to provide advice on R&D projects and develop projects pursued within firms (Markus and Kathryn 2008). The objective of exploring a high-risk concept on behalf of a firm-outside the firm's main stream activities is generally initiated by both academia and industry. On the other hand, the objective of generating knowledge in general carrying out research on topics of broad interest to a firm is often mainly the interest of academics. Indeed, firms prefer to initiate collaborative contract research that is short-term, with specific and foreseeable results (Schmoch 1999). Furthermore, such research is normally characterized by one-way knowledge transfer from academia to industry.

Based on an analysis of 46 university–business collaborations in several European countries, showed that industrial partners usually prefer research with lower risk and with higher potential for developing concrete applications within a reasonable timeframe (Carayol 2003). Even when firms decide to go for risky research, the rationale for this still lies in the potential applicability of the research into their product or process development. By contrast, academics often try to exploit synergies between industrial partners' and their own research agendas. Carayol (2003) suggested that the balance may depend on the distance between the academics' own research and their industrial partners' research objectives. These contributions suggest that projects with and without industrial involvement may have different objectives. Consequently, we would expect projects with industrial involvement to be more industrially relevant (Lee and Miozzo 2015).

3 Results: Research Training in the University-Business Collaboration

Doctoral training has always been one of the vital aspects of higher education. Doctoral researchers are, on the one hand, producing academic knowledge, and, on the other hand, receiving research training as competent researchers. Their research training and hence competences are often defined through the doctoral research projects (Lee and Miozzo 2015).

The academic research with industrial involvement might affect the nature of academics' research on fostering university-business collaboration and impact on users affects doctoral students' projects. In SME, academics that have greater industry involvement also support more post-graduate students (Bozeman and Boardman 2013). Doctorates are special types of academic personnel as many will go to industry after graduation, enabling direct knowledge transfer from academia to industry (Mangematin 2001). This means that doctoral research training must meet the dual challenge of preparing future academics and industrial scientists at the same time.

European universities are increasingly embracing an open innovation model whose ambition is to make a more productive use of knowledge, technology and resources, increasingly involving the industry in doctoral education (Santos 2016). In addition, involving the industry in doctoral research training, a doctoral research training was defined in order to achieve the reinforcement of the university business collaboration.

At the same time, one of the objective was to promote entrepreneurial education and training in business skills to technical Ph.D. students within the faculty of engineering. The latter objective involved the creation of a new Business Model and a fostering of new holistic innovation champions through ongoing training. In fact, the training itinerary program was designed to include the participation of Ph.D. students from different European universities. The participants' objective was to develop their business ideas in relation to their technical doctoral thesis.

In general, Ph.D. students in their final stages of their Ph.Ds, Post Doctorate students and researchers have very good technical background in specific knowledge areas. However, there is a distinct lack of the utilization of these skills beyond the classroom setting. In order to overcome this limitation, they must improve their skills in innovation, communication, designing business models according to customer needs and markets. In this vein, the methodology offers a concrete framework for creating and training innovation champions through industry, in order to develop such skills and knowledge that will provide those participants with the capability to boost innovation, generate new business ideas and put the subsequent deliverable on the market.

The concept of itinerary is a plan for a training journey. Within this process a general itinerary will be suggested. Depending on the need of the industry and the background or experiences of the future Doctor, this general itinerary may be customized to the participant/project. It is conceived as learning by doing path. Besides, while participants are working in the agreed project within the industry, they have to create new ideas that will lead to innovation. Also participants are expected to create the associated value proposition and business model. The creation of a business model is an evolution process as seen in the following Fig. 1.

Along with the activities of the above process there will be specific moments to learn and share. Figure 2 illustrates the whole training itinerary process which consists in 3 thematic workshops. The integration of the three workshops relays on the next figure.

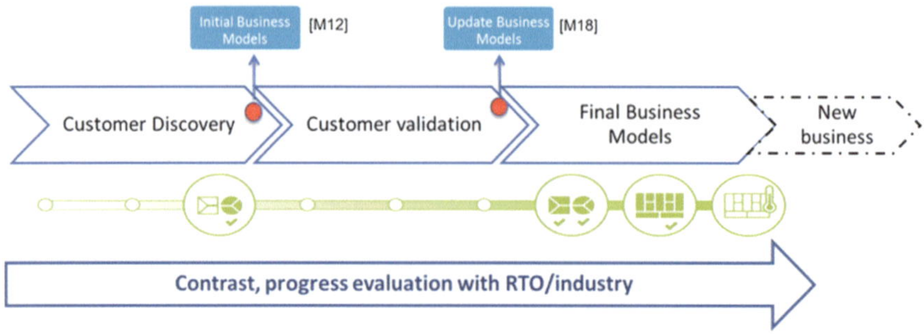

Fig. 1 Structure of the itinerary

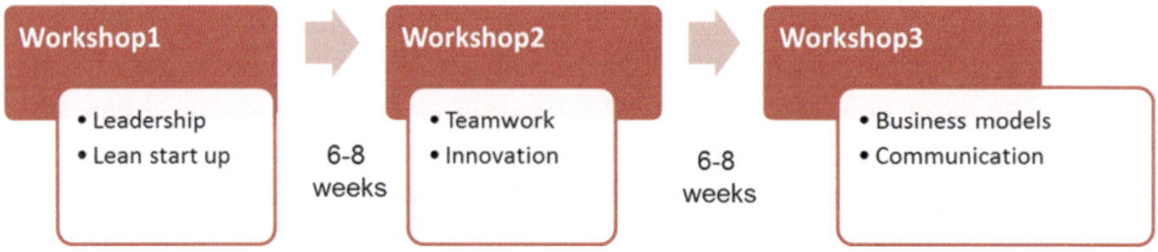

Fig. 2 Workshops in the itinerary

The objectives of the workshops are:

- To create an international community among participants. During these workshop participants will have face to face discussions about their technical projects and how to implement the new topics learn in the workshop to their projects and to their day-to day activities.
- To learn, share experiences and create their own concepts.
- To widen their network, as the workshop will be in different organisations.
- To follow up of activities: Participants will have to elaborate a plan in relation with the topic address in the WS and coaches will assist and follow up the activities.

Regarding the topics of the workshops:

WS1: Leadership and lean start up: The main objective is to work on personal development, Individual and team leadership and teamwork.
WS2: Team work, Innovation and Customer development: Industry proposes lines of research to participants at the beginning. It can happen that solutions have been already generated or not. The aim is to work on possible solutions or in the project they are developing for the industry.
WS3: Business models, marketing and communication: This workshop it to understand marketing, relation with customers, new marketing channels, branding, communication and pitching together with the ability to create business models.

4 Conclusions

The results obtained show the great commitment of participating Ph.D. candidates, as well as the fact all Ph.D. candidates continue with the ongoing improvement of their business model design approach. The results have been successful both for participants and industry-academy mentors due to its practical approach and the industry collaboration scheme developed.

This paper explains the training itinerary and workshops scheme process, of this experience based on student's experience.

This pilot experience has reached its main objective, which was to promote the interaction within industry and university. This project has been a first step, and there is need and desire from both parts to follow this way and to go on searching for new opportunities and experiences that can lead to fruitful relationships.

Acknowledgements We would like to thank the EIT KIC—Raw Materials for their support in the development of T-DORE project, as well as to all participating agents.

References

Bozeman B, Boardman C (2013) Academic faculty in university research centers: neither capitalism's slaves nor teaching fugitives. J High Educ 84:88–120

Carayol N (2003) Objectives, agreements and matching in science–industry collaborations: reassembling the pieces of the puzzle. Res Policy 32:887–908

Etzkowitz H (2003) Research groups as "quasi-firms": the invention of the entrepreneurial university. Res Policy 32:109–121

Etzkowitz H, Leydesdorff L (2000) The dynamics of innovation: from national systems and "mode 2" to a Triple Helix of university-industry-government relations. Res Policy 29:109–123

European Commission (2001) Building an innovative economy in Europe. In: Commission, E (ed). Luxembourg

Grimaldi R, Kenney M, Siegel DS, Wright M (2011) 30 years after Bayh–Dole: reassessing academic entrepreneurship. Res Policy 40:1045–1057

Kim HH (2008) The influence of entrepreneurial activities on teaching at Universities in the United States. Georgia Institute of Technology

Klofsten M, Jones-Evans D (2000) Comparing academic entrepreneurship in Europe—the case of Sweden and Ireland. Small Bus Econ 14:299–309

Lee YS (2000) The sustainability of university-industry research collaboration: an empirical assessment. J Technol Transf 25:111–133

Lee H-F, Miozzo M (2015) How does working on university–industry collaborative projects affect science and engineering doctorates' careers? Evidence from a UK research-based university. J Technol Transf 40:293–317

Louis KS, Blumenthal D, Gluck ME, Stoto MA (1989) Entrepreneurs in academe: an exploration of behaviors among life scientists. Adm Sci Q 34:110–131

Mangematin V (2001) Individual careers and collective research: is there a paradox? Int J Technol Manag 22:670–675

Markus P, Kathryn W (2008) Engaging the scholar: three types of academic consulting and their impact on universities and industry. Res Policy 37:1884–1891

Meyer M (2003) Academic entrepreneurs or entrepreneurial academics? Research–based ventures and public support mechanisms. R&D Manag 33:107–115
Meyer-Krahmer F, Schmoch U (1998) Science-based technologies: university–industry interactions in four fields. Res Policy 27:835–851
Osiri JK, McCarty MM, Jessup L (2013) Entrepreneurial culture in institutions of higher education: impact on academic entrepreneurship. J Entrepreneurship Educ 16:1–11
Philpott K, Dooley L, O'Reilly C, Lupton G (2011) The entrepreneurial university: examining the underlying academic tensions. Technovation 31:161–170
Santos P (2016) Moving the universities to the «Third Mission» in Europe, new impulses and challenges in doctoral education. Foro de Educación 14:107
Schmoch U (1999) Interaction of universities and industrial enterprises in Germany and the United States-a comparison. Ind Innov 6:51–68
Wright M, Clarysse B, Mustar P, Lockett A (2008) Academic entrepreneurship in Europe. Edward Elgar Publishing Limited, United Kindom

Business Model Innovation from a Technology Perspective: A Review

Dorleta Ibarra, Juan Ignacio Igartua and Jaione Ganzarain

Abstract The following article presents a brief literature review conducted to extend our knowledge of what Industry 4.0 means and its impact on business models. Based on the results of the research three challenges related to the digital transformation of the companies are suggested such as: a service-oriented approach, a network-oriented approach and a user-driven approach. Furthermore, the implications on Business Model Innovation based on the identified issues are presented. As a result, four different ways to innovate in business models based on different degrees of innovation are proposed to embrace digital transformation.

Keywords Business model · Business model innovation · Technology
Industry 4.0

1 Introduction

The increasing fusion of industrial production and information and communication technologies has brought the so-called Industry 4.0, also known as Industrial Internet of Things (IIoT), into the manufacturing world (Arnold et al. 2016). This phenomenon is making it possible to connect information, objects and people due to the convergence of the physical world and the virtual world (cyberspace) in the form of Cyber-Physical Systems (CPS) and therefore, it is enabling the transformation of factories into smart environments (Thoben et al. 2017; Kagermann et al. 2013).

D. Ibarra (✉) · J. I. Igartua · J. Ganzarain
Mechanical and Industrial Production Department, Faculty of Engineering,
Mondragon University, Loramendi 4, 20500 Mondragon, Spain
e-mail: dibarra@mondragon.edu

J. I. Igartua
e-mail: jigartua@mondragon.edu

J. Ganzarain
e-mail: jganzarain@mondragon.edu

Á. Ortiz et al. (eds.), *Engineering Digital Transformation*, Lecture Notes in Management and Industrial Engineering,
https://doi.org/10.1007/978-3-319-96005-0_5

In addition, the business world's rapid digitalization is breaking down the traditional barriers of the industry and many academics and practitioners are emphasizing the need to rethink the existing business models (Gerlitz, 2016). Besides, with the rise of the IIoT the world will face a new era of innovation and change. Thus, some authors, are already considering it the most powerful driver of innovation over the next few decades triggering the next wave of innovation (Kaggermann 2015).

Consequently, the discussion and initiatives that promote the digital transformation of factories are increasing between researchers, industries and policy makers around the world. Germany was the first to refer to the digitalization of the industry publicly as "Industrie 4.0" in 2011. The term has been expanded to the Anglo-Saxon world as "Industry 4.0". Furthermore, Unites States focuses on "Smart Manufacturing" as do Japan and Korea. General Electrics popularized the "Industrial Internet" concept. Finally, other commonly related terms found in the literature are "intelligent manufacturing", "advanced manufacturing", "Integrated Industry", "Smart Industry" and "Smart Factory" (Thoben et al. 2017; Hermann et al. 2016).

Hence, in response to the increasing interest of the topic and the many concepts related to it, the need for a deeper understanding of the Industry 4.0 term and its relation with the business model has been identified. Hence, the aim of this paper is to present a brief literature review conducted to address the following goals: To have a clear vision of what Industry 4.0 means and its impacts on business models.

2 Research Methodology

The literature survey was conducted in articles indexed in Web of Science, Engineering Village, Scopus and Business Source Premier Databases, as they are the referents in our field, for the purpose of carrying out a critical analysis of collected data and the presentation of some conclusions and future research opportunities.

The method of analysis used is the one proposed by Becheikh et al. (2006). In order to do this, the steps related to the establishment of the inclusion criteria and the strategy for the selection of potential studies are particularly important. In our case, the criteria used are the following: (1) the content must show the impact of Industry 4.0 from a managerial approach, (2) the research must illustrate opportunities and threats related to the implementation of the Industry 4.0 approach in business models and (3) the papers should not be focused on concrete technological implementations into specific cases.

On the one hand, the time period of the literature review was defined between 2011 and the present, since the term "Industry 4.0" became publicly known in 2011. Furthermore, only conference papers and journal articles related to business and management were taken into account to avoid too technical an approach.

On the other hand, to select the keywords within the search, the synonym terms mentioned above were taken into account. The keywords included in the title, abstract and/or full text were combined using the Boolean operators "AND" and "OR" in the following form: "Business model" AND "Industry 4.0" OR "Industrie 4.0" OR

Table 1 Features, challenges and requirements related to the Industry 4.0

Main features of the Industry 4.0	Main issues affecting traditional business model	Main requirements to face digital transformation
Interoperability (connection and communication between humans and smart factories through the IIoT technologies) Virtualization Decentralization (of decision making) Real-time capability Service orientation Modularity	Networking and reduction of barriers Flexibility and personalization of the product and production system Individualized mass production Local production Low price Smart goods and services Fragmentation of the value chain Globalization and decentralization of production V-H integrated production systems Thoben et al. (2017) Automation Human ingenuity	Standardization (of systems, platforms, protocols, connections, etc.) Work organization Availability of products New business models Know-how protections Availability of skilled workers Research investment Professional development Legal frameworks

"smart manufacturing" OR "advanced manufacturing" OR "Integrated Industry" OR "smart Industry" OR "smart Factory" OR "Industrial Internet" OR "fourth industrial revolution" OR "Intelligent Manufacturing".

In the end, 229 papers were collected. After removing duplicates, a first quick content check was conducted by reading titles and abstracts to check whether the content of the articles was aligned with the criteria mentioned above, excluding the ones that did not meet them.

Bearing in mind all the criteria set, a total of 22 papers were taken into account. In addition, after reading through the lists of references in the reviewed articles, 4 papers that had been missed in the search and provided meaningful insights were included as well. Finally, a total of 26 articles were encompassed.

3 Results of the Literature Review

After reviewing the different approaches from several authors, it has been concluded that notwithstanding the lack of a common definition for the term Industry 4.0, in short, we are able to say that it results in real-time capable, intelligent, horizontal and vertical connection of people, machines, objects, and ICT systems to dynamically manage complex systems (Arnold et al. 2016).

Simultaneously, there are some related issues, features and requirements that seem to be important in the research field. Table 1 summarizes the findings related to this issue (Maslaric et al. 2016; Thoben et al. 2017).

Additionally, this fusion of the physical and the digital world has brought different challenges to the industry. Those challenges are depicted in detail in Sect. 3.1 and later they are related to the Business Model and Business Model Innovation concept in Sect. 3.2.

3.1 Challenges Suggested from the Literature Review

This chapter suggests the challenges identified in the literature review. Each one is described from a different approach despite the fact that they all are interrelated between them. Thus, three proposals such as service-oriented approach, network-oriented approach and user-driven approach can be appreciated below.

3.1.1 A Service-Oriented Approach

The need to rethink the optimal mix of product and service business has been identified, since the digital part of a hybrid solution is always a service. Besides, Business Model patterns which were limited to digital industries until now, are becoming relevant to traditional good sales industries. Consequently, the Industry 4.0 is pushing companies to a change from product to service mindset (Fleisch et al. 2014; Iivari et al. 2016). Moreover, many researchers have suggested that manufacturing firms in developed economies should expand their role in the value chain by extending their products with services so they do not have to compete solely on manufacturing cost. The result is the so-called product-service system (PSS) concept, a framework describing the integrated development, realization, and offering of specific product-service bundles as a solution for the customer. As a result, suppliers, customers, and other partners become part of a networked ecosystem around the CPS.

3.1.2 A Network-Oriented Approach

The horizontal and vertical integration of the value chain and the related interoperability expands firms' traditional boundaries due to the organization and the stakeholders' network. New actors arise and the role of existing ones is changing. As a consequence, new ways of creating and offering value through ecosystems that goes beyond individual value chains are arising. Accordingly, traditional manufacturing companies oriented to product sales, feel increasingly compelled to revise their existing Business Models in response to new competitive dynamics and to tap into those Industry 4.0 inspired opportunities (Iivari et al. 2016).

3.1.3 A User-Driven Approach

This context opens up inroads to make manufacturing more responsive to user-driven design and to align it better with customer value creation processes and contexts (Ehret and Wirtz 2017). From this approach, companies need to develop new capabilities in both learning more about their customers (using digital capabilities to obtain information about customers, promoting evidence-based decision making, developing integral customer experiences, etc.) and becoming more of an ecosystem beyond individual value chains (become great at building partnerships with new stakeholders). Thus, the Industry 4.0 provides opportunities to create new and more flexible value propositions to respond to customer demands such as the provision of individualized products and even batch-size-line production (Arnold et al. 2016). Moreover, an enhanced customer orientation is shown by the expansion of innovative service offering.

3.2 Implications of the Identified Challenges in Business Models

To describe the impact of the challenges mentioned before on the Business Model elements, we make use of the elements which constitute the Business Model Canvas defined by Osterwalder and Pigneur (2010). According to the authors, a Business Model "describes the rationale of how an organization creates, delivers, and captures value". Therefore, the table below summarizes how Industry 4.0 may influence the nine elements constituting the Business Model (Arnold et al. 2016; Laudien and Daxböck 2016; Weinberger et al. 2016).

According to Osterwalder and Pigneur (2010), the customer segments are the different groups of customers the company is targeting. The value proposition seeks to solve customer problems and satisfy customer needs by means of products and services. Value propositions are delivered to customers through communication distribution and sales channels and, as a consequence, a relationship is established. Key resources and activities are the assets and performances needed to develop the value propositions. The key partnerships are the stakeholders that are outside the enterprise. Finally, the revenue streams and the cost structure describes how firm captures economic value and the related costs.

Based on the findings described in Table 2, four ways to conduct the digital transformation in manufacturing companies have been identified according to the innovation degree applied.

1. Internal and External Process Optimization: This transformation represents an incremental innovation that optimizes the current business without involving big changes. New IIoT technologies are introduced just to optimize the value creation architecture (key resources and activities) due to increasing efficiency and improve performance (reducing costs, time and failures, employee training,

Table 2 Influence of the Industry 4.0 in the business model elements

Business model elements	Opportunities through Industry 4.0
Customer segments	Changing roles through networking and co-creation New segment identification due to new product and service offers and value networks
Value propositions	More flexible offers: individualized mass production, product customization, etc. Customer oriented innovative offers through goods-services hybrid solutions New smart and connected product and services
Channels	New channels based on online platforms that provide more transparent and open relationships allowing the interaction and participation of customers and collaborators
Customer relationships	More direct, closer, efficient and long term relationships (due to more flexible offers and service-orientation)
Revenue streams	New revenue models: dynamic pricing, pay-per-use, performance-based revenues, etc.
Key resources	New physical, human and intellectual resources are needed Better maintenance and traceability of resources Use optimization of resources
Key activities	New value creation activities: monitoring, collection and interpretation of data, decentralized processes control, connected objects, pattern-based decision-making, etc. Quality and efficiency improvement in the manufacturing processes, internal and external logistics, marketing, etc. Optimization of production in costs, quality, time, efficiency, reliability, etc. Employee training (from operators to controllers and problem solvers)
Key partnerships	New key players to achieve new required skills and resources Integration of clients and partners in the value chain
Cost structure	Cost optimization due to more efficient processes and use of resources Cost redistribution due to decentralization of activities in new value networks

etc.). This could be the first step for traditional manufacturing companies to embrace the Industry 4.0 without addressing high risks.
2. Customer Interface Improvement: This other incremental innovation is focused on improvement of the customer interface (customer segment, channels and customer relationship). By the introduction of new IIoT technologies, new ways of interaction through new or improved touchpoints are created, allowing a better understanding of customers' needs and greater customer experiences. Once internal and external processes are optimized, this could be the next investment to add more value to the traditional business.
3. New Ecosystems and Value Networks: This model proposes a radical innovation of the current Business Model where new IIoT technologies are introduced so it links the focal firm's value creation process with the stakeholders' processes. Moving from value chains to ecosystems and, as a consequence, increasing stakeholders' knowledge, requires a radical change in many of the Business Model elements (key activities, channels and relationships) with both customer and partnerships. As a result, new ways of capturing value should be defined too.
4. New Business Model 4.0: In contrast to the previous cases, this type proposes a completely new Business Model based on new IIoT technologies that allow companies to offer innovative and smart goods and services (e.g. remote installation of maintenance or activation of product upgrades). Thus, a disruptive innovation that provides the change of almost all the elements of the Business Model is needed. These kinds of new Business Models could be implemented parallel to the current Business Model. In this way, the company could experiment with the new Business Model while the old one still provides revenues.

4 Conclusions and Future Research

As a result of the conducted literature review, the goals defined at the beginning have been fulfilled. We have expanded our understanding about the Industry 4.0 as well as the possible implications for Business Models and Business Model Innovation.

On the one hand, a set of features, issues and requirements have been identified and three different challenges have been suggested to bring firms closer to the Industry 4.0 phenomenon from different approaches such as service orientation, networked ecosystems and customer orientation.

On the other hand, the implications of the Industry 4.0 in Business Model elements have been identified enabling the identification of different ways to transform Business Models. Firstly, the improvement of traditional Business Models due to an incremental innovation of both, value creation architecture and customer interface has been defined. Secondly, the diversification of the current Business Model through the reconfiguration of value networked ecosystems has been described as a radical innovation. Finally, a new Business Model typology based on the *smartization* of the value proposition has been proposed.

To conclude, the need for further research on the topic has been detected to provide a deeper understanding of the process of Business Model Innovation and archetypes derived from the introduction of IIoT in the manufacturing world. Besides, the role of value in new ecosystems and how it is recognized and generated should be analyzed too in order to develop different ways to support the Business Model discovery-driven process. Lastly, a wider survey of methodological approaches, techniques and tools to help firms to deal with the opportunities and threats of the Industry 4.0 should be covered in future research.

Acknowledgements We would like to thank the Basque Government and InterregPOCTEFA for their support in the development of TRANSFERINN project, as well as to all participating agents.

References

Arnold C, Kiel D, Voigt KI (2016) How the industrial internet of things changes business models in different manufacturing industries. Int J Innov Manag 20

Becheikh N, Landry R, Amara N (2006) Lessons from innovation empirical studies in the manufacturing sector: a systematic review of the literature from 1993–2003. Technovation 26:644–664

Ehret M, Wirtz J (2017) Unlocking value from machines: business models and the industrial internet of things. J Mark Manag 33:111–130

Fleisch E, Weinberger M, Wortmann F (2014) Geschäftsmodelle im Internet der Dinge. HMD Praxis der Wirtschaftsinformatik 51:812–826

Gerlitz L (2016) DESIGN management as a domain of smart and sustainable enterprise: business modelling for innovation and smart growth in Industry 4.0. Entrepreneurship Sustain Issues 3:244–268

Hermann M, Pentek T, Otto B (2016) Design principles for Industrie 4.0 scenarios. In: Proceedings of the annual Hawaii international conference on system sciences, pp 3928–3937

Iivari MM, Ahokangas P, Komi M, Tihinen M, Valtanen K (2016) Toward ecosystemic business models in the context of industrial internet. J Bus Models 4:42–59

Kagermann PW, Wahlster W, Helbig D (2013) Recommendations for implementing the strategic initiative INDUSTRIE 4.0. Acatech-National Academy of Science and Engineering, Frankfurt

Kaggermann H (2015) Change through digitization-value creation in the age of the Industry 4.0. Manag Permanent Change 23–45

Laudien SM, Daxböck B (2016) The influence of the industrial internet of things on business model design: a qualitative-empirical analysis. Int J Innov Manag 20:1640014 (28 p.)

Maslarić M, Nikoličić S, Mirčetić D (2016) Logistics response to the Industry 4.0: the physical internet. Open Eng 6:511–517

Osterwalder A, Pigneur Y (2010) Business model generation: a handbook for visionaries, game changers, and challengers. Wiley

Thoben KD, Wiesner SA, Wuest T (2017) "Industrie 4.0" and smart manufacturing—a review of research issues and application examples. Int J Autom Technol 11:4–16

Weinberger M, Bilgeri D, Fleisch E (2016) IoT business models in an industrial context. At-Automatisierungstechnik 64:699–706

The Impact of Real Estate Taxations on the Inter-municipal Migration Which Influences the Housing Construction Dynamics

Petra Janež and Marija Bogataj

Abstract This study investigates the Tiebout-Tullock hypothesis of "voting with one's feet", using NUTS5-level data for the year of the first real estate appraisal in Slovenia, 2011. The general approach is developed and the case study for Slovenia is presented. This analysis differs from previous related studies by Tullock, Cebula and other authors because the income tax burden in Slovenia is equal in all spatial units. However, the housing property tax burdens, which have not been introduced yet, but will be in the near future, are supposed to be different in different municipalities. We have found that the existence of local property taxes has strongly influenced migration patterns over time, therefore it can also influence the dynamics of needed real estate construction. From the study, we can conclude that the flows to the areas of smaller percentage of total population are more sensitive to changing housing property burdens. This finding means that spatial planners, who are planning the future construction of housing, and investors in constructions in the municipalities need to adjust their forecasts regarding the supply and demand of housing units also to the future real estate taxation policies, whereas the local authorities have to be aware how their taxation policy might influence the growth or decline of their urban area. The numerical results are limited by parameters of the gravity model on the NUTS5-level in Slovenia.

Keywords Property tax · Migration pattern · Housing construction dynamics
Spatial interaction · Gravity model

P. Janež (✉) · M. Bogataj
Faculty of Organisation Studies, Ulica talcev 3, 8000 Novo Mesto, Slovenia
e-mail: petra.janez.1@gmail.com

M. Bogataj
e-mail: marija.bogataj@guest.arnes.si

M. Bogataj
CERRISK, Vrtača 9, 1000 Ljubljana, Slovenia

Á. Ortiz et al. (eds.), *Engineering Digital Transformation*, Lecture Notes in Management and Industrial Engineering,
https://doi.org/10.1007/978-3-319-96005-0_6

1 Introduction

According to Tiebout (1956, p. 418), migrants move to the community whose local government best satisfies their set of preferences. Tullock (1971, p. 917) further extended this hypothesis by stating that *»ceteris paribus, the individual deciding where to live will take into account the private effects upon himself of the bundle of government services and taxes«*. It means that the consumer-voter evaluates both the government goods and services and the tax burden at potential locations of choice. This conclusion was further investigated by Cebula (2009), studying the net in-migration rate determinants for the period 2000–2005, using state-level data for the United States. He provided a broader interpretation of "taxes" as implicitly stressed by Tullock. While he included both local property tax burdens and state income tax burdens, our study is limited by the fact that the income tax burdens are equally allocated in all studied areas (municipalities of Slovenia), and the local property taxes will be introduced in near future and could differ among municipalities.

Our early studies have empirically addressed economic and environmental determinants of internal migration in Slovenia (Bogataj et al. 1995; Bogataj and Drobne 1997; Drobne 2003, 2005). In the United States, these topics have been elaborated by Cebula and many other authors (Cebula and Belton 1994; Cebula 2005, 2009; Cebula and Alexander 2006; Davies et al. 2001). Later the theory was further developed by Bogataj and Drobne (2005) and empirically addressed by Drobne and Bogataj (2005, 2009, 2011, 2013), Drobne et al. (2008a, b, 2013), Drobne (2014), Drobne and Lakner (2014). In a further theoretical and empirical study of Slovenian migration and commuting flows, fiscal factors were also studied in detail by Janež et al. (2016, 2017). Here we shall present that migration flows to the municipalities with decreased real property taxation tend to increase if other indicators which influence migration are unchanged.

2 The Model

2.1 The Normalised Spatial Interaction Model

We can analyse migration flows between municipalities and therefore the growth or decline of certain urban areas, using the Normalised Spatial Interaction Model—N-SIM, developed by Drobne and Bogataj (2011) in any economic area consisting of many smaller spatial units: the impact of factors influencing stickiness in the origin and the impact of factors influencing attractiveness in destination (attractiveness) on

$$M_{ij} = kK\left(d(t)_{ij}\right)^{\beta} \prod_{i} K(g)_i^{\gamma(g)} \prod_{j} K(g)_j^{\alpha(g)}$$
$$i = 1, 2, \ldots, n; \quad j = 1, 2, \ldots, n; \quad i \neq j \qquad (1)$$

where M_{ij} is the intensity of migration flow (number of migrants per year) from municipality i to municipality j; k is the proportionality constant of the model; $K(d(t)_{ij})$ is the coefficient of the travel time by car between the municipality of origin i and municipality of destination j; $K(g)_i$ and $K(g)_j$ are the coefficients of the analysed independent variables g, such as the number of inhabitants in the municipality in the origin i or in the destination j, the employment rate or the its growth in the municipality of origin and destination, gross or net earnings in the area of origin or destination, environmental factors, factors of safety and many others. The coefficients of variables are the ratios between the value of variables in the analysed spatial unit and the average of the values in the analysed spatial area. Among these values, we are focused on the values of real estate which have been found to be an important indicator of attractiveness and stickiness of spatial units in many analysed spatial areas.

2.2 The Market Value of a Real Estate Property as a Capitalization of Rent

A property tax is a levy on the value of a property. This value is often calculated as a capitalization of a potential rent from a property, or in this case housing units. Many countries, regions and local communities levy property tax on real estate based upon land use and value. This is usually a major source of revenue for local governments. Taxation of potential real estate rents influences the actual market value of the property while the rent is often considered less dependent on taxes, being mostly dependent on the distribution of incomes of families who need to rent property (McDonald and McMillen 2011). Therefore, areas with higher income could have higher rents. Rent variation depends more on income variation than on taxation. This assumption will be included in our further study.

The market value of a residential property V is determined by the present discounted value of the stream of rents, whereby the discount factor includes real estate taxes see Eq. (2). For the sake of simplicity, let us presume that this is a ground rent (perpetual annuity), but similarly, we could also consider the time limitation in the real estate use. If R stands for the annual amount of ground rent, r is the profitability rate and the annual real estate tax is expressed as the percentage u of market value V. Here the usability of a residential unit goes to infinity, therefore the relationship between the market value and rent can be written as:

$$V = \frac{(R - u \cdot V)}{r} \rightarrow V \cdot r = R - u \cdot V \rightarrow V = \frac{R}{r + u}, \tag{2}$$

where $r + u$ is the capitalization rate. A more detailed explanation of this assumption is explained by McDonald and McMillen (2011).

2.3 The Variation of Migration Flows

Based on our previous experience (Janež et al. 2016), the following factors have been found to be very significant: $K\left(d(t)_{ij}\right)$—the coefficient of distances; $K(INH_{\circ})$—the coefficient of the number of inhabitants in the spatial unit; $K(EMP_{\circ})$—the coefficient of employment in the spatial unit; $K(GPI_{\circ})$—the coefficient of average gross personal income per capita in the municipality; $K(MR_{\circ})$—the coefficient of the municipal revenue per capita in the municipality regarding average of the total studied area (state); $K(URA_{\circ})$—The coefficient of the usable residential area expressed in square metres per capita in the municipality regarding the average spatial standard in the state. All these factors are assumed to be unchanged while $K(V_{\circ})$—the coefficient of the current average net market price per square metre of residential area in the municipality will be variable, depending on the changing taxation u to u^*.

In this case, the spatial interaction model (1) can be written as:

$$\begin{aligned} M_{ij} = k\ & K\left(d(t)_{ij}\right)^{\beta} K(INH)_i^{\gamma(INH)} K(INH)_j^{\alpha(INH)} K(EMP)_i^{\gamma(EMP)} \\ & \cdot K(EMP)_j^{\alpha(EMP)} K(GPI)_i^{\gamma(GPI)} K(GPI)_j^{\alpha(GPI)} K(V)_i^{\gamma(V)} \\ & \cdot K(V)_j^{\alpha(V)} K(MR)_i^{\gamma(MR)} K(MR)_j^{\alpha(MR)} \\ & \cdot K(URA)_i^{\gamma(URA)} K(URA)_j^{\alpha(URA)} \end{aligned} \quad (3)$$

We assume that only the municipality j introduces higher taxation. In case that $V_j = \frac{R_j}{r+u}$ the value decreases to $V_j^* = \frac{R_j}{r+u^*}$.

If $p_j\%$ of all inhabitants live in this municipality j and $\left(100 - p_j\right)\%$ of inhabitants live in other spatial units, then the new average is:

$$\bar{V}^* = (p_j V_j^* + (100 - p_j)\bar{V}/100.$$

Therefore, we can write for $\forall i$:

$$\begin{aligned} M_{ij}^*/M_{ij} &= \left(\frac{V_j^*}{V_j}\frac{\bar{V}}{\bar{V}^*}\right)^{\alpha(V)} = \left\{\frac{r+u}{r+u^*}\frac{100\ \bar{V}}{\frac{p_j(r+u)}{r+u^*} + \left(100 - p_j\right)\bar{V}}\right\}^{\alpha(V)} \\ &= \left\{\frac{r+u}{r+u^*}\frac{100}{\frac{p_j(r+u)}{r+u^*} + \left(100 - p_j\right)}\right\}^{\alpha(V)} \end{aligned} \quad (4)$$

The important remarks in Eq. (4) are the following:

(1) The percentage of population living in the spatial unit j is always much smaller than 100 and always positive, therefore when taxes increase, the denominator $\frac{p_j(r+u)}{r+u^*} + \left(100 - p_j\right)$ decreases and $\frac{100}{\frac{p_j(r+u)}{r+u^*}+\left(100-p_j\right)}$ increases over 1, while $\frac{r+u}{r+u^*}$ decreases below 1, but there are no extreme values inside the defined interval of $\frac{r+u}{r+u^*}$ for the real cases.

Table 1 The parameters of the migration flow model between Slovenian municipalities in 2011

R	0.641		Observations	43,890	
R^2	0.411		ANOVA stat. F	2,354.13	
Adjusted R^2	0.411		ANOVA sign. p	<0.001	
Standard error	2.636				
Parameters	Est. value	t	Parameters	Est. value	t
Constant	2.12E−21	−34.519	$\alpha(GPI_j)$	0.400	0.122
$\beta(d(t)_{ij})$	−2.388	−105.620	$\gamma(V_i)$	0.875	0.125
$\gamma(INH_i)$	1.297	67.555	$\alpha(V_j)$	**0.787**	**0.125**
$\alpha(INH_j)$	1.257	65.452	$\gamma(RM_i)$	1.420	0.066
$\gamma(EMP_i)$	−0.356	−5.856	$\alpha(RM_j)$	1.375	0.066
$\alpha(EMP_j)$	−0.345	−5.674	$\gamma(URA_i)$	0.094	0.040
$\gamma(GPI_i)$	−0.501	4.106	$\alpha(URA_j)$	0.233	0.040

(2) If $\alpha(V)$ is positive then M^*_{ij}/M_{ij} always decreases when u^* increases on the acceptable interval in real life.
(3) If $\alpha(V)$ is negative then M^*_{ij}/M_{ij} always increases when u^* increases on the acceptable interval.

The main finding of this study is that the migration variable (expressed now in a simple form of the coefficient M^*_{ij}/M_{ij}) is an increasing function of the property tax burden only in the conditions where $\alpha(V) < 0$.

3 Numerical Example

The model calibration (1) was performed for Slovenia using the SPSS 23.0 computer software and the Ordinary Least Squares Regression model. The effects on internal migrations between Slovenian municipalities were estimated in the regression analysis. The spatial interaction model (1) has the following parameters, as listed in Table 1.

From Table 2 we can conclude that the flows to the areas of smaller percentage of total population are more sensitive to the changes in the housing property tax burden.

Here we can see that $\alpha(V_j) > 0$, therefore with increasing u^*, because p_j is small and always much less than 100, the flow will always decrease.

The case for r = 2%, u = 0.09 and $\alpha(V_j) = 0.787$ is given in Table 2.

For example, if housing tax burden increased from 0.09% of property value per year (as would be the case at all other municipalities) to 0.225% only at municipality j,

Table 2 Matrix of coefficient of internal migrations M_{ij}^{*}/M_{ij} in case of changing the real estate tax burden on interval [0.075, 0.225], at basic value of u = 0.09 by changing p_j on interval [0.001–0.3]

p_j							
u*	(r+u)/(r+u*)	0.001	0.01	0.05	0.1	0.2	0.3
0.07500	1.00723	1.00568	1.00568	1.00568	1.00568	1.00567	1.00567
0.09000	1.00000	1.00000	1.00000	1.00000	1.00000	1.00000	1.00000
0.12000	0.98585	0.98885	0.98885	0.98885	0.98886	0.98887	0.98888
0.15000	0.97209	0.97797	0.97797	0.97798	0.97799	0.97801	0.97804
0.18000	0.95872	0.96736	0.96737	0.96738	0.96740	0.96743	0.96746
0.20000	0.95000	0.95193	0.96044	0.96046	0.96047	0.96051	0.96055
0.22500	0.93933	0.95193	0.95194	0.95196	0.95198	0.95202	0.95207

the migration inflow to this municipality would decrease by more than 4.8% if this was a very small municipality and by a little less than 4.8% in communities where the percentage of inhabitants was over 15% of all citizens in Slovenia (the biggest is Ljubljana: 14%). Such reduction would reduce the dynamics of construction of housing units.

4 Conclusion

In this study we have investigated the Tiebout-Tullock hypothesis of "voting with one's feet." According to Tiebout, the migrants move to the community whose local government best satisfies their set of preferences. Tullock further extended this hypothesis by stating that »*ceteris paribus, the individual deciding where to live will take into account the private effects upon himself of the bundle of government services and taxes*«. It means that the consumer-voter evaluates both the government goods and services and the tax burden at potential locations of choice. We started our investigation with the hypothesis that when we reduce the real estate taxes, the in-migration flow to such a municipality will increase. But we cannot simply generalise this hypothesis. The dynamics depend on the value of $\alpha\left(V_j\right)$ in the normalised spatial interaction model. If this value is positive, the hypothesis can always be accepted. The sensitivity of flows also depends on size—the number of inhabitants who live in incoming spatial units.

Using NUTS5-level data for the year of the first real estate appraisal in Slovenia, 2011, the Tiebout-Tullock hypothesis of "voting with one's feet." was analysed by developing the normalised spatial interaction model NSIM for Slovenia. This analysis differs from previous related studies by Tullock, Cebula and other authors because the income tax burden in Slovenia is equal in all spatial units. However, the housing property tax burdens, which will be introduced in Slovenia in the near future, are supposed to be different in different municipalities. We have found that the existence of a local property tax has strongly influenced migration patterns over time, therefore it can also influence the dynamics of needed real estate construction. Different states could have different values of parameters in the NSIM model, but it is assumed that the power $\alpha(V_j)$ is always positive. It would be advisable to study further the influence on $\alpha(V_j)$. What if the power of attractiveness regarding the value of real estate is negative? In this case the Tiebout-Tullock hypothesis cannot simply be accepted as it is. Knowing more about the behaviour of power $\alpha(V_j)$ would help to better evaluate the dynamics of housing demand at changing tax policies.

References

Bogataj M, Drobne S (1997) The influence of investments in highways on gravity and interaction in Slovenia. In: Proceedings, the 4th international symposium on operational research in Slovenia, Ljubljana: Slovenian Society Informatika, SOR, pp 55–60

Bogataj M, Drobne S (2005) Does the improvement of roads increase the daily commuting? Numerical analysis of Slovenian interregional flows. Selected decision support models for production and public policy problems, (SDI-SOR series, no. 3). Slovenian Society Informatika, Section of Operational Research, Ljubljana, pp 185–206

Bogataj M, Drobne S, Bogataj L (1995) The influence of investment and fiscal policy on growth of spatial structure. Suvremeni promet 15:239–245

Cebula RJ (2005) Internal migration determinants: recent evidence. Int Adv Econ Res 11(2):267–274

Cebula RJ (2009) Migration and the Tiebout-Tullock hypothesis revisited. Am J Econ Sociol 68(2):541–551

Cebula RJ, Alexander GM (2006) Determinants of net interstate migration, 2000–2004. J Reg Anal Policy 27(2):116–123

Cebula RJ, Belton WJ (1994) Voting with one's feet: an analysis of public welfare and migration of the American Indian. Am J Econ Sociol 53(3):273–280

Davies S, Greenwood MJ, Li H (2001) A conditional logit approach to U.S. state-to-state migration. J Reg Sci 41(2):337–357

Drobne S (2003) Modelling accessibility fields in Slovene municipalities. In: Zadnik Stirn L, Drobne S (eds) Proceedings of the 6th international symposium on operational research in Slovenia, SOR '03 proceedings. Slovenia Society Informatika (SDI), SOR, Ljubljana, pp 89–96

Drobne S (2005). Do administrative boundaries fit accessibility fields in Slovenia? In: Cygas D, Fröhner KD (eds) Environmental engineering, the 6th international conference, Selected papers. University Press Technika, Vilnius, pp 537–542

Drobne S (2014) Impact of the recession on the attractiveness of urban and rural areas of Slovenia. Geodetski vestnik 58(1):103–139

Drobne S, Bogataj M (2005) Intermunicipal gravity model of Slovenia. In: Proceedings of the 8th international symposium on operational research in Slovenia, SOR '05 Proceedings. Slovenian Society Informatika, Ljubljana, pp 207–212

Drobne S, Bogataj M (2009) Razvojna ogroženost regij in stalne selitve. In: Nared J, Perko D (eds) Razvojni izzivi Slovenije, (Regionalni razvoj, 2), ZRC, Ljubljana, pp 285–294

Drobne S, Bogataj M (2011) Accessibility and flow of human resources between Slovenian regions. Mathematical economics, operational research and logistics, serial no. 11. Faculty of Civil and Geodetic Engineering, & Šempeter pri Gorici: MEDIFAS, Ljubljana

Drobne S, Bogataj M (2013) Vpliv recesije na parametre kakovosti regionalnih središč in njihovo privlačnost. Revija za univerzalno odličnost 2(2):A25–A42

Drobne S, Lakner M (2014) Which distance-decay function for migration and which one for commuting? A case study of Slovenia. Croatian Oper Res Rev 5(2):259–272
Drobne S, Bogataj M, Bogataj L (2008) Spatial interactions influenced by European corridors and the shift of the Schengen border regime. In: KOI 2006 proceedings, the 11th international conference on OR. Croatian Operational Research Society, Zagreb, pp 185–198
Drobne S, Bogataj M, Lisec A (2008) The influence of accessibility to inter-regional commuting flows in Slovenia. In: Bernard L (ed) Taking geoinformation science one step further, Girona, pp 1–12
Drobne S, Rajar T, Lisec A (2013) Dynamics of migration and commuting to the urban centres of Slovenia, 2000–2011. Geodetski vestnik 58(2):313–353
Janež P, Bogataj M, Drobne S (2016) Impact of the real estate taxation and municipal revenue on dynamics of internal migration: case study for city municipal of Ljubljana. Geodetski vestnik 60(4):644–684
Janež P, Bogataj M, Drobne S (2017) Forecasting dynamics of daily commuting to work to other municipality in the case of changing taxation policies. In: Proceedings CIO-ICIEOM-IIE-AIM (IJC 2016), San Sebastián, Spain (to appear)
McDonald JF, McMillen DP (2011) Urban economics and real estate: theory and practice. George Hoffman Publisher, Danvers, MA
Tiebout CM (1956) A pure theory of local expenditures. J Polit Econ 64(1):416–424
Tullock G (1971) Public expenditures as public goods. J Polit Econ 79(5):913–918

Developing a Pre-scale for Evaluating Supervisors' Directive Style in Continuous Improvement Environments

Lucía Lomas Echeveste, Álvaro Lleó de Nalda, Elisabeth Viles Díez and Daniel Jurburg Melnik

Abstract The level of workers' commitment and participation has a huge impact on the success of continuous improvement initiatives. One of the non-written duties of middle managers is creating a trust-culture that enhances commitment and participation, which can be achieved with the appropriate directive style. In this paper, we present a pre-scale of four dimensions and fifty-five middle managers' behaviors that encourage employee commitment towards and participation in continuous improvement activities. The list has been developed putting together the opinions of operators and supervisors, as well as businessmen, university professors and consultants.

Keywords Middle manager behaviors · Participation · Continuous improvement Commitment · Group dynamics

1 Introduction

Continuous improvement (CI) can be defined as 'the inter-related group of planned, organized and systematic processes of constant change across the entire organization, focused on engaging everyone inside the organization into achieving greater business productivity, quality, safety, ergonomics and competitiveness' (Jurburg et al. 2016).

In order to attain operational excellence, many companies decide to implement continuous improvement systems. Even though these activities focus on constant

L. L. Echeveste (✉) · Á. L. de Nalda · E. V. Díez
Industrial Management Department, Tecnun – University of Navarra, San Sebastian, Spain
e-mail: llomas@tecnun.es

Á. L. de Nalda
e-mail: alleo@tecnun.es

E. V. Díez
e-mail: eviles@tecnun.es

D. J. Melnik
CINOI, University of Montevideo, Montevideo, Uruguay
e-mail: djurburg1@um.edu.uy

Á. Ortiz et al. (eds.), *Engineering Digital Transformation*, Lecture Notes in Management and Industrial Engineering,
https://doi.org/10.1007/978-3-319-96005-0_7

improvement and innovation, which make organizations more competitive, sustaining CI systems over time is not easy (Rapp and Eklund 2007). In particular, the lack of employee commitment towards CI is regarded as one of the main factors hindering the success of CI systems (Assen 2016).

Existing literature shows two possible ways of improving the level of employee commitment that stand out. One could be creating and organizational context (based on a series of organizational variables affecting the design of daily CI management systems) that promotes employees' intention to participate in CI activities (Jurburg et al. 2016). The second approach involves understanding which are the adequate managerial practices that strengthen individual and collective participation (Lam et al. 2015). Both approaches somehow share the belief that certain behaviors and actions occurring between employees and their supervisor (also regarded as middle managers in most cases) are determinant to achieve employee commitment with CI. However, not many authors have investigated about which specific managerial practices could foster employee commitment towards the continuous improvement system.

Therefore, this papers' aim is to provide deeper understanding of how supervisors should behave in order to encourage employee participation. In this study, we put together the opinion of both people experienced in continuous improvement, as well as experts coming from the business, university and consultancy world. As a result, we present a pre-scale with four dimensions and 55 different middle manager specific behaviors that enhance employee participation and commitment.

2 Context and Purpose

Given that one of the main issues regarding CI sustainability is the importance of achieving the active involvement of everyone within the organization (Garcia-Arca and Prado-Prado 2011), we undertook an in-depth study focused on the human role in CI activities (Jørgensen et al. 2008).

CI literature identifies different organizational factors that are related to employee participation and commitment to CI activities (Jurburg et al. 2016). In particular, some of these factors relate with interpersonal relations in the work environment such as trust, social influence, organizational support or workplace satisfaction.

When analyzing key enablers of CI success, there are many authors that list top manager support as one of the most important factors (Jaca et al. 2012). However, beyond top managers, given the key role that middle managers play in any organizational structure, understanding the importance they have when promoting employee commitment towards CI activities is worth studying (Holmemo and Ingvaldsen 2015). Connected with these managers' behaviors, one factor that directly predicts employee commitment is the level of trust generated by their supervisor (Dirks and Ferrin 2002; Song et al. 2009). It seems valuable, to have a list of specific middle managers' conducts for reflecting, evaluating and developing a directive style that potentiates operators' participation and commitment to CI tasks.

3 Methodology

In order to obtain the most accurate group of behaviors (assuring content validity of the final list of behaviors), the research was structured based on a series of group dynamics aimed at gathering the voice of the different key players, in this case: shop floor workers, middle managers, consultants, experienced practitioners and academics. Taking this objective into consideration, a first round based on a concept mapping technique with shop floor workers and middle managers was conducted, followed by a second round based on an internal reflection of the research team, and a final round based on an Expert Panel took place.

3.1 Round 1: The Concept Mapping with Shop Floor Workers and Middle Managers

First, two different concept mappings were carried: one with middle managers and another one with operators. This technique was used in order to make a particular concept explicit and operational based on the personal knowledge of a group of participants (Windsor 2013). Once the group was together, we presented the following question: *what specific supervisor trustworthy behaviors foster worker commitment to and participation in continuous improvement activities?*

With the responses, a list of statements was created, and then each participant was asked to classify them into different groups according to his/her personal experience, and to evaluate the importance of each affirmation within that group.

Afterwards, using a multidimensional scaling and cluster analysis, the perceptual maps that delimited the concept being studied could be determined. This way, it was possible to collect the participants' opinions in a structured way by reducing the inherent subjectivity of the group moderator (Bigne et al. 2002).

3.2 Round 2: Reflection Process

The Concept Mapping helped arise one list of behaviors from the operators' point of view and another one from middle managers' point of view. On the one hand, operators grouped 51 different behaviors into four dimensions. On the other hand, middle managers agreed on a list of 87 behaviors grouped into five dimensions.

Based on these results, and with the aim of answering the investigations' objective, the research team needed to group all items into one unified list of managerial trustworthy behaviors. Following an internal reflection process (similar to a focus group dynamic), the research team decided to maintain the original dimensions created by the supervisors, because it was a more detailed classification, and the statements of the operators' list could be easily redistributed within the middle managers' dimensions. The research group also worked towards eliminating very similar and redundant elements from both lists.

The outcome of this phase was a list of 64 behaviors distributed in five dimensions that condensed all the ideas, feelings and opinions previously expressed by shop floor workers and middle managers. Special care was taken to include only clear and unambiguous statements.

3.3 Round 3: Expert Panel

Given that the research objective was to arrive to an agreed list of managerial trustworthy behaviors to improve employees' participation and commitment towards CI activities, the research team decided to gather the voice of all relevant players in this topic. During Round 1 and Round 2, the voice of operators and middle managers was obtained, mediated by the experience knowledge of the research team. In Round 3, the target was to harvest the experienced opinion and knowledge from other key players such as HR and CI consultants, experienced practitioners and managers, and academics related with industrial management research areas. To achieve this objective, an expert panel was formed, and a structured questionnaire was developed.

Expert panels are recommended when a specialized or very specific opinion is required for a certain topic (Melnyk et al. 2009). In general, these panels are formed by a variety of experts (in this case consultants, practitioners and academics) who are engaged based on various fields of expertise (in this case HR, CI and industrial management) to express their opinions and make recommendations about a complex topic.

We selected a first sample of 20 experts and 12 of them accepted our invitation. Afterwards, we provided them with a questionnaire with the 64 behaviors, and three questions were asked regarding the list.

- Assess from 0 (not important) to 10 (very important) the weight of each of the behaviors within each of the five dimensions it belongs to
- Would you change this behavior to another existing group? Do you think a new group should be created, and move this item to this new group?
- Would you eliminate this statement? Give reasons for elimination

An additional field was created for experts to be able to provide any additional information or opinion about the list of behaviors or about the research itself and to add behaviors in case they thought something was missing.

4 Results

After conducting all the different phases of the research, the results can be found in Table 1. According to Table 1, the most important supervisor dimension regards the

Human qualities, which ensure that the person is educated, approachable, fair and consistent. It is also important for the middle manager to invest time and effort in training and developing people in order to increase their autonomy, which is reflected in the Training and development group.

Equally important to this group is the Technical and managerial competence of the supervisor, ensuring the technical competence and making sure the he/she knows how to manage a continuous improvement project. Last, but not least, it is important that middle manager behave in a way that promotes Team-building, having the ability to build a team, integrating himself/herself and getting involved within the group.

Given the scoring of each dimension, we can conclude that in order to have good leaders that foster commitment among employees, companies must not only promote the human qualities of their leaders or make sure they are competent, but also teach them how to train and develop their workers, as well as make sure they are team players and know how to build a team.

5 Discussion and Conclusion

Based on a multiple step procedure to gather the voice of key players' experience and experts' knowledge, this paper identifies 55 managerial trustworthy behaviors that help supervisors encourage their employees and make them more committed towards participating in continuous improvement activities within a company.

Comparing the results with existing management literature, Butler (1991) had already identified ten managerial trustworthy behaviors (availability, competence, consistency, discreetness, fairness, integrity, loyalty, openness, promise fulfilment and receptivity). Most of Butler's behaviors (except for competence) belong to our dimension of Human qualities. Based on our results, this research extends Butler's view of the human qualities needed to be a good leader, by highlighting the importance of also being calm and thankful.

Regarding Training and development, on one hand, Nyhan (2000) already talked about empowerment and feedback. Yet, our study also includes other variables such as forgiveness and error management, which had not been included in previous research related with trustworthy managerial behaviors in continuous improvement.

The behaviors in Technical and managerial competence, are aligned with the findings of Dirks and Ferrin (2002) that talked about competence and communication. However, our research also emphasizes the importance of supporting the promotion of the people with the best performance.

Finally, some authors had mentioned the importance of being receptive (Butler 1991) and caring (Dirks and Ferrin 2002), items included in our Team Builder dimension, but there is scarce literature about the importance of spending time in the workshop with operators, a behaviour highlighted as necessary in this study.

Moving forward, our study not only groups all the behaviors in four different dimensions, but also identifies two major abilities that every good leader must have. On the one hand, managers need to be task focused and professionally competent,

Table 1 Final list of managerial trustworthy behaviors, with scoring of the Delphi study

Human qualities	9.04
My supervisor…	
Looks for a solution whenever there is a problem, instead of looking for the one to blame	10.00
Apologizes for his/her own mistakes and is consistent	9.64
Fulfils the commitments undertaken towards the staff	9.60
Is a straightforward person, and is sincere	9.45
Knows how to maintain worker-boss confidentiality	9.09
Communicates with diplomacy, tact, and patience; knows how to treat each person	9.09
Acts towards people the way my supervisor wants people to act towards him/her	9.09
Treats people with humanity, paying attention to peoples' personal issues	9.00
Is grateful	8.91
Is available, approachable and communication with him/her is direct	8.80
Knows how to separate personal and professional life	8.73
Bases his/her opinions on facts	8.64
Shows calm and composure; he/she is a positive person	8.36
Knows how to say "no", whenever he/she thinks it is necessary	8.09
Training and development	8.83
My supervisor…	
Appreciates well done work, and teaches how to do things well whenever they are done wrong	9.40
Educates others about error management (each person is responsible for his/her own errors) and knows how to forgive them	9.33
Seeks workers' feedback and is ok with being assessed	9.20

(continued)

Table 1 (continued)

Gets involved in workers' professional development, challenging them and helping them grow	8.90
Teaches and shares his/her knowledge and experience	8.90
Looks after his/her team's interests when dealing with management and communicates the teams' proposals	8.80
Delegates providing autonomy, but assuming his/her corresponsibility part	8.10
Puts a lot of effort into peoples' initial training	8.00
Technical and managerial competence	8.79
My supervisor…	
Instils strength and believes in the project	9.60
Listens to proposals, keeps workers informed, and responds to the proposals	9.40
Does not create unachievable high-standards	9.20
Is ready to solve difficult situations whenever they come up	9.11
Cares about workers having respectable working conditions and the appropriate tools	9.10
Has team management skills and is organized	9.00
Informs workers of goals, expected results and the importance of the orders he/she gives	8.90
Instils safety at work	8.90
Gives quick answers to problems; is effective, fast and efficient	8.80
Is a serious person at work	8.80
Supports promoting the people with the best performance	8.70
Anticipates future needs before they come up	8.70
Insists on the importance of continuous improvement and is committed to it	8.22
Communicates information about the company, clients and suppliers	8.10
Is technically competent (understands the machines and tasks)	7.30

(continued)

Table 1 (continued)

Team builder	8.77
My supervisor…	
Does not step on other workers	9.78
Communicates successes and failures (internal and external recognition) to the team, and makes them collective	9.67
Listens to everyone's opinion, and appreciates each participation (shows that everyone's opinion counts)	9.33
Stands shoulder to shoulder, gives and not only expects to receive	9.14
Corrects behaviors acting against the group	9.10
Is a role model	9.10
Values everyone's work	9.00
Gets involved (collaborates, participates) in others' ideas	9.00
Knows everyone's strengths and weaknesses	8.90
Supports workers both at work and in their personal life (we all have bad days)	8.80
Leads rather than orders	8.80
Fosters creativity, letting people do things by themselves	8.78
Shows interest in how workers feel when the work is accomplished	8.33
Is in the garage along with the workers, side by side	8.30
Does not allow behaviors that go against the organization's culture (principles)	8.22
Is not afraid of communicating to the team their limitations	8.00
Knows the relationship between team-members	7.80
Promotes camaraderie and does not create rivalry within the group	7.80

expressed by the Technical and managerial competence and Training and development dimensions. On the other hand, good managers also need to develop their most personal or human side, expressed by the Human qualities and the Team builder dimensions.

Further research steps would be to conduct an empirical study for testing this measurement instrument. The validated scale would be useful for managers, in order to analyze and develop a managerial style that promotes employee participation and commitment in continuous improvement initiatives.

References

Assen MF (2016) Exploring the impact of higher management leadership styles on Lean management. Total Qual Manage Bus Excellence, Published online: 10 Nov 2016. https://doi.org/10.1080/14783363.2016.1254543

Bigne JE, Aldas-Manzano J, Küster I, Vila N (2002) The concept mapping approach in marketing: an application in the travel agencies sector. Qual Market Res Int J 5(2):87–95

Butler JK Jr (1991) Toward understanding and measuring conditions of trust: evolution of a conditions of trust inventory. J Manage 17(3):643–663

Dirks KT, Ferrin DL (2002) Trust in leadership: meta-analytic findings and implications for research and practice. J Appl Psychol 85:611–628

Garcia-Arca J, Prado-Prado JC (2011) Systematic personnel participation for logistics improvement: a case study. Human Factors Ergon Manuf Serv Ind 21(2):209–223

Holmemo MDQ, Ingvaldsen JA (2015) Bypassing the dinosaurs? How middle managers become the missing link in lean implementation. Total Qual Manage Bus Excellence, Published online: 26 Aug 2015. https://doi.org/10.1080/14783363.2015.1075876

Jaca C, Viles E, Mateo R, Santos J (2012) Components of sustainable improvement systems: theory and practice. TQM J 24(2):142–154

Jørgensen F, Hyland P, Kofoed LB (2008) Examining the role of human resource management in continuous improvement. Int J Technol Manage 42(1/2):127–142

Jurburg D, Viles E, Tanco M, Mateo R (2016) What motivates employees to participate in continuous improvement activities? Total Qual Manage Bus Excellence, Accepted for publication. https://doi.org/10.1080/14783363.2016.1150170

Lam M, O'Donnell M, Robertson D (2015) Achieving employee commitment for continuous improvement initiatives. Int J Oper Prod Manage 35(2):201–215

Melnyk SA, Lummus RR, Vokurka RJ, Burns LJ, Sandor J (2009) Mapping the future of supply chain management: a Delphi study. Int J Prod Res 47(16):4629–4653

Nyhan RC (2000) Changing the paradigm: trust and its role in public sector organizations. Am Rev Public Adm 30(1):87–109

Rapp C, Eklund J (2007) Sustainable development of a suggestion system: factors influencing improvement activities in a confectionary company. Human Factors Ergon Manuf Serv Ind 17(1):79–94

Song JH, Kim HM, Kolb JA (2009) The effect of learning organization culture on the relationship between interpersonal trust and organizational commitment. Human Resour Dev Q 20(2):147–167

Windsor LC (2013) Using concept mapping in community-based participatory research a mixed methods approach. J Mixed Methods Res 7(3):274–293

Mandatory Convertible Bonds as an Efficient Method of Issuing Capital

Angel Huerga and Carlos Rodríguez-Monroy

Abstract Mandatory Convertibles Notes (MCNs) mean only a small fraction of all the securities issued by corporate or financial institutions, however, they represent nearly a 30% in volume of the convertible securities issued every year. MCNs share characteristics of equity and debt securities but rating agencies assign them a high equity component and are commonly treated as equity by accounting standards. Despite the high facial coupon that MCNs seem to pay, a deeper analysis shows that the cost of MCN can be lower than the cost of issuing hybrid or subordinated debt and in some cases similar to the cost of issuing senior debt. Mandatory convertibles were profusely issued by financial institutions amid the global crisis as a means to increase capital and could be considered as a predecessor of some types of AT1 and Contingent Convertible. The academic literature about Mandatory Convertibles is scarce and we consider necessary to shed some light on a type of security that can be very useful for the real economy.

Keywords Mandatory convertible note · Capital · Dividend protection
Call-spread · Cost of capital

1 Introduction

Mandatory convertibles Notes (MCN) are hybrid securities that share characteristics of both debt and equity. They are designed and documented as a bond, pay coupons regularly but upon redemption or at maturity are mandatorily converted into a fixed or limited number of common shares and no cash or other security is delivered. This unique feature makes this security to behave mostly like common shares but with

A. Huerga · C. Rodríguez-Monroy (✉)
Dpto. de Organización, Administración de Empresas y Estadística. ETS de Ingenieros Industriales, Universidad Politécnica de Madrid (UPM), Madrid, Spain
e-mail: crmonroy@etsii.upm.es

A. Huerga
e-mail: angel.huerga@gmail.com

Á. Ortiz et al. (eds.), *Engineering Digital Transformation*, Lecture Notes in Management and Industrial Engineering,
https://doi.org/10.1007/978-3-319-96005-0_8

specific characteristics and potentially with lower cost. MCNs pay a regular coupon that in general is higher than the dividend yield of the underlying shares, but the issuer can elect to defer the coupons at its sole discretion. Mandatory convertibles are junior to other debt securities and senior only to common equity. Mandatory convertibles do not have voting rights. Issuing a mandatory convertible can be considered as a deferred capital increase. Historically the scarce academic literature has focused on the information asymmetry implications as a rationale for issuing MCNs and only a few authors have studied the theoretical pricing model and its relationship and consistency with the observed market prices.

2 Objectives

This paper is the first document of an ongoing academic research that aims to study the potential advantages for corporate and financial institutions of issuing MCNs compared to other forms of capital. One of the main objectives of this study is to describe and to shed some academic light on a type of security about which the number of academic studies is very limited.

3 Methods

This paper researches the academic literature about MCNs and the notes issued, describes its main characteristics from the issuer point of view and also studies the potential advantages for issuers. Additionally, this is the starting point of a research that aims to develop an empirical analysis using the price data series and each MCNs characteristic using market data providers (Bloomberg, Reuters) to confirm the potential cost advantages for issuers of this form of capital.

4 Size of the Market

According to Bloomberg in 2008 the market of MCNs peaked to its higher recorded level at USD24.5bn. The issuance of MCNs represented globally USD1.8bn in 1996 and the market grew steadily until 2008. The latest financial crisis meant a significant reduction of the amount issued due to the lower appetite of investors and their preference for cash, treasuries and other safe-haven investments. The market recovered in 2013 and in 2016 the USD volume issued reached USD21bn. The total size of the convertible securities outstanding, mandatory and non-mandatory, was USD392bn on average from 2013 to 2016 (Bank of International Settlements, BIS 2016), and the new issues amounted to USD58bn, a mere 9% of the total debt issued globally. The issuance of MCNs represents on average 29% of all the convertible bonds

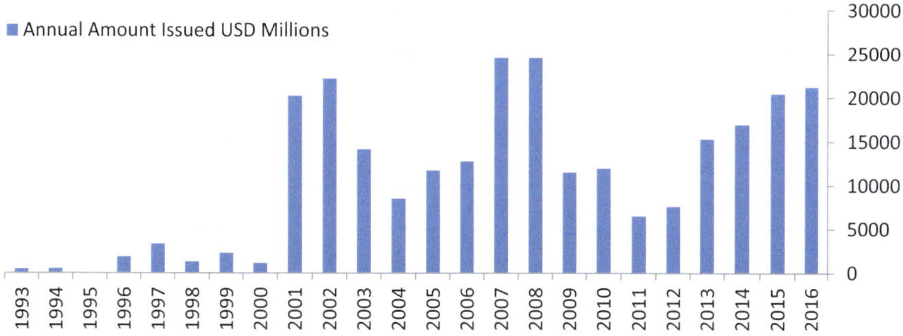

Fig. 1 Annual issuance of mandatory convertible securities. *Source* Bloomberg

issued annually. Mandatory convertibles tend to have shorter maturities (typically 3 years) than standard convertible bonds (5–7 years) therefore the annual issuance tends to be higher. Nevertheless, the academic literature related to standard convertible bonds is disproportionally larger than the one studying mandatory convertibles (Rodríguez-Monroy and Huerga 2014) (Fig. 1).

5 Pricing Model

There are several types of MCNs, (Arzac 1997), but all of them share some characteristics: a predetermined coupon regularly paid and common stock conversion at redemption with one or two conversion ratios that transform the initial notional into a limited number of shares. Due to market appetite, historical and fiscal reasons, the typical MCN has two conversion triggers, hence two conversion ratios, the lower conversion ratio and the upper conversion ratio. The number of shares delivered at redemption is given by (Fig. 2):

$$If\ S \le Trigger_{Lower}\quad Number\ of\ Shares = Amount\ Issued \cdot Conversion\ Ratio_{Lower}$$

$$If\ S \ge Trigger_{Upper}\quad Number\ of\ Shares = Amount\ Issued \cdot Conversion\ Ratio_{Upper}$$

$$If\ Trigger_{Lower} < S < Trigger_{Upper}\quad Number\ of\ Shares = \frac{Amount\ Issued}{S}$$

Being,

$$S = Stock\ Price\ at\ Redemption$$

$$Conversion\ Ratio_{Lower} = \frac{Amount\ Issued}{Trigger_{Lower}}$$

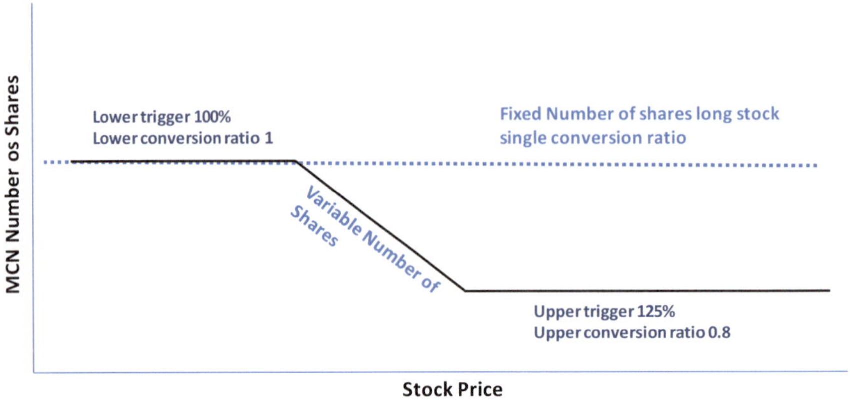

Fig. 2 Number of shares delivered at maturity versus stock price. *Source* Authors

$$Conversion\ Ratio_{Upper} = \frac{Amount\ Issued}{Trigger_{Upper}}$$

$$Conversion\ Ratio_{Lower} > Conversion\ Ratio_{Upper}$$

$$Trigger_{Lower} < Trigger_{Upper}$$

The payoff for an investor in MCNs is substantially different to the payoff of a standard convertible since there is no downside protection, and it is also different to the total return of the common stock. The return is closer to a prepaid forward share purchase agreement. One feature of MCNs is that the at higher stock prices the mandatory convertible delivers fewer shares and therefore less dilution for existing shareholders than a straightforward share sale and represents a lower total cost for the issuer (Fig. 3).

A MCN can be synthetically replicated as a prepaid forward share sale agreement plus a strip of fixed coupons, plus a sold equity call option and a purchased equity call option at a higher strike on a lower notional, a ratio call-spread (Arzak 1997):

$$\begin{aligned} Price = {} & PV_{riskfree}(N) + PV_{risky}(C) - Conversion\ Ratio_{Lower} \cdot Call_{Lowerstk} \\ & + Conversion\ Ratio_{Upper} \cdot Call_{Upperstk} \end{aligned}$$

The present value of the amount issued must be discounted using the risk-free rate because the issuer can always deliver shares even in a distressed situation; however, the coupons must be discounted using the issuer's credit curve (Amman and Seiz 2006). Since the credit component is limited to the present value of the strip of coupons, standard options valuation as Black-Scholes and Merton or numerical

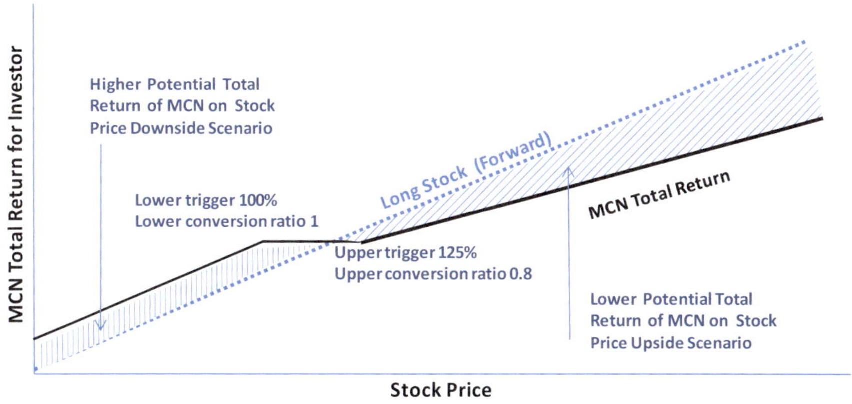

Fig. 3 Total return MCN versus total return common shares. *Source* Authors

models well tested in options valuations can be applied to calculate the value of MCNs (Black and Scholes 1973; Merton 1973). Dividends may affect the pricing of MCNs and forecasting future dividends is never an easy task. To avoid this issue, recent MCNs include full dividend protection clauses to make them more appealing for new investors as convertible arbitrage funds and easier to hedge for them and for investment banks or even for the same issuers (Zimmermann 2016). Standard dividend protection clauses read as follows for both the upper and lower conversion ratios, where S_{cum} is the price of the underlying stock the day before the actual dividend payment:

$$Conversion\ Ratio_{New} = \frac{S_{cum}}{S_{cum} - div} Conversion\ Ratio_{Old}$$

Increasing the conversion ratio is equivalent to reducing conversion triggers and it adjusts the mandatory convertible by the value reduction on the underlying stock that represents the dividend payment. Full dividend protection clauses are in the center of an academic discussion that can be explored in future research.

6 Mandatory Convertible Notes Greeks Profiles

The Greeks are defined as the sensitivities of derivatives and investment securities to the change of some observed underlying parameter and are necessary to manage and understand the financial risk of investment portfolios or securities. The more representative Greeks are the Delta, the Gamma, the Vega and the Theta. Mandatory convertibles have Greek parameters that differ greatly from traditional convertibles and other securities (Calamos 2011). The Delta can be defined as the change of the price of the mandatory convertible linked to the unitary change of the price of the

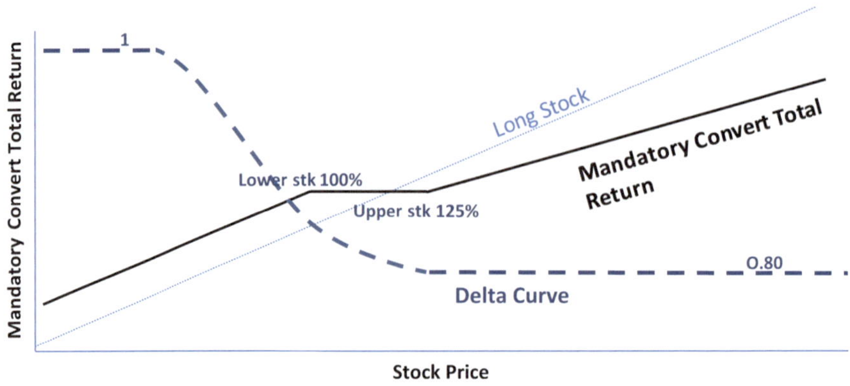

Fig. 4 Delta of a mandatory convertible security. *Source* Authors

underlying stock. The Delta is close to one when the stock price is below the lower trigger, the Delta decreases between the lower and the upper strike, and stabilizes close to the upper trigger (Fig. 4).

The Gamma of a security can be defined as the changes in the Delta of MCN per unitary change in the underlying stock. Gamma management is important as a means to manage the Delta of a portfolio and to avoid abrupt changes in the price. Portfolio hedging techniques and options are used to manage Gamma.

7 Accounting for Mandatory Convertibles

A MCN is an instrument that includes an unconditional obligation requiring the issuer to redeem the security delivering a specified number of the underlying stock at a specific date. Under the International Financial Reporting Standard (IFRS) accounting rules, if the MCN represents the obligation to deliver a fixed amount of shares, the instrument can be bifurcated as an equity component and a debt component. The initial carrying amount of the debt component is the present value of the coupons that the issuer is obliged to pay, discounted at the credit risk curve of the issuer (Ramirez 2011). The equity component is calculated as the difference of the issuer proceeds and the debt component. Most mandatory convertibles deliver two fixed amounts of shares, the lower ratio and the upper ratio and a gap where the number of shares is variable. Therefore, issuers tend to bifurcate the instrument into a fixed parity mandatory convertible and a sold call-spread that is accounted at mark to market.

8 Potential Motivations for Issuing Mandatory Convertibles

A few studies research the motivations of the issuers to access the MCN market and the effects of the announcement of MCNs on the stock prices, and some other papers apply the asymmetry information theories to the issuance of convertible bonds (Chemmanur et al. 2014; Wang 2017). The commercial and academic literature highlights the following advantages:

1. Firms issue a mandatory convertible because of its rating agencies treatment as equity. Additionally, mandatory convertible securities are accounted partially or totally as equity.
2. Huckins (1999) shows in his research that MCNs allow companies to increase capital delaying the equity dilution to redemption.
3. Private placements or accelerated book building stock sales imply a substantial price discount that can be avoided by issuing MCNs.
4. Issuers of mandatory convertibles tap an investor base that is different to the straight debt or equity investor base.
5. The lower slope of the mandatory convertible ratio at higher stock prices can represent a lower dilution and a lower cost for the issuer compared to straight equity, thanks to the ratio call-spread purchased by the issuer.

Additionally, our research determines another motivation for issuing MCNs, since the total cost for the issuer can be lower than the cost of hybrid debt and sometimes in line with the cost of senior debt. The real cost for the issuer is related to the cost of a long term repurchase agreement of the underlying shares.

9 Preliminary Results of the Research

MCNs investors traditionally compare the coupon of the note with the dividend yield of the common stock at issuance. Preference for a constant and high dividend stream are the main driving aspect in the long-only investors decision making process, but the ratio call-spread embedded in the note can have an important weight in the coupon, depending of the volatility of the underlying shares. Arbitrage investors can hedge the MCN with a short equity position and a purchased call-spread. Stripping out the cost of the call-spread and bid-ask spreads, the minimum required coupon for the buyer would be the cost of a long term stock borrow agreement plus its liquidity cost. Therefore for the issuer, the cost of the instrument is mainly linked to the availability and cost of long term stock borrow and the liquidity of the stock and not fully correlated with its own credit spread, credit rating or credit availability.

There are presently 71 public mandatory convertible securities outstanding. The notes have been issued from 2014 to 2017 with an average maturity of 3 years. Out of the 71 notes, 51 include dividend protection mechanisms that facilitate the access to the arbitrage investor base. The average running cost of the embedded ratio call-spread is 3.7% annually. Under the above assumptions, if we want to compare the cost of issuing a MCN with the cost of issuing senior or subordinated debt, the spread paid by the issuer can be broken down as:

$$MCN\ Spread = Nominal\ Coupon - Call\text{-}spread\ running\ cost \\ - Mid\ Swap\ Rate\ at\ issuance$$

Our analysis of the outstanding MCN shows that the average MCN cost is 2.3%.

10 Conclusions

The present paper represents the starting point of an ongoing academic research that aims to shed some light on a barely known type of security that is becoming increasingly popular among issuers, corporate and financial institutions, as a means to increase capital. Some studies show that MCNs can offer substantial advantages for securities issuers. This paper shows that the real price paid by issuers of MCN is relatively low compared to other forms of capital, subordinated debt and in some cases comparable to senior debt. The investor community tends to react positively to the announcement of this type of security.

References

Amman M, Seiz R (2006) Pricing and hedging mandatory convertible bonds. J Deriv 13(3): 30–46

Arzac ER (1997) PERCS, DECS, and other mandatory convertibles. J Appl Corp Finan 10(1):54–63

Bank for International Settlements (2016) BIS quarterly review December 2015. Available at http://www.bis.org

Black F, Scholes M (1973) The pricing of options and corporate liabilities. J Polit Econ 81:637–654

Calamos N (2011) Convertible arbitrage: insights and techniques for successful hedging. The Wiley Finance Series

Chemmanur T, Nandy D, Yan A, Jiao J (2014) A theory of mandatory convertibles? J Bank Finan 42:352–370

Huckins NW (1999) An examination of mandatorily convertible preferred stock. Finan Rev 34:89–108

Merton R (1973) Theory of rational option pricing. Bell J Econ Manage Sci 4(1): 141–183 (Spring)

Ramirez J (2011) Handbook of corporate derivatives and equity capital markets. The Wiley Finance Series

Rodríguez-Monroy C, Huerga A (2014) Los instrumentos convertibles en acciones para la financiación empresarial en un entorno económico de restricción del crédito bancario. Economía Industrial (392): 89–100

Wang S (2017) A theory of mandatory convertibles: distinct features for large repeated financing. Eur J Finan, 1–18

Zimmermann P (2016) The fallacy of fully dividend protected stock options and convertible bonds. J Deriv 23(3): 61–72 (Spring)

Culture and Environment as Antecedents of Technological Entrepreneurship

Simona Parente, Marianna Perfili, Gustavo Morales-Alonso and Michele Grimaldi

Abstract The present research points out how engineering students from two different countries, and therefore subject to different cultural and role models, show different entrepreneurial intentions. A questionnaire was submitted to engineering students divided into two groups on the basis of their country origin: Italy and Spain. The sample in use is for students attending Engineering Schools, as engineers are considered as possible founders of future innovative start-ups. The results of the survey show that Italian students have higher entrepreneurial intentions than Spanish ones. This is attributed to their different cultural origins and in particular, the sub-dimensions of individualism and masculinity from Hofstede's Cultural Dimension Theory. Among the results of the research, it has been proved that students whose parents are entrepreneurs present higher entrepreneurial intention than those whose parents are not self-employed.

Keywords Entrepreneurial intention · Cultural dimensions · Parental role models · Career choice · Start-up

S. Parente (✉) · M. Perfili · M. Grimaldi
Dpto. of Civil and Mechanical Engineering, University of Cassino and Southern Lazio, Via G. Di Biasio 43, 03043 Cassino, FR, Italy
e-mail: simona.parentee@gmail.com

M. Perfili
e-mail: marianna.perfili@gmail.com

M. Grimaldi
e-mail: m.grimaldi@unicas.it

G. Morales-Alonso
Dpto. de Ingeniería de Organización Administración de Empresas y Estadística. ETSI Industriales, Universidad Politécnica de Madrid, C/José Gutèirrez Abascal 2, 28006 Madrid, Spain
e-mail: gustavo.morales@upm.es

Á. Ortiz et al. (eds.), *Engineering Digital Transformation*, Lecture Notes in Management and Industrial Engineering,
https://doi.org/10.1007/978-3-319-96005-0_9

1 Introduction

Entrepreneurship "is the practice of starting new organizations or revitalizing mature organizations, particularly new businesses generally in response to identified opportunities" (Eroglu and Piçak 2011). The entrepreneur is the single most important player in a modern economy; he carries out three basic functions that are absolutely essential to the operation of any company: the capitalist or financial function; the managerial function; and the booster function (Cuevas 1994).

Literature suggests that entrepreneurship contributes to an economic performance by introducing innovations, creating change, creating competition and enhancing rivalry, and thus encourages economic wealth and spending power (Holmgren and From 2005). Particular emphasis has been done in the last years to point out the link between the creation of new technology-based firms (NTBFs) and economic growth (Audretsch 1995; Carree and Thurik 2003). The creation of a new firm requires an individual, the entrepreneur, willing to risk his time and money on the venture; he is the detonating and sine qua non-condition for economic growth (Morales-Alonso et al. 2016).

On the other hand, human capital, in which education plays a significant role, is one of the most important factors for NTBFs' growth (Colombo et al. 2004; Colombo and Grilli 2005).That is to say, in order to successfully create and lead an NTBF, it is necessary to have an entrepreneur with (1) high technical skills (i.e. an engineer) and (ii) high entrepreneurial intentions. For this reason, understanding the determinants of engineering students' entrepreneurial intentions is essential (Pablo-Lerchundi et al. 2015).

The present research focuses on the comparison on the entrepreneurial intention of engineering master students of Università di Cassino e del Lazio Meridionale and Universidad Politécnica de Madrid. Two main topics have been addressed: entrepreneur's personal characteristics or traits and the influence of contextual aspects in entrepreneurship.

The paper is organized as follows. In the next section, the state of the art is summarized and the research question and hypotheses of the study are established. Later, there is a section in which the methodology used to obtain the results is described. In the following section, the obtained results are described and discussed, while the last section gathers the conclusions of this work.

2 Research Framework

2.1 Cultural Dimensions

Intentions and behaviours of an individual are determined by contextual and internal factors. Among the latter, the culture in which the individual has been raised plays a dominant role (Schwarzt 1990). Hence, it is with good reason that culture is termed as "software of the mind" by researcher Hofstede (1980). Hofstede presents, in fact, a set of cultural dimensions that address individual's behaviours and intentions. Out of the whole set, collectivism/individualism and feminity/masculinity have been identified as the most relevant for this study.

Individualism describes tendencies to orient values and actions towards independence, competition, and oneself or one's immediate family (Tiessen 1997). On the other hand, collectivists perceive themselves as interdependent members of an "in-group," a collection of people perceived as sharing the same fate (Triandis 1993) so they tend to act cooperatively in their group's interest (Hofstede 1980; Hofstede et al. 1991; Triandis 1993).

Hofstede and Hofstede (2001) puts individualism and collectivism at opposite ends of a continuum, which the majority of researchers accept (Morris et al. 1994). People in individualist cultures tend to be more autonomous and independent than people in collectivist cultures, they view uncertainty in the external environment more optimistically than collectivist individuals and they are more likely to involve themselves in situations that collectivist individuals perceive as being extremely risky (Morris et al. 1994).

On the other hand, masculinity refers to the overall "toughness" and competitiveness of a society (Steensma et al. 2000), being for this reason related to the desirability of achievement; while femininity relates to interpersonal relationships.

In the above-mentioned measures, Italy has a level of individualism of 76%, while Spain ranks at 51%, being the percentage of masculinity of 70% for Italy and 42% for Spain. So Spain could be considered a collectivistic and feminine country in comparison with the rest of the European countries, and in particular, with Italy.

In conformity with the above, and considering that in general, researchers have hypothesized that entrepreneurship is facilitated by cultures that are high in individualism, low in uncertainty avoidance, low in power-distance, and high in masculinity (Hayton et al. 2002), it is appropriate to establish the first two hypotheses:

H1a. Italian students have high entrepreneurial intention because they are more individualistic and masculine.
H1b. Spanish students have low entrepreneurial intention because they are more collectivistic and feminine.

2.2 *Influence of Parental Occupation*

A role model is a common reference to individuals who set examples to be emulated by others and who may stimulate or inspire other individuals to make certain (career) decisions and achieve certain goals (Basow and Howe 1980; Wright). The relevance of role models for entrepreneurs is evident in the popular business press that is littered with stories of, and references to, entrepreneurial endeavours and successes that have influenced other entrepreneurs (Bosma et al. 2012).

Parental occupation plays a key role in the development of potential entrepreneurial intention in their children, self-employed parents being a motivator, while civil servants a discouraging factor for entrepreneurship (Morales-Alonso et al. 2016). The family, particularly the father or mother, plays the most powerful role in establishing the desirability and credibility of entrepreneurial action for the individual (Shapero and Sokol 1982). For this reason, it is hypothesized:

H2. Students who have fathers entrepreneurs have a higher entrepreneurial intention.
H3. Students who have mothers entrepreneurs have a higher entrepreneurial intention.

3 Methodology

The methodology used in the present research follows a quantitative approach. A questionnaire has been submitted to a sample of engineering master degree students from two different universities, namely "Università di Cassino e del Lazio Meridionale" (Italy), and "Universidad Politécnica de Madrid" (Spain). The questionnaire was filled at the beginning of a teaching session during the first semester of academic year 2016/2017.

The main topics treated in the survey are cultural dimensions, entrepreneurial intention as defined by the Theory of Planned Behaviour (Ajzen 1991), and barriers to entrepreneurship. A set of control variables has been used to explain the demographic and cultural aspects of each student.

The students who took part in these surveys belong to master degrees of Mechanical, Civil and Industrial Engineering and their number is of 154 for Italy and of 620 for Spain. Regarding genders, in the Italian sample it results a percentage of 69.7% of men and 30.3% of women, while for the Spanish one there is a percentage of 62% of men and 38% of women. Of the two final samples, 33.1% of Italian students were studying civil engineering, 30.3% were industrial engineering and 36.6% mechanical engineering; on the other side, 26.6% of Spanish students were civil engineering, 37.2% industrial engineering and 36.2% mechanical engineering.

The survey is composed of 38 items and some demographic questions, divided into 4 modules. The first one is composed of a set of standard segmentation questions, such as student's age, gender, nationality, as well as nationality, education and employment status of student's parents. The second module contains questions related to Hofstede's cultural dimensions theory, in particular to the items of collectivism/individualism, femininity/masculinity and risk aversion. The third module relies on Liñán and Chen's (2009) questionnaire on students' purpose for starting their own firms. The fourth and last part contains propositions of the institutional economic theory (North 2005) and also some influences from social capital research (Anderson and Jack 2002; Liñán and Santos 2007).

All these items are evaluated on a five-point Likert scale, where 1 = total disagree and 5 = total agree; exception for module 3 that uses a seven-point Likert scale, where 1 = total disagree and 7 = total agree.

Table 1 Student's t-test for entrepreneurial intention, collectivism/individualism and femininity/masculinity

		N	Mean (1–7)	df	T
Entrepreneurial intention	Italy	145	3.34	704	2.36*
	Spain	561	2.85		
Collectivism/individualism	Italy	145	3.40	704	2.36*
	Spain	561	3.15		
Femininity/masculinity	Italy	145	3.33	704	2.30*
	Spain	561	3.06		

***$p<0.001$; **$p<0.01$; and *$p<0.05$

4 Results and Discussion

Considering the whole sample of 706 respondents designed valid and meaningful for the study, descriptive analysis was conducted on two sub-samples of 145 students from Italy and 561 from Spain. It was detected an average value of entrepreneurial intention equal to 3.34 for Italy and 2.85 for Spain, with a relative value of standard deviation of $\sigma = 1.86$ and $\sigma = 2.29$, respectively.

The average values obtained for collectivism/individualism are 3.4 ($\sigma = 0.549$) in Italy and 3.15 ($\sigma = 1.23$) in Spain, while for femininity/masculinity are 3.33 in Italy ($\sigma = 0.675$) and 3.06 in Spain ($\sigma = 1.33$). Besides, a gender effect was also assessed, resulting in that men in both countries have a higher entrepreneurial intention than women [Italy: 3.45 (men), 3.09 (women) and Spain: 3.09 (men), 2.48 (women)].

The non-parametric Kolmogorov-Smirnov test was used to verify the normality of all items considered in the analysis, presenting all of them a normal distribution with a significance level of $p<0.001$. On the other hand, a reliability analysis was conducted on the entire questionnaire, which shows an excellent internal consistency with Cronbach's alpha value equal to $\alpha = 0.972$. Moreover, the accuracy found among the six items of EI was excellent ($\alpha = 0.933$), for collectivism/individualism is $\alpha = 0.767$ and for femininity/masculinity $\alpha = 0.669$.

Hypotheses have been checked with the Student's t-test for difference between average values. Table 1 summarizes the results from comparing average values of EI between Italian and Spanish respondents. As it can be seen, Italians present a higher entrepreneurial intention, being this difference statistically significant. When the cultural values are dealt with, it is also found that Italians are more individualistic and rank higher in Masculinity than Spaniards (see also Table 1), being the again statistically significant. Therefore, the Hypothesis H1a and H1b are positively verified regarding both cultural values: collectivism/individualism and femininity/masculinity.

Table 2 Student's t-test for entrepreneurial intention

		N	Mean (1–7)	df	t
Fathers with own business	Yes	155	3.53	698	3.73***
	No	545	2.80		
Mothers with own business	Yes	70	3.58	695	2.45*
	No	697	2.91		

***$p < 0.001$; **$p < 0.01$; and *$p < 0.05$

These results are congruent with previous findings in the literature (Hofstede 1980; Shane 1992). Mcgrath et al. (1992) establish that a good entrepreneur should have a high level of individualism and a high level of masculinity.

Regarding the influence of parents' employment, both Hypotheses 2 and 3 are confirmed over the whole sample, as shown in Table 2. That is, children from self-employed parents show higher entrepreneurial intention and rank higher on entrepreneurial attitudes than the offspring of non-entrepreneurs. Many researchers have declared that vocational outcome depends so much from parental support and encouragement, with the result that the attitudes and behaviours that young people adopt toward work are the consequence of what their parents say. Parents provide their influence to children through comments, beliefs and interactions, such as conversations and verbal and nonverbal reactions, influencing what children think, say and perceive about several careers. The reason for the correlation between parents' influence and children's choice is that parents possess a unique position to influence the future behaviour of their sons/daughters (Morales-Alonso et al. 2016). In fact, if an individual feels strongly identified with someone in his social context, he will learn by the observation of actions and others themselves, to act and think in the same way.

5 Conclusions

The present research sheds light on how cultural differences mediate on the entrepreneurial intention of an individual. Besides, the effect of exposure to positive and negative role models for entrepreneurship has also been sought after.

To this end, a questionnaire has been developed and responded by a set of 774 students from two different countries: Italy and Spain.

As illustrated by the results, it emerges that Italians have a higher entrepreneurial intention than Spaniards, because they are more individualistic and masculine than Spanish students, which are more collectivistic and feminine.

The second outcome shows that students are more prone to entrepreneurship when exposed to positive (i.e. entrepreneurs) in their close environment (father and mother).

Contrarily, students whose father and/or mother is not an entrepreneur, respond as if subjected to a negative role model to entrepreneurship. That is, exposure to positive role models foster entrepreneurial intention, while this is hindered when subjected to the influence of negative role models. The relevance of this result is that it is independent of the cultural origin of the individual.

References

Ajzen I (1991) The theory of planned behavior. Organ Behav Hum Decis Process 50(2):179–211

Anderson AR, Jack SL (2002) The articulation of social capital in entrepreneurial networks: a glue or a lubricant? Entrepreneurship Reg Dev 14(3):193–210

Audretsch DB (1995) Innovation and industry evolution. MIT Press, Cambridge

Basow SA, Howe KG (1980) Role-model influence: effects of sex and sex-role attitude in college students. Psychol Women Q 4(4):558–572

Bosma N, Hessels J, Schutjens V, Van Praag M, Verheul I (2012) Entrepreneurship and role models. J Econ Psychol 33(2):410–424

Carree MA, Thurik AR (2003) The impact of entrepreneurship on economic growth. In: Handbook of entrepreneurship research. Springer, US, pp 437–471

Colombo MG, Grilli L (2005) Founders' human capital and the growth of new technology-based firms: a competence-based view. Res Policy 34(6):795–816

Colombo MG, Delmastro M, Grilli L (2004) Entrepreneurs' human capital and the start-up size of new technology-based firms. Int J Ind Organ 22(8):1183–1211

Cuevas JG (1994) Towards a taxonomy of entrepreneurial theories. Int Small Bus J 12(4):77–88

Eroglu O, Piçak M (2011) Entrepreneurship, national culture and Turkey. Int J Bus Soc Sci 2(16)

Hayton JC, George G, Zahra SA (2002) National culture and entrepreneurship: a review of behavioral research. Entrepreneurship Theor Pract 26(4):33

Hofstede G (1980) Motivation, leadership, and organization: do American theories apply abroad? Organ Dyn 9(1):42–63

Hofstede GH, Hofstede G (2001) Culture's consequences: comparing values, behaviors, institutions and organizations across nations. Sage, London

Hofstede G, Hofstede GJ, Minkov M (1991) Cultures and organizations: software of the mind, vol 2. McGraw-Hill, London

Holmgren C, From J (2005) Taylorism of the mind: entrepreneurship education from a perspective of educational research. Eur Educ Res J 4(4):382–390

Liñán F, Chen YW (2009) Development and cross-cultural application of a specific instrument to measure entrepreneurial intentions. Entrepreneurship Theor Pract 33(3):593–617

Liñán F, Santos FJ (2007) Does social capital affect entrepreneurial intentions? Int Adv Econ Res 13(4):443–453

McGrath RG, Macmillan IC, Scheinberg S (1992) Elitists, risk-takers, and rugged individualists? An exploratory analysis of cultural differences between entrepreneurs and non-entrepreneurs. J Bus Ventur 7(2):115–135

Morales-Alonso G, Pablo-Lerchundi I, Núñez-Del-Río MC (2016a) Entrepreneurial intention of engineering students and associated influence of contextual factors/Intención emprendedora de los estudiantes de ingeniería e influencia de factores contextuales. Revista de Psicología Social 31(1):75–108

Morales-Alonso G, Pablo-Lerchundi I, Vargas-Perez AM (2016) An empirical study on the antecedents of knowledge intensive entrepreneurship. Int J Innov Technol Manage 1640011

Morris MH, Davis DL, Allen JW (1994) Fostering corporate entrepreneurship: cross-cultural comparisons of the importance of individualism versus collectivism. J Int Bus Stud 25(1):65–89

North D (2005) Understanding the process of economic change. SPE, Storia del pensiero economico. Fascicolo 2(2): 1000–1004

Pablo-Lerchundi I, Morales-Alonso G, González-Tirados RM (2015) Influences of parental occupation on occupational choices and professional values. J Bus Res 68(7):1645–1649

Schwartz SH (1990) Individualism-collectivism critique and proposed refinements. J Cross Cult Psychol 21(2):139–157

Shane SA (1992) Why do some societies invent more than others? J Bus Ventur 7(1):29–46

Shapero A, Sokol L (1982) The social dimensions of entrepreneurship. Encyclopedia of entrepreneurship, pp 72–90

Steensma HK, Marino L, Weaver KM, Dickson PH (2000) The influence of national culture on the formation of technology alliances by entrepreneurial firms. Acad Manage J 43(5):951–997

Tiessen JH (1997) Individualism, collectivism, and entrepreneurship: a framework for international comparative research. J Bus Ventur 12(5):367–384

Triandis HC (1993) Collectivism and individualism as cultural syndromes. Cross Cult Res 27(3–4):155–180

Drivers' Perception of the Major Advantages of Electric Vehicles

María del Carmen Pardo-Ferreira, Antonio López-Arquillos, María Martínez Rojas, Manuel Suarez-Cebador and Juan Carlos Rubio-Romero

Abstract Governments support the use of electric. These vehicles have environmental, economic and technical advantages. However, drivers of electric vehicles perceive the environmental benefits as most important advantages.

Keywords Electric vehicle · Drivers' perception · Environment benefits

1 Introduction

The White Paper on Transport spells out among its goals to achieve CO_2-free city logistics in major urban centers by 2030. Therefore, it proposes to reduce the use of 'conventionally fueled' cars in urban transport by 2030 and to phase them out of cities by 2050. Consequently, amount of electric vehicles in the cities is increasing. The advantages of these vehicles include different aspects as environmental, since they present zero exhaust emissions, economic aspects, since their efficiency reduces energetic costs and they can be charged during cheap electric tariffs and technical aspects, since they have low noise engine and fast acceleration.

2 Objectives

The study aimed to determine what are the major advantages of electric vehicles from driver's perspective.

M. del C. Pardo-Ferreira (✉) · A. López-Arquillos · M. M. Rojas
M. Suarez-Cebador · J. C. Rubio-Romero
Dpto. de Economía y Administración de Empresas, Universidad de Málaga, Málaga, Spain
e-mail: carmenpf@uma.es

Á. Ortiz et al. (eds.), *Engineering Digital Transformation*, Lecture Notes in Management and Industrial Engineering,
https://doi.org/10.1007/978-3-319-96005-0_10

3 Methods

To achieve the objective, a questionnaire was administered to 95 experienced electric vehicle drivers. They indicated the degree of advantage on a scale of 1 (no advantage) to 100 (major advantage). Most of these drivers, who participated in the study, had driven electric and hybrid vehicles for more than two years while working. Companies collaborating in this study had hybrid and electric vehicles in their fleet, both motorcycles and cars. The questionnaire was filled out on paper and the answers were subsequently processed. This questionnaire was drawn from the study by Labeye et al. (2016) in Paris and original questionnaire was designed by German research team that worked on the MINI E study (Cocron et al. 2011).

4 Results

The following Table 1 shows the average of the degree of advantage of each item.

Table 1 Drivers' evaluation of the major advantages of the electric vehicles

In your opinion, what are the major advantages of electric vehicles?	Degree of advantage
Fewer localized carbon emissions while driving	96
Support of developing the availability of renewable energies	94
Fewer carbon emission in general if the electricity is produced by renewable energy	92
Less dependence on fossil fuels such as oil	91
Fewer carbon emission in general (incl. electricity generation), when charged with 'green power' (low CO_2)	90
Possibility to charge during peak hours (cheap)	85
Lower costs (energy costs) than for a conventional combustion engine vehicle	83
Quieter inside the car	83
Feeling less guilty about driving a car	78
Fast acceleration of the vehicle	74
Less noise outside	73

5 Conclusion

For the drivers the major advantages of electric vehicles are environment benefits. These advantages are located in the first five positions, highlighting 'fewer localized carbon emissions while driving' and 'support of developing the availability of renewable energies'. Secondly, advantages related to savings and lower costs are located. Finally, drivers gave less importance to other advantages such as silent feature, less feeling of guilt and fast acceleration.

Acknowledgements We wish to thank the General Direction of Traffic of Spain for financing project SPIP2015-01765, of which this paper forms a part, and also to the University of Malaga for its support through the Research Own Plan.

References

Cocron P, Bühler F, Neumann I, Franke T, Krems JF, Schwalm M, Keinath A (2011) Methods of evaluating electric vehicles from a user's perspective—the MINI E field trial in Berlin. IET Intell Transp Syst 5(2):127–133

Labeye E, Hugot M, Brusque C, Regan MA (2016) The electric vehicle: a new driving experience involving specific skills and rules. Transp Res Part F: Traffic Psychol Behav 37:27–40

A Study on Consumer Behaviour in the Sharing Economy

Marianna Perfili, Simona Parente, Michele Grimaldi and Gustavo Morales-Alonso

Abstract Sharing economy is a socio-economic model, built around the sharing of human and physical resources, which give consumers the possibility of accessing to product and assets without having the individual ownership. The present research has the purpose of analysing consumers' intentions to join in with sharing economy system, and, in particular, of focusing on how this participation is influenced by Hofstede's cultural dimensions and Materialism. Data for this study were collected by distributing a questionnaire among engineering students from two technical universities, respectively in Italy and Spain, in order to observe the different behaviour about sharing intention. The obtained results point out that cultural dimensions like Collectivism and Femininity foster sharing intention and that, therefore, consumers' attitude to join in with sharing economy, whereas Individualism and Masculinity inhibit sharing intention. It is possible to say the same thing about Materialism that, as demonstrated by the results of the research, has a negative impact on sharing intention, considered as one of the principal inhibitors of sharing.

Keywords Sharing intention · Culture · Materialism · Shared good

M. Perfili (✉) · S. Parente · M. Grimaldi
Department of Civil and Mechanical Engineering, University of Cassino and Southern Lazio, Via G. Di Biasio 43, 03043 Cassino, FR, Italy
e-mail: marianna.perfili@gmail.com

S. Parente
e-mail: simona.parentee@gmail.com

M. Grimaldi
e-mail: m.grimaldi@unicas.it

G. Morales-Alonso
Dpto. de Ingeniería de Organización Administración de Empresas y Estadística. ETSI Industriales, Universidad Politécnica de Madrid, C/José Gutiérrez Abascal 2, 28006 Madrid, Spain
e-mail: gustavo.morales@upm.es

Á. Ortiz et al. (eds.), *Engineering Digital Transformation*, Lecture Notes in Management and Industrial Engineering,
https://doi.org/10.1007/978-3-319-96005-0_11

1 Introduction

The sharing economy is a phenomenon that has had a rapid growth over the last five years, especially after the global economic recession of 2008. Network and communication technologies such as Internet have started to challenge traditional thinking about how resources can and should be offered and consumed (Cohen and Kietzmann 2014). The sharing economy marks an important shift in the behaviour of individuals (Bonciu et al. 2016), and points to a new stage in the evolution of the consumer role (Perren and Grauerholz 2015), which assumes peculiar characteristics both in the consumption and in the creation of products and services.

However, there is lack of consensus around sharing economy definition (Borcuch 2016). Terms like collaborative economy, collaborative consumptions, and sharing economy have been used interchangeably (Botsman 2013). On one hand, "collaborative economy" is usually referred to as a series of systems which have the function to release the value of assets and resources that not used to best of their ability level. On the other hand, Botsman (2013) defines "collaborative consumption" as an economic model based on sharing, swapping, trading, or renting products and services enabling access over ownership, while "sharing economy" is defined as economic models based on sharing underutilized assets from spaces to skills to stuff for monetary or non-monetary benefits, that it can take place only in a peer-to-peer (P2P) market and not in the business-to-consumer (B2C).

Regardless of the term used, the sharing economy has brought changes in the consumers' value system. Indeed, the typical values of the capitalist mentality such as the Individualism, the personal well-being, the cult of the ego, is leaving space to completely different values. The community, the relationships, less tolerance for waste, a strong and growing desire for sociality and belonging, service orientation, environmental responsibilities, will of sharing and participation, are now trends that characterize the collaborative consumer profile. Sharing intention is a culturally learned behaviour as well as possession and ownership (Furby 1976). The current research has been conducted in order to verify if culture dimensions and materialistic attitudes can influence positively or negatively the consumers' intention to participate in the world of sharing economy.

The paper is organized as follows. In the next section, the state of the art is summarized and the research question and hypotheses of the study are established. Later, there is a section in which the methodology used to obtain the results is described. In the following section, the obtained results are described and discussed, while the last section gathers the conclusions of this work.

2 Research Framework

2.1 Influence of Culture Over Sharing Intentions

Hofstede (1980) defines a set of six cultural dimensions that allow distinguishing different national cultures across countries. Out of the six factors, two of them have been selected as possible mediators in consumer's interest towards the sharing economy.

The first dimension considered is the Individualism/Collectivism, in which the definition of self, expressed in terms of "I" and "We" is involved (Triandis and Harry 1995), Members of individualistic cultures see themselves as independent of others. Individualism can be seen as the emotional independence from "groups, organizations, or other" (Hofstede 1980). On the contrary, members of collectivist cultures see themselves as interdependent members of an "in-group/community", a collection of people perceived as sharing the same fate (Triandis 1993), so they tend to act cooperatively in their group's interest (Hofstede 1980; Kagitcibasi and Berry 1989; Triandis 1993) and give priority to the goals of the groups (Hofstede 2001).

A collectivist consumer is prone to collaborating and compromising in group activities and sharing in group outcomes (Hui et al. 1991). "Sharing Resources", whether tangible or intangible, is seen as a characteristic of collectivism (Hui and Triandis 1986; Sinha and Verma 1987). In this sense, the act of sharing is one of distribution, and it is an active practice to have something in common with someone (John 2013), so the "real" sharing experience is consuming together, sharing or exchanging assets and resources like products, time, space, skills, food, money from and/or with peers (Botsman and Rogers 2011). In this sense, sharing can be considered a form of consumption that promotes cooperation between people, awakens a sense of community and reinforces the sense of commitment, in which the social aspect becomes more relevant and important (Vaquero and Calle 2013).

The second dimension under study is Masculinity/Femininity, where masculinity stands for a society in which social gender roles are clearly distinct: men are supposed to be assertive, tough and focused on material success; women are supposed to be modest, tender and concerned with the quality of life. Contrariwise femininity stands for a society in which social gender roles overlap: both men and women are supposed to be modest, tender and concerned with the quality of life (Hofstede 2001). Cultures with high masculinity emphasize work goals, assertiveness, heroism and materialism, while feminine cultures stress personal goals, such as employment security, human relationships, concern for others, and nurturing relationships (Hofstede 1980). While masculinity can be represented by a competitive society that view the world in terms of winners and losers, femininity refers to the environment of cooperation where people feel secure to share their knowledge and resource with others, where alliances are typically viewed as "win-win" situations.

If sharing is about openness and mutuality, there is a sense in which it can be perceived as reflecting a set of values that can be seen as feminine (John 2013). Jaggar (1992) in fact defines "interdependence, community, connection, sharing" as feminine attributes, a set of characteristic that may be associated to sharing economy.

Italy and Spain are the two countries under consideration in this study, and they show distinct values for the abovementioned measures. Italians rank higher in individualism (76%) than Spaniards (51%). At the same time, Italians are more masculine (70%) than Spaniards (42%). For Italians, competing and winning are important things in their life; they show their success with new car, travels, big houses, etc. On the contrary Spaniards are less competitive. They are educated in search of harmony, relationships and cooperation with others, also with weak or needy people.

Hence, in accordance with the literature and Hofstede's studies, it is possible to state that Italians are individualistic and masculine, while Spaniards are collectivist and feminine, thus reacting differently over sharing economy. For this reason it is hypothesized:

H1a: "Italians are individualistic and masculine. They have a low Sharing Intention"
H1b: "Spaniards are collectivistic and feminine. They have a high Sharing Intention"

2.2 *Influence of Materialism Over Sharing Intentions*

Materialism has been identified as playing a key role in the consumers' sharing decision (Akbar et al. 2016). It is defined as a personality trait reflecting the importance a person places on material possessions (Belk 1985); it refers to how important material good are to a person's life (Goldsmith and Clark 2012). Materialism, in essence, is an orientation among individuals who find their greatest personal satisfaction from owning valuable objects, and who have relatively little interest in humanistic value such as altruism or community feeling.

However, in accordance with Belk (1984), materialism is a multi-dimensional concept, which consists of three sub-dimensions: Possessiveness, Non-generosity and Envy. In this research we rely on the previous study of Akbar et al. (2016), in which only the first two sub-dimensions are accounted for when dealing with sharing intentions. Possessiveness represents the inclination and tendency to retain control or ownership of one's possessions (Belk 1983), it is the excessive desire to acquire materialistic things, people, and memories. On the other hand, non-generosity stands for an unwillingness to give possessions to or share possessions with others (Belk 1984); it is to value things more than people (Belk and Ger 1994), and appeared to represent the interpersonal components of materialism (O'Guinn and Faber 1989).

As abovementioned, Italians are considered more individualistic than Spaniards and for this reason, more prone to materialism. Therefore the following hypotheses are made:

H2a: "Italians are materialistic, so they have a high Possessiveness and a low Sharing Intention"
H2b: "Spaniards are not materialistic, so they have a lower Possessiveness than Italians, and have a high Sharing intention"
H3a: "Italians are materialistic, so they have a high Non-Generosity and a low Sharing Intention"

H3b: "Spaniards are not materialistic, so they have a lower Possessiveness than Italians, and have a high Sharing intention".

3 Methodology

A quantitative approach has been selected for this research, with the use of a questionnaire submitted to a sample of engineering master degree students from two different universities, namely "Università di Cassino e del Lazio Meridionale" (Italy), and "Universidad Politécnica de Madrid" (Spain). The respondents of the survey are enrolled in master degrees of Mechanical, Civil and Industrial Engineering, being their number of 144 for Italy and of 119 for Spain. Similar results for gender distribution have been found for both samples: Italy (men: 71.5%, women: 28.5%) and Spain (men: 71.4%, women: 28.6%).

The items that form the survey are divided in three modules. The first one is devoted to demographic identification of respondent (age, gender, nationality and employment status of student's parents). The set of cultural dimensions (individualism/collectivism, masculinity/femininity and risk aversion) are gathered in the second module. Last, the survey developed by Akbar et al. (2016) for analysing the link between materialism and sharing intentions has been used as the third module. All items are measured in a five point Likert scale.

4 Results and Discussion

The first variable measured in this study is General Sharing Intention (GSI), for which an average of 2.64 has been found for Italy ($\sigma = 0.944$) and 3.20 for Spain ($\sigma = 0.884$). The second variable is materialism, which is composed of two subdimensions: Possessiveness, with an average value of 4.34 ($\sigma = 0.852$) for Italy and 3.71 ($\sigma = 0.940$) for Spain; and Non-Generosity, which averages 2.50 ($\sigma = 0.953$) for Italy and 2.15 ($\sigma = 0.709$) for Spain. The Cronbach alpha for these variables is 0.78, 0.90 and 0.90 for GSI, Possessiveness and Non-Generosity, respectively, proving a high internal consistency in the questionnaire.

Prior to the statistical analyses, the non-parametric Kolmogorov-Smirnov test has been used to verify the normality of all items considered in the analysis, presenting all of them a normal distribution with a significance level of $p < 0.001$.

Hypotheses have been checked with the Student's t-test for difference between average values between both samples. Table 1 summarizes the results obtained. As it can be seen, H1 is confirmed in its two propositions: Italians, which belong to an individualistic and masculine culture, have a lower GSI than Spaniards (collectivist and feminine), being this difference statistically significant.

This result implies that an individualistic and masculine behaviour, as that of Italian consumer, does not foster participating in the sharing economy. In Italy there

Table 1 Student's t-test for the variables measured in this research

		N	Mean (1–5)	df	T
General sharing intention	Italy	144	2.64	261	−4.95***
	Spain	119	3.20		
Possessiveness	Italy	144	4.34	261	5.62***
	Spain	119	3.72		
Non-Generosity	Italy	144	2.50	261	3.35***
	Spain	119	2.15		

***$p < 0.001$; **$p < 0.01$; and *$p < 0.05$

are some prevailing individualistic feelings among consumers, which clash with those typical of the sharing economy and therefore slow the spread of the same phenomenon. Contrarily, Spaniards do not feel comfortable with competiveness, preferring to cooperate and give attentions to needs and preference of others rather than competing to distinguish oneself. In fact, Triandis et al. (1984) described the Spanish as entertaining, warm, pleasant, and friendly people who show respect for others and try to establish harmony in their interpersonal relationships, can be seen also in the work-place as teamwork. They adopt a pattern of social relationships that has been described as communal sharing and that instils a sense of belonging to a social group, that is frequently giving presents, behaving altruistically, being generous, and perceiving relationships as being eternal (Fiske 1992).

With respect to H2a, H2b, H3a and H3b, they are all positively verified. That is, Italians rank higher in both sub-dimensions of materialism. This is linked with the previous result obtained for GSI, as previously proved by Akbar et al. (2016). Not in vain, Rifkin (2014) argues that "Materialists are less empathetic not only against other people but also toward other creatures and the natural environment. In their view, the nature is just a tool, a resource to be exploited, instead of a community to preserve. For them the environment, as well as relationships with others, should be considered only in terms of utility and market value and never for its own sake".

5 Conclusions

The determinants of consumer behaviour when dealing with sharing economy have been dealt within this research. Particular emphasis has been made on the influence of culture and materialism.

Regarding cultural dimensions, results obtained point out that of individualism/collectivism and masculinity/femininity influence individual behaviour and propensity to sharing economy. Specifically, individualism and masculinity are negatively correlated with sharing intention.

With respect to materialism, its influence on sharing intention has been confirmed in this study. Individuals with high materialistic disposition do not show a high sharing intention, in consequence of their personal satisfaction in owning valuable objects rather than sharing them.

References

Akbar P, Mai R, Hoffmann S (2016) When do materialistic consumers join commercial sharing systems. J Bus Res
Belk RW (1983) Worldly possessions: issues and criticisms. NA-Adv Consum Res 10
Belk RW (1984) Three scales to measure constructs related to materialism: reliability, validity, and relationships to measures of happiness. NA-Adv Consum Res 11
Belk RW (1985) Materialism: trait aspects of living in the material world. J Consum Res 12(3):265–280
Belk RW, Ger G (1994) Problems of marketization in Romania and Turkey. Res Consum Behav 7
Bonciu F, Bâlgăr AC (2016) Sharing economy as a contributor to sustainable growth. An EU perspective. Rom J Eur Aff
Borcuch A (2016) The sharing economy: understanding and challenges
Botsman R (2013) The sharing economy lacks a shared definition. Fast Company
Botsman R, Rogers R (2011) What's mine is yours: how collaborative consumption is changing the way we live. Collins, London
Cohen B, Kietzmann J (2014) Ride on! Mobility business models for the sharing economy. Organ Environ 27(3):279–296
Fiske AP (1992) The four elementary forms of sociality: framework for a unified theory of social relations. Psychol Rev
Furby L (1976) The socialization of possession and ownership among children in three cultural groups: Israeli kibbutz, Israeli city, and American. Piagetian Res Compilation Commentary 8:95–127
Goldsmith RE, Clark RA (2012) Materialism, status consumption, and consumer independence. J Soc Psychol 152(1):43–60
Hofstede G (1980) Culture's consequences: international differences in work-related values, Abridged edn. Sage Publications, Newbury Park, CA
Hofstede G (2001) Culture's consequences: comparing values, behaviours, institutions and organizations across nations, 2nd edn. Sage Publications, Thousand Oaks, CA
Hui CH, Triandis HC (1986) Individualism-collectivism a study of cross-cultural researchers. J Cross Cult Psychol 17(2):225–248
Hui CH, Triandis HC, Yee C (1991) Cultural differences in reward allocation: is collectivism the explanation? Br J Soc Psychol 30(2): 145–157
Jaggar AM (1992) Feminist ethics
John AN (2013) The social logics of sharing. The communication review. Routledge, Abingdon
Kagitcibasi C, Berry JW (1989) Cross-cultural psychology: current research and trends. Ann Rev Psychol 40(1):493–531
O'Guinn TC, Faber RJ (1989) Compulsive buying: a phenomenological exploration. J Consum Res
Perren R, Grauerholz L (2015) Collaborative consumption. Int Encycl Soc Behav Sci: 139–144
Rifkin J (2014) The zero marginal cost society: the internet of things, the collaborative commons, and the eclipse of capitalism. Macmillan, Basingstoke
Sinha JBP, Verma J (1987) Structure of collectivism

Triandis HC (1993) Collectivism and individualism as cultural syndromes. Cross Cult Res 27(3–4):155–180
Triandis HC (1995) Individualism & collectivism. Westview press
Triandis HC et al (1984) Simpatia as a cultural script of Hispanics. J Pers Soc Psychol 47(6):1363

Levels of Application of Public R&D&I Policy Models

Alexander Piñero, Carlos Rodríguez-Monroy and Miguel Ángel Peláez García

Abstract With the analysis by researchers and institutions regarding experiences in designing public R&D&I policies, to strengthen the capacity of innovation in the industrial sector, international cooperation was possible to understand the difficulties and recommendations for linking and integrating SMIs with the other members of the Innovation System. With this information, four levels of implementation of public policy models in R&D&I are presented. Among the proposed alternatives level 4 is the one recommended since it presents the best model of integration, coordination and governmental evaluation in the design of public R&D&I policies.

Keywords Public policy for R&D&I · Capacity of innovation · Innovation system · SMIs

1 Introduction

The development of Research, Technological Development and Innovation activities (R&D&I) have become a priority for the government administration that designs public policy R&D&I, to understand that innovation represents the alternative to promoting the economic, social and sustainable development of a country. Benavente et al. (2012) consider it essential that government policies have an important role in

A. Piñero
Universidad Técnica de Manabi, Portoviejo, Ecuador
e-mail: alexanderpinero12@gmail.com

C. Rodríguez-Monroy (✉)
Dpto. de Organización, Administración de Empresas y Estadística. ETS Ingenieros Industriales, Universidad Politécnica de Madrid (UPM), Madrid, Spain
e-mail: crmonroy@etsii.upm.es

M. Á. Peláez García
Dpto.de Organización, Administración de Empresas y Estadística, ETS Ingenieros Industriales, UPM, Madrid, Spain
e-mail: miguelangel.pelaez@upm.es

Á. Ortiz et al. (eds.), *Engineering Digital Transformation*, Lecture Notes in Management and Industrial Engineering,
https://doi.org/10.1007/978-3-319-96005-0_12

establishing a suitable environment for companies to improve their level and disposition to innovation. These initiatives have led to other studies such as Ben (2012), in the evolution of public policies oriented collaboration of interactive processes in R&D&I with a focus on Innovation System (IS).

As Hobday et al. propose (2012), in several countries enhancing the contribution of design innovation policy for national economic growth is becoming a key objective of government policy.

The Organisation for Economic Co-operation and development (OECD 2008) believes that for a better coordination of public policy it is critical that the public administrations improve their understanding of the innovative capacity in business, which is crucial for designing innovation policies adapted to their needs.

However, we have knowledge and experience that the vast majority of Small and Medium Industries (SMIs), require the support of an IS as a support structure to undertake and develop R&D&I that generate products to meet the needs and opportunities for innovation in the market.

2 Objectives

This paper aims to develop a proposal application-level public policy models in R&D&I to develop the capacity of innovation in small and medium industries.

3 Methods

The study represents a documentary and applied type research. This work is initiated from the reflections and recommendations made by different researchers and international cooperation organisations regarding the design of public R&D&I policy to promote the demand for innovation in the market. Among other studies, those performed by the Technopolis-group (2011, 2012) for the European Commission, OECD (2010a, b), OECD (2011), OECD (2012), OECD (2013), OECD (2014) and OECD (2015) have to be cited.

Besides the interest to improve the design and implementation of public policies of R&D&I in the countries of the European Union can be shown in the following studies: European Commission (2010, 2011, 2014); COTEC Foundation in Spain (2012, 2013, 2015) which recommended promoting a favorable environment for industries to significantly increase investment and carry out R&D&I activities through the coordination and cooperation with other IS actors.

Subsequently, the recommendations of the theories of the innovation process proposed by Kline and Rosenberg (1986), Rothwell (1994) and of the Innovation System provided by Freeman (1987), Lundvall (1992) and Nelson (1993) become the basic guidelines for the design of various levels of implementation of public policy models in R&D&I in order to promote demand and opportunities for industrial innovation.

4 Results

As a result of this study, different levels of implementation of proposed models of public policy in R&D&I for the development of the innovation capacity, specifically in the SMIs are designed. Each level is represented in a figure and its present situation and the role of the government institutions in the implementation of public R&D&I policy is analysed. (See Figs. 1, 2, 3 and 4).

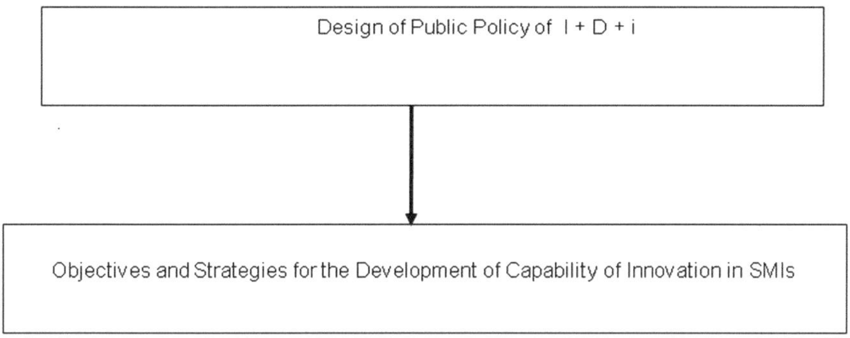

Fig. 1 Level 1 model of public R&D&I policy. *Source* Authors

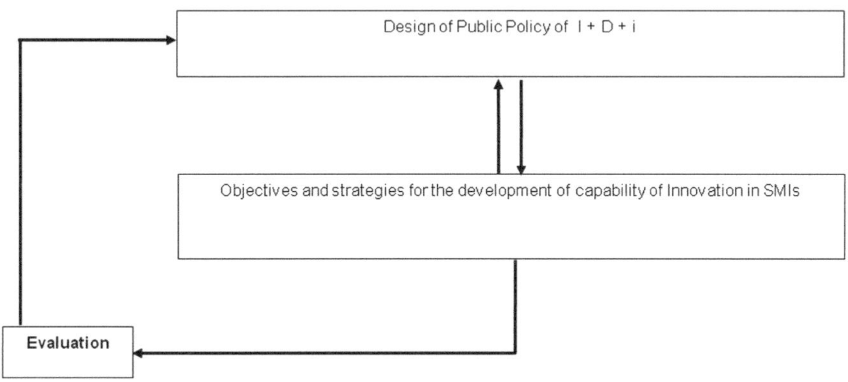

Fig. 2 Level 2 model of public R&D&I policy. *Source* Authors

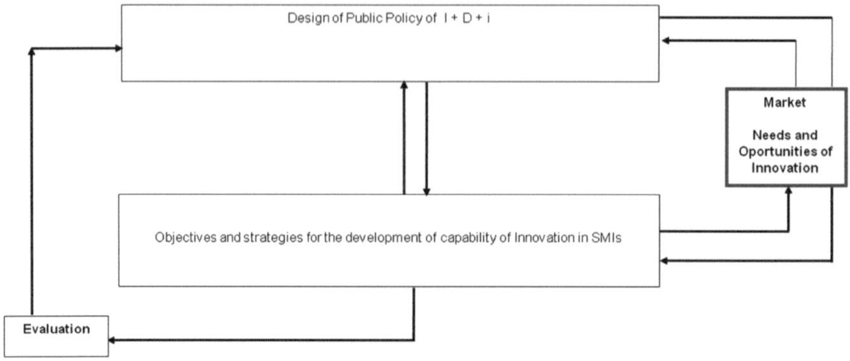

Fig. 3 Level 3 model of public R&D&I policy

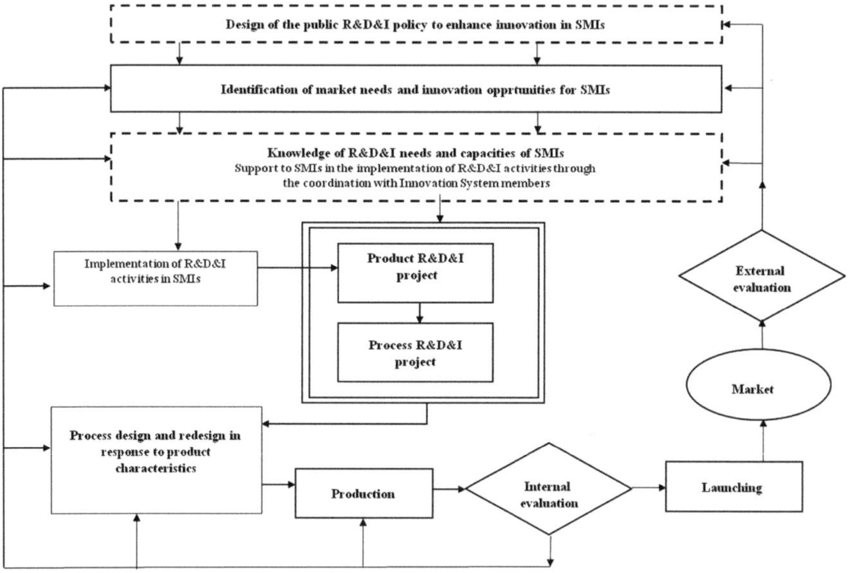

Fig. 4 Level 4 model of public R&D&I policy. *Source* Authors

4.1 Role of Public R&D&I Policy in Level 1

As shown in Fig. 1, there is no communication or evaluation of the public R&D&I policy in collaboration with SMIs.

In the Level 1 model public R&D&I policy guidelines are designed without prior information on the needs and capabilities of R&D&I of the SMIs, and about innovation opportunities in the market. No evaluation of the implementation of strategies or compliance with the objectives and targets is performed.

4.2 Role of Public R&D&I Policy in Level 2

In Fig. 2, there is communication and evaluation of the public R&D&I policy in collaboration with SMIs.

In the Level 2 model, public R&D&I policy guidelines are designed with prior information on the needs and capabilities of R&D&I of the SMIs. An evaluation of the implementation of strategies or compliance with the objectives and targets of the public policy is performed. There is no knowledge about the demand for products and about innovation opportunities in the market.

4.3 Role of Public R&D&I Policy in Level 3

In Fig. 3, there is communication and collaboration with SMIs and information about the market is available.

In the Level 3 model, public R&D&I policy guidelines are designed with prior information on the needs and capabilities of R&D&I of the SMIs. The government institutions have knowledge of the demand for products and services of public and private companies. They promote and create a favourable environment for innovation in the market through public and private purchasing. Ongoing communication with the SMIs is performed and an evaluation of the implementation of strategies and compliance with the goals and objectives of the policy is made.

4.4 Role of Public R&D&I Policy in Level 4

In Fig. 4 there is communication and collaboration with SMIs and information about the market is available. In addition, a support structure for SMEs is created to respond better to market opportunities through coordination with members of the Innovation System.

In the Level 4 model shown in Fig. 4 there is communication, linking of public R&D&I policy with SMIs and the government institution has knowledge of the demands and opportunities for product innovation in the market. Further evaluation of public policy R&D&I at the internal level of SMIs or at the external level in relation to product placement on the market is performed in order to detect possible obstacles in the achievement of the planned objectives so that SMIs can be endowed with better capacities for innovation. Without the stages of internal and external evaluation, there is the risk of not introducing timely corrective actions to ensure compliance and learning of public R&D&I policy. If necessary, the government institution establishes alliances and negotiations with actors on the IS to reinforce those areas that are deficient in SMIs in order to provide effective support in the innovation process.

5 Considerations for Each of the Proposed Application Levels of the Public Policy Model for R&D&I

Most errors in the implementation of public R&D&I policy are presented assuming that the SMIs have in R&D&I capabilities, to start on their own and successfully complete the product innovation without a previous study of the needs of SMIs and without having accurate information on market demands. This situation results in loss of resources, effort and time. The levels of a public policy model for R&D&I found, in this case, are those of Levels 1 and 2.

Level 3, as described in Fig. 3, applies to SMIs that have learning experience and capabilities of R&D&I for developing new products. The government institution is aware of market needs and a permanent communication and evaluation are performed with the SMIs to measure compliance with objectives and policy goals. At this level the relationship between the government institution and the SMIs it is possible to respond to market demands.

Most SMIs have deficiencies and obstacles that affect their capacity for R&D&I, plus the difficulty for researching market needs. Therefore, the intervention of public R&D&I policy and coordination with IS members, to support the R&D&I activities in the SMIs is necessary. Level 4, depicted as a model of public R&D&I policy in Fig. 4, is better suited to this scenario by allowing to stimulate demand and create a favorable environment for innovation in the market. Finally, internal and external evaluation of the progress of R&D&I in SMIs is considered a priority.

6 Conclusions

At the different levels of implementation of models of public in R&D&I policy it is necessary to promote market opening and create favorable conditions for innovation, in addition to supporting the SMIs to participate in both public procurement and purchases by large private companies, in order to achieve the development of new or improved products that meet the demands and market opportunities. Hence, the market becomes a source of essential information for the design of public R&D&I policy, which mobilizes the energy and vision of the SMIs towards innovation.

In the international context it is evident in government institutions that there is a constant effort to improve the implementation of a public R&D&I policy, aimed at consolidating a culture of cooperation and establishing interrelations between the different actors that make up the innovation system, creating favorable conditions for the SMIs such as: promotion of opportunities for innovation in the market, supporting a legal framework to promote and facilitate interest in innovation, funding for R&D&I, accessibility to scientific and technological infrastructure centers and R&D laboratories of universities and technology parks among others in order to strengthen the innovation process in the SMIs.

Among the different levels of public R&D&I, policy models the recommended one is level 4. The model raises the relationship, linking and assessment between the government institution and the IS members in order to promote and support R&D&I in SMIs. Therefore, it is essential for the responsible government agency to design and coordinate the public R&D&I policy and be aware of the needs, constraints and capabilities of innovation in SMIs and also know the strengths and weaknesses of the relation of SMIs with other players in the SI. With this prior knowledge, it may be possible to design in an adequate way the instruments, objectives and strategies of public R&D&I policy so that SMIs manage to respond to opportunities for innovation of new products or processes on the market.

References

Ben M (2012) The evolution of science policy and innovation studies. Res Policy 41(7):1219–1239

Benavente D, Dutta S, Wunsch-Vincent S (2012) The global innovation index 2012: stronger innovation linkages for global growth. Chapter 1. http://www.globalinnovationindex.org/gii/main/fullreport/index.html. INSEAD and WIPO 2012. ISBN: 978-2-9522210-2-3. France

Comisión Europea (2010) Europa 2020. Una Estrategia para un Crecimiento Inteligente, Sostenible e Integrador. Brussels, Belgium

Comisión Europea (2011) Estado de la Unión por la innovación 2011. Informe de la Comisión al Parlamento Europeo, al Consejo, al Comité Económico y Social Europeo y al Comité de las Regiones. COM (2011) 849. Brussels, Belgium

Comisión Europea (2014) Comprender las políticas de la Unión Europea: Investigación e innovación. Brussels, Belgium

Freeman C (1987) Technology, policy and economic performance: lessons from Japan. Pinter Publishers, London

Fundación COTEC (2012) Tecnología e Innovación en España. Informe Cotec 2012. Fundación Cotec para la Innovación Tecnológica. Madrid. Spain. http://www.cotec.es

Fundación COTEC (2013) Un decálogo de retos de la innovación para la competitividad de España. 2013. Fundación Cotec para la Innovación Tecnológica. Madrid. Spain. http://www.cotec.es

Fundación COTEC (2015) Tecnología e Innovación en España. Informe Cotec 2015. Fundación Cotec para la Innovación Tecnológica. Madrid. Spain. http://www.cotec.es

Hobday M, Boddington A, Grantham A (2012) Policies for design and policies for innovation: contrasting perspectives and remaining challenges. Technovation 32(5):272–281

Kline S, Rosenberg N (1986) An overview of innovation, the positive sum strategy: harnessing technology for economic growth. In Landau R, Rosenberg N (eds) National Academy Press, Washington. D.C. pp 275–305

Lundvall B (1992) National systems of innovation: towards a theory of innovation and interactive learning. Pinter Publishers, London

Nelson R (1993) National innovation systems. Oxford UP, Oxford

OECD (2008) Science, Technology and Industry Outlook 2008: Highlights. OECD Publishing. Organisation for Economic Co-operation and Development (OECD), París

OECD (2010a) Innovation strategy. Ministerial report on the OECD innovation strategy. Innovation to strengthen growth and address global and social challenges. Key Findings. OECD Publishing, Paris

OECD (2010b) The OECD innovation strategy: getting a head start on tomorrow. ISBN. 978-92-64-084704. OECD Publishing, Paris

OECD (2011) Demand-side innovation policies. OECD Publishing, Paris

OECD (2012) Science, Technology and Industry Outlook 2012. OECD Publishing. Organisation for Economic Co-operation and Development (OECD), París
OECD (2013) Innovation in Southeast Asia. OECD Publishing, Paris
OECD (2014) OECD science, technology and industry outlook 2014. OECD Publishing, Paris
OECD (2015) Innovation policies for inclusive growth. OECD Publishing, Paris
Rothwell R (1994) Towards the fifth-generation innovation process. Int Mark Rev 11(1):7–31
Technopolis-group (2011) Trends and challenges in demand- side innovation policies in Europe. www.technopolis-group.com
Technopolis-group (2012) regional innovation monitor. Thematic Paper 3. Demand-side innovation policies at regional level. www.technopolis-group.com

Limitations and Pitfalls of the Brain That Prevent Us from Thinking

José Luis Portela and Carlos Rodríguez-Monroy

Abstract When we think or make decisions, we must not forget that there are several internal "pitfalls" that cause our brain to make decisions incorrectly. Many authors have written on various pitfalls that may affect us. In this paper, five of them are identified and they are: The use of patterns when thinking, the relativity of things, the anchor effect, loss aversion, value attribution and diagnosis bias.

Keywords Thought · Brain · Patterns · Business innovation

1 Introduction

Everyday knowledge, theories and procedures are being discovered and shared by people and organizations. The key, therefore, is not usually knowledge. The key lies in how managers make decisions: in other words, how these managers are able to think in a different way. This research work focuses on identifying the "limitations and pitfalls", both external and internal, which influence us to think and make decisions.

It is a well-known fact that the brain makes certain errors in decision making. While it is true that it is a known fact that decisions are made by the unconscious brain, to date little is known about exactly how it does this. And while it is true that we cannot know for certain how it does this, it is also true that, if these errors are repetitive, we can establish a cause-and-effect relationship and that is what it is all about. The key is to recognize that when we are irrational, we are predictably irrational, which would lead to thinking that our irrationality always occurs in the same way and, therefore, we should identify these behavioral patterns.

J. L. Portela
Operations Department IE Business School, Calle María de Molina 11, Madrid, Spain
e-mail: jlportela@faculty.ie.edu

C. Rodríguez-Monroy (✉)
Higher Technical School of Industrial Engineers of Madrid, Technical University of Madrid, C/José Gutiérrez Abascal 2, 28006 Madrid, Spain
e-mail: crmonroy@etsii.upm.es

Á. Ortiz et al. (eds.), *Engineering Digital Transformation*, Lecture Notes in Management and Industrial Engineering,
https://doi.org/10.1007/978-3-319-96005-0_13

2 Use of Patterns

The way in which the brain works is by connecting neurons. A single neuron can create between 10,000 and 15,000 connections.

If the entire brain has an average of 100,000 million neurons, the average number of synapses in a human brain is a total of: one trillion synapses (100,000 million neurons average times 10,000 connections), a one followed by 15 zeros.

The first-time connection of neurons through synapses consumes more energy than the use of these connections once they have been connected. Thus, for a baby, energy consumption by the brain reaches 60% of the total energy instead of 20%, which occurs in adults. This is due to the fact that it is specifically during these early stages that the bulk of new connections in the brain is being formed.

Neurologist Marcus Raichle E. published an article in Science (2006) called 'The Brain's Dark Energy'. "It is a complex problem," says Pascual-Leone, "allow me to give you some figures to give you an idea of the magnitude of the issue. The brain is approximately 2% of the weight of the human body. However, it consumes 20% of the energy of the human body".

Making a decision requires some calculations. The brain is always trying to save the maximum amount of energy. The way it does this is through the establishment of objectives, the restriction of communications and the creation of models that it uses to "pattern" behaviors and facts.

The brain has to make approximately 4000 decisions per day. The best way to do it, using the lowest possible energy consumption, is to use synapses which are already established instead of having to create new ones from scratch.

What does this mean?

According to Edward de Bono (1994), the mind creates patterns, because, once created, they can be reused in the process of new thoughts and there is no need to be creating new connections from scratch.

The dominant ideas prevent us from creating new situations as they mark the journey of our thoughts and as a result we always think in a similar way.

What happens is that when something different from what was experienced reaches the brain for the first time, the first thing it does is to see if it fits into some of the patterns it has. If an adult, who played with cars as a child, picks up an iPad for the first time and drops it, their brain knows perfectly what the outcome will be, because as a child when they dropped the car, it fell. What the brain does here is to relate that object with other previous ones and says "the iPad is the same as the little cars you had as a child".

The first thing that the brain does, therefore, is to see whether things of the past are similar to assign one to it. Once we have this list of ideas of the past, we select the one we deem best. In other words, we do not think, we simply bring thoughts from the past to be used in the present.

3 The Relativity of Things

Dan Ariely, in his book "Predictably Irrational" speaks of the way in which the brain is able to compare things. As Ariely indicates "the brain is not able to assign an absolute value to an object unless it compares it with another".

As a result, we humans rarely choose things in absolute terms. We do not have a meter of internal value that tell us how much things value. To be able to give a value to something, we have to look at the advantages or disadvantages of one thing in relation to another, and in this way we will be able to estimate its value.

The example that Dan Ariely in his book "Predictably Irrational" reflects this situation.

On the website www.economist.com there were three subscription options. The options were as follows:

(1) Subscription to the on-line access for only $59.
(2) A subscription to the printed version for only $125.
(3) Subscription to on-line access and printed version for $125.

In the case, he explains, 84% chose option 3. Why is this?

In choosing the best option, the brain tries to use the minimum amount of energy, or, to put it another way, it tries to think as little as possible. Of the three options, comparing the first and the third, or the first and last is much more complicated than comparing the second and the third, since these are much closer and in addition there is the fact that one is clearly better than the other.

4 The Anchor Effect

The naturalist Konrad Lorenz discovered that the gosling (offspring of goose), upon hatching from the egg, sticks to the first moving object that it finds, which naturally tends to be its mother.

Lorenz discovered it in one of his experiments, because he was the first thing the young geese saw and since that moment they have followed him loyally everywhere.

What Lorenz demonstrated is that the goslings' initial decisions not only make decisions based on what is available in the environment, but stick to their decision once it is made. Lorenz called this natural phenomenon "imprinting".

Does our brain act the same as that of these goslings? Could our first impressions and data also produce an imprint? This phenomenon is known in behavioral economics as an "anchor".

This happens when we observe an object that we had never seen before for the first time and our brain "anchors" the first price that we see and thus, almost magically, is linked to it, conditioning future prices of the product to this first observed price.

Drazen Prelec, Professor of MIT's Sloan Management School, did an experiment to explain the phenomenon of locking in prices. He pulled out a bottle Cotes du Rhone Jaboulet Parallel of 1998.

In the class, there were 55 students. That day Drazen George Loewenstein and Dan Ariley made a request of the students. They were asked to record the last two digits of their social security numbers on a piece of paper. What he was trying to show is what is called "arbitrary coherence".

The idea was to demonstrate that the best of having an arbitrary number in one's head would create an anchor in the head that would determine what the students would be willing to pay.

Drazen grabbed another bottle, a Hermitage Jaboulet La Chapelle of 1996 with a ranking of 92 points in Wine Advocate magazine. The previous bottle that we had talked about had a score of 86.

There were also four articles, a wireless trackball (Logitech TrackMan Marble FX), a Wireless Keyboard and Mouse set, (Logitech iTouch), a design book (the perfect package: how to increase value by using graphic design) and a box of half a kilo of Belgian chocolates from Neuhaus.

Forms were distributed where all the articles appeared. Then, next to each article, they were to write the last two digits of their social security numbers, for example 23, and indicate that this would be the price in dollars of the article.

Then, they simply have to say whether or not they would buy the item for that price.

In addition to this there was another box in which they were to enter the maximum amount they would pay for each of these items. This would be like a bid in an auction for each article. Once they received all articles, the winner was given the article once the amount had been paid.

The question is thinking about whether the social security number influenced, or not, the form in which each one of the students answered.

After analyzing the data, the conclusions were devastating. Students with higher digits (80–99) were the ones that made higher bids, while those who had lower digits (1–20) were the ones that made lower bids.

The top quintile, for example, pushing an average of $56 for the wireless keyboard, while the bottom quintile had bid only an average of $16.

At the end, students whose social security numbers were in the top quintile made bids that were between 216 and 346% higher than those of students whose last two digits of the social security were in the bottom quintile.

The data showed that the relative prices between the various products, comparing them was logical. They paid more for the best wine than for the worst, for example.

What is significant is that once the students were willing to pay a certain price for a product, their willingness to pay for other articles in the same category happened to be in related with the first price, the anchor. This is termed "arbitrary coherence".

The anchor effect was also mentioned by Daniel Kahneman, who won the Nobel in economic science, in his book "thinking fast and slow". In one experiment, experienced German Judges were inclined to give a shoplifter a longer sentence if they had just rolled a pair of dice loaded to give a high number.

5 Loss Aversion, Value Assignment and Diagnostic Bias

In the book "he irrational impulse" of Ori Brafman and Rom Brafman about the effect of loss and how it affected both the time of making decisions, but also speaks to us of how loss aversion affects us when making decisions.

The best way to explain it is to start with an example that he narrates in his very informative book.

Macarthur Job and Matthew Tesch described the chain of events that led to the air collision of Tenerife in Air Disasters: Volume 1, Aerospace Publications, Sydney, Australia, 1994, pp. 165–180.

The KLM Flight 4805 was a Jumbo 747 to the path that Amsterdam airport of Las Palmas in the Canary Islands.

The flight was quiet and received an urgent message from international air traffic control. A terrorist bomb had exploded in the flower shop of the airport and so the airport was closed. It was noted that it had to land at the Tenerife airport.

When the plane landed, Jacob realized that if he was there for a very long time, he would have to pass through the compulsory rest period, so that they could take off later than half past six in the evening.

Two hours after it occurred to him to refuel in order to gain time and the news that Las Palmas had finally opened, came but it was already too late to stop the process of refuelling, which lasted for 35 min. But this was not all, suddenly a dense layer of fog began to descend on the track, after which Van Zanten realized that if he continued staying there, he could not take off.

In those moments, what then seems like something out of an abnormal situation, he accelerated the engines and headed toward the track.

"Wait a moment, said of Van Zanten co-driver bewildered. We do not have authorization for takeoff.

I know, "replied the commander while brakeing—Come on, ask for it."

The co-pilot turned on the radio and received the permission to fly—the adoption of the flight plan, but the tower did not say anything about the crucial authorization take-off

However, Van Zanten turn the throttles to full power and began to take off, with the fatality of crossing a Pan Am 747 in middle of the track, crashing into it. A total of 584 people lost their lives.

How was it possible that Van Zanten has been trying to get off the ground without flight authorization?

Our behavior and decision-making is influenced by a set of underlying psychological flows much more powerful and penetrating than most think

What were the internal hidden forces that influenced this decision?

- **Loss aversion**. Our tendency to do everything possible to avoid potential losses. We experience pain associated with a loss much more vividly than the joy of getting a profit.
- **Value attribution**: Our inclination to imbue a person or thing with certain qualities based on the initial perception of its value and not on objective data.

- **Diagnostic bias**: Our blindness to all evidence that contradicts our initial assessment of a person or situation.

If we go back to the air accident, we realize the error that the pilot committed who talked with the commander who had no authorization to fly. The fact of attributing a supposed value of always being right, not just because one is the commander but to because one is the best commander of KLM, attached to the diagnosis bias that made the brain hide very clear and accurate information that they had no authorization to fly, it turned out that the pilot did nothing to stop the plane, despite being an obligation.

6 Conclusions

As we have seen there are many "pitfalls" into which our brain can fall when making a decision. The key is to get to know them in advance for when we are thinking to not fall into any of them.

We should avoid using learned patterns whenever we are thinking. To do this we must check if the premises from which they originate, are learned or reasoned. Another way to avoid falling into learned patterns, is to erase them although these seem necessary for the decision. In other words, we must check our patterns, to be sure that they are true or valid in the current context.

When we are comparing between several options and we have chosen one, we need to review them again to reflect on whether we have eliminated some of them simply by the fact that it is not comparable with anything.

To avoid the anchoring effect, we have to try to eliminate all potential anchors that we have about the object that we are thinking about. That is to say, if we thought that something had a certain value, we should reflect on whether this value is true or it is just there because of the anchoring effect.

Finally, when we have made a decision, we have to check whether we have made it because of the fear of the losing something. At the same time, we should also check whether we have attached some value to someone and we have believed what he or she has said, without even questioning it.

In summary, to be sure that the choices we make are correct, we should systematically check all these aspects to be sure that we have not fallen in any of the 'pitfalls' of our brain.

References

Ariely D (2008) las trampas del deseo. Ariel
Brafman O, Brafman R (2009) Irrational impulse, SLU. Espasa Libros
De Bono E (1994) Creative thinking. power of lateral thinking for the creation of new ideas. McQuaig Group, Inc

Job M, Tesch M (1994) Air disasters: (1994) volume 1. Aerospace Publications, Fyshwick, Australia, pp 165–180
Kahneman D (2001) Thinking fast and slow. Farrar, Straus and Giroux
Lorenz KZ (1935) The companion in the bird's world. J für Ornithologie 83(2–3):137–215, 289–413 (reprint 1965, In: Human and Animal Behaviour, vol 1)
Raichle ME (2006) Published in Science. The Brain's dark energy. N. p., 2017. Print

Lessons Learned in Assessment of Technology Maturity

Marisela Rodríguez Salvador, René Lezama Nicolás, Rosa María Río Belver and Alejandro Rodríguez Andara

Abstract Maturity is a stage in the technology life cycle where technology is highly competitive and it can (safely) be integrated into products or processes. This article analyzes the most important methods to assess maturity and proposes a method based on their best features. On the one hand, Future-oriented Technology Analysis, an approach that incorporates different decision-taking techniques. On the other hand, Technology Readiness Levels, a quantitative scale used to measure technology's progress. The proposed approach uses logistic-growth fitting to match technology publication trends to the Technology Readiness Level scale. Three technologies at different maturity levels were successfully tested.

Keywords Technology maturity assessment · Technology readiness level
Future-oriented technology analysis · Logistic-growth function

---

M. Rodríguez Salvador · R. Lezama Nicolás
Tecnologico de Monterrey, Escuela de Ingeniería y Ciencias, Ave. Eugenio Garza Sada 2501, 64849 Monterrey, N.L., Mexico
e-mail: marisrod@itesm.mx

R. Lezama Nicolás
e-mail: rene.lz.nicolas@gmail.com

R. M. Río Belver (✉) · A. Rodríguez Andara
Foresight, Technology and Management (FTM) Group,
Industrial Organization and Management Engineering Department, University College of Engineering of Vitoria-Gasteiz, University of the Basque Country UPV/EHU, Basque Country, Leioa, Spain
e-mail: rosamaria.rio@ehu.eus

A. Rodríguez Andara
e-mail: alejandro.rodriguez@ehu.eus

Á. Ortiz et al. (eds.), *Engineering Digital Transformation*, Lecture Notes in Management and Industrial Engineering,
https://doi.org/10.1007/978-3-319-96005-0_14

1 Introduction

The acquisition and development of emerging technologies has demonstrated to be an effective booster for expanding new technological capabilities, increasing strategic options, gaining improved efficiency and responding to the competitive environment (Mortara and Ford 2012). However, emerging technologies might be a double edge sword if they are not assessed wisely. As new technological components intertwine with others in an integrated system, likelihood of error grows geometrically.

The technology life cycle determines how technology evolves: from being a scientific concept to a fully functional system. Maturity, as defined by Ernst (1997), is the stage where "technology becomes integrated in products or processes and maintains its high competitive impact". Thus, to diminish the risks when acquiring or developing technology, decision takers need to keenly evaluate how mature it is to their specific context.

There are currently two highly accepted approaches for technology maturity assessment. Chronologically introduced, the first being Future-oriented Technology Analysis (FTA) whose maturity assessment tools—which are mainly qualitative—stem from Technology Forecasting in the 1950s. For many years, these methods were deployed almost exclusively by professional forecasters. The second is Technology Readiness Level (TRL), a nine-level scale that quantitatively describes "increasing levels of technical maturity based on demonstrations of capabilities" (US Government Accountability Office 2016). It was developed by NASA in the 1970s. However, it became globally known in the 1990s, when it boomed as an unprecedented tool for assessing technology maturity following the publication of Mankins' (1995) white paper on "Technology Readiness Levels". Its main strength is the ability to classify technology maturity on a standardized numerical scale. During 1990s and 2000s, TRL's methodologies outshined FTA's. Nevertheless, as TRLs were tested in further fields besides government agencies, serious limitations and critiques appeared. Many TRL-hybrid methods came out as an attempt to counteract them (EARTO 2014), yet few of them were powerful enough to broaden the initial TRLs limitations. In the current decade, FTA methods have made breakthroughs for unsolved challenges by TRL methodologies; especially for predicting when a specific level of maturity will be reached (Gao et al. 2013).

2 Approaches of Technology Maturity Assessment

The following sections will delve further into state-of-the-art methodologies to assess technology maturity. First, the key TRL-based methodologies—developed, improved and published from the 1970s to the early 2000s—will be explained. Then, FTA tools that began in the 1950s and have been updated as recent as 2010s will be reviewed. Lastly, a method where both approaches are linked is proposed.

2.1 Technology Readiness Level (TRL)

The TRL scale was developed by NASA as a "systematic tool that enables assessments of the maturity of a particular technology and the consistent comparison of maturity between different types of technology" (Mankins 2009). Table 1 shows TRLs original definition by Mankins (1995), maturity stages for non-government organizations by the European Association of Research and Technology Organisations (EARTO 2014) and the system fidelity corresponding to each level by the US Department of Energy (DOE 2009).

In the early 2000s, the US Department of Defense (DoD) developed the Technology Readiness Assessment (TRA), i.e. "a systematic, evidenced based process that evaluates the maturity of hardware and software technologies" with the aim to assess the maturity of critical technology elements, namely, a technology element wherein the system depends on it to satisfy operational requirements (US Government Accountability Office 2016). TRA determines TRLs through review teams of subject matter experts, although there are other ways. For instance, there are official reports—such as Gartner's Hype Cycles—that are constructed by experts' surveys and that aim to determine the ideal time for technology adoption (Fenn et al. 2015). In addition, a calculator was developed by Nolte et al. (2003) on a Microsoft Excel spreadsheet as an attempt to create a standard repeatable method to determine TRLs.

TRAs and other TRL-based techniques have spread to non-government organizations, although with significant adaptation. EARTO (2014) has spotted several limitations such as "lack of attention in case of maturity setbacks", "unqualified approach for assessing multiple technologies in a system context" or "focus on product development, rather than manufacturability, commercialization and organizational changes" that confirm the need for the TRLs to be adapted to each organization's specific needs. Solutions to these issues have been provided, however they have often turned out to be uneasy to implement.

2.2 Future-Oriented Technology Analysis (FTA)

Proposed as an attempt to unite future-oriented technology analysis methodologies (such as Technology Foresight, Technology Forecasting, Technology Intelligence, Technology Roadmapping or Technology Assessment), FTA is an approach that incorporates the best practices to anticipate future technology developments (Cagnin et al. 2008). It helps decision makers to anticipate and deal with transformations (Cagnin et al. 2013). The following sections will examine the most notable FTA methods regarding technology maturity assessment.

Table 1 Technology readiness level (TRL) summary

Maturity cluster (EARTO 2014)	TRL	TRLS original definition (Mankins 1995)	System fidelity (US DOE 2009)
Invention	1	Basic principles observed and reported	System exists on paper (no hardware system)
	2	Technology concept and/or application formulated	
Concept validation	3	Analytical and experimental critical function and/or characteristic proof-of-concept	System matches a piece or pieces of the final application
	4	Component and/or breadboard validation in laboratory environment	
Prototyping and incubation	5	Component and/or breadboard validation in relevant environment	System matches final application in almost all respects
Pilot production and demonstration	6	System/subsystem model or prototype demonstration in a relevant environment (ground or space)	
	7	System prototype demonstration in a space environment	
Initial market introduction	8	Actual system completed and "flight qualified" through test and demonstration (ground or space)	System matches final applications in all respects
Market expansion	9	Actual system "flight proven" through successful mission operations	

2.2.1 Technology Forecasting

Technology forecasting methods were an early attempt to assess technology maturity. Its aim has been to determine the most feasible alternatives in technology development (Roper et al. 2011). It includes a range of quantitative and qualitative procedures; some examples are environment monitoring, expert opinion, trend extrapolation, analogies or causal models. These methods are ideal when evaluating one

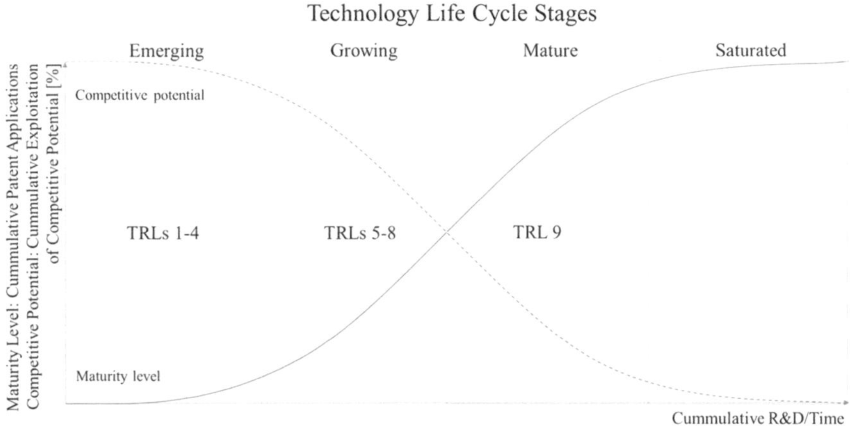

Fig. 1 LCA by means of maturity level and competitive potential. Adapted from Reinhart and Schindler (2010) and Ernst (1997)

technology against other. However, they fall short when assessing the overall level of risk that comes with it. As new methods were developed, attempts were made to overcome this restriction.

2.2.2 Life Cycle Analysis (LCA)

LCA has arisen as an attempt to manage TRL's inability to forecast. Technology can be classified, according to its life cycle stage, as emerging, growing, mature or saturated (Ernst 1997). Figure 1 shows technology life cycle and how TRLs would match to it. When a technology concept is conceived and validated (TRLs 1–4), it belongs to the emerging stage. When it is prototyped and launched (TRLs 5–8), it pertains to the growing stage. Finally, when the technology flourishes (TRL 9), it is considered mature. The saturated stage does not include TRLs, since this stage—where competitive potential is lost—goes beyond TRL intended assessment.

LCA looks for ways to match technology according to the life cycle stages. However, it typically attempts it through qualitative methods. Having no repeatable frameworks, forecasting advances were almost impossible, and the limits of each stage were not well defined—which dimmed objectivity from the assessment (Reinhart and Schindler 2010).

2.3 TRL-FTA Hybrid Model

TRL and FTA methodologies have proven to be symbiotic. The TRLs offer a solid maturity index, however the means to determine them rely almost exclusively on expert assessment. The FTA broadens the techniques for determining maturity, yet it is usually based on LCA, whose stages and transitions are not as well established as TRLs.

In 1997, Watts and Porter published an attempt to link the technology life cycle stages to specific bibliometric indicators. They suggested that fitting growth logistic curves might help visualize the life cycle. This approach has been used to estimate technology substitution by means of market penetration; namely the last transitions along the life cycle (Daim et al. 2010; Paixão-Garcez and Coulter-Wright 2010). However, there are few studies that take Watts and Porter (1997) original proposal to fit logistic curve functions to science and technology indicators. This might have been because growth logistic curves require an upper limit known as "carrying capacity", which in market penetration can reach up to 100%. Nevertheless, science and technology indicators (such as papers or patents) do not reach such limit. Wong and Goh (2010) pointed out that science and technology carrying capacity tends to self-propagate.

The logistic function (also known as Fisher Pry model or Pearl curve) adapted to the proposed method is:

$$r_t = \frac{K}{1 + ae^{-bt}} \tag{1}$$

The variable r_t is the amount of records given in a time t from each database respective life cycle stage. K is the carrying capacity, which will be $\max\{r_t + 1\}$. The variable a is the initial stage of diffusion; and b is the velocity of diffusion. These last two coefficients are obtained by linearizing the function.

Our proposal to link TRLs with life cycle stages is shown in Table 2. Equation 1 is applied on *Typical Sources,* and a coefficient of determination (R^2) is extracted from this function to determine the development level of each stage. As a rule of thumb, if $R^2 \geq 0.80$, then the goodness of fit can be accepted and the TRLs attached to it can be considered overcome.

The Technology Policy & Assessment Center at Georgia Tech (1999) provides a useful MS Excel macro to fit the model and obtain the value of R^2. Table 3 shows the application of our model to three different technologies.

This maturity assessment approach has relevant implications for decision takers who regard technology. It offers a reliable and objective way to determine technology readiness and may therefore aid strategic planning. Thus, by analyzing the maturity assessment applied in our model, important insights might be revealed. S*mart dust* is at an embryonic stage and may interest academic research. *Breast cancer detection* is halfway towards maturity, therefore product prototyping is encouraged to establish a competitive advantage. *Augmented reality* is in a development stage where strategic patents have already been issued; licensing them would be recommended to guarantee market share early on.

Table 2 TRL-FTA hybrid model

TRL	R&D stage	Typical sources	Life cycle stage
1	Basic research	Science citation index	Emerging
2			
3			
4	Applied research	EiCompendex/INSPEC/medline	
5			
6	Development	Patent scope/USPTO/Espacenet/PatSeer	Growing
7			
8	Application	Factiva/LexisNexis	
9			Mature

Adapted from Watts and Porter (1997), who indicated that a certain amount of "hits" would rise and peak in each *Typical Source* as each *R&D Stage* was attained

Table 3 TRL-FTA hybrid model example on 3D printing technologies at different maturity levels

Technology	Basic research (science citation index)[a] TRL 1–3	Applied research (INSPEC/medline)[a] TRL 4–5	Development (PatSeer)[b] TRL 6–7	Application (Factiva)[c] TRL 8–9	TRL
	Growth logistic curve R^2				
Smart dust	0.04	0.02	0.2	0.01	1–3
Breast cancer Detection	0.86	0.81	0.64	0.32	4–5
Augmented reality	0.82	0.83	0.82	0.64	6–7

Each query consisted on the technology term from the first column
[a]The query was applied on the field "Topic"
[b]The query was applied on the fields "Title" and "Abstract"
[c]The query was applied on "Free Text Search" and filtered by "Corporate/Industrial News"

3 Conclusions

This research reviewed the most transcendental technology maturity assessment tools and classified them in two categories: TRL and FTA. After a state-of-the-art review, a method was proposed to match the benefits of both approaches: the quantitatively determination of TRLs to the robustness of the FTA. Logistic curve fitting, which has been typically used for predicting, was arranged to depict the goodness of fit of each life cycle stage. Three technologies at different maturity levels were successfully tested. Results showed that maturity can clearly be positioned within a numbered scale (TRL) based on a bibliometric indicator (publication trend). Future lines of research should focus on narrowing the TRL output, since the proposed model divides the 9-level scale into a 4-level one. Finally, the validation of technology experts is always recommended to accompany the use of this model.

References

Cagnin C, Keenan M, Johnston R, Scapolo F, Barre R (2008) Future-oriented technology analysis: strategic intelligence for an innovative economy. Springer, Berlin

Cagnin C, Havas A, Saritas O (2013) Future-oriented technology analysis: its potential to address disruptive transformations. Technol Forecast Soc Change 80(3):379–385

Daim T, Ruida G, Martin H, Gerdsri P (2010) Forecasting emerging technologies: use of bibliometrics and patent analysis. Technol Forecast Soc Change 73(8):981–1012

EARTO (2014) The TRL scale as a research & innovation policy tool, EARTO recommendations. Available online at: http://www.earto.eu/fileadmin/content/03_Publications/The_TRL_Scale_as_a_R_I_Policy_Tool_-_EARTO_Recommendations_-_Final.pdf

Ernst H (1997) The use of patent data for technological forecasting: the diffusion of CNC-technology in the machine tool industry. Small Bus Econ 9(4):361–381

Fenn J, Raskino M, Burton B (2015) Understanding gartner's hype cycles. Report no. G00251964. Gartner, Inc

Gao L, Porter AL, Wang J, Fang S, Zhang X, Ma T, Wang W, Huang L (2013) Technology life cycle analysis method based on patent documents. Technol Forecast Soc Change 80(3):398–407

Georgia Institute of Technology (1999) Growth curve fitting excel macro. Accessed on 10th May 2017. Available at: https://tpac.gatech.edu/sites/default/files/projectfiles/DSMod_trend_macro-Solar_cells_CUM_data.xls

Mankins JC (1995) Technology readiness levels. White Paper. Available online at: https://www.hq.nasa.gov/office/codeq/trl/trl.pdf

Mankins JC (2009) Technology readiness assessments: a retrospective. Acta Astronaut 65(9):1216–1223

Mortara L, Ford S (2012) Technology acquisition: a guided approach to technology acquisition and protection decisions. University of Cambridge, Institute for Manufacturing, Cambridge

Nolte WL, Kennedy BC, Dziegiel RJ (2003) Technology readiness level calculator. In: NDIA systems engineering conference, vol 20. Available online at: www.dtic.mil/ndia/2003systems/nolte.ppt

Paixão-Garcez M, Coulter-Wright JT (2010) Estudo de modelos de previsão tecnológica aplicados à substituição de embalagens de refrigerantes para o mercado brasileiro. Revista de Administração. 45(3):255–270

Reinhart G, Schindler S (2010) A strategic evaluation approach for defining the maturity of manufacturing technologies. World Acad Sci Eng Technol 47(4):730–735

Roper AT, Cunningham SW, Porter AL, Mason TW, Rossini FA, Banks J (2011) Forecasting and management of technology, 2nd ed. Wiley

US Department of Energy (2009) Technology readiness assessment guide. Guide no. DOE G 413.3-4. Available online at: https://www.directives.doe.gov/

US Government Accountability Office (2016) Technology readiness assessment guide: best practices for evaluating the readiness of technology for use in acquisition programs and projects. Report no. GAO-16-410G. Available online at: www.gao.gov/products/GAO-16-410G

Watts R, Porter A (1997) Innovation forecasting. Technol Forecast Soc Change 56(1):25–47

Wong CY, Goh KL (2010) Modeling the behaviour of science and technology: self-propagating growth in the diffusion process. Scientometrics 84(3):669–686

Assessing Worldwide Research About Performance Measurement for SMEs: 2006–2016

Ximena Rojas-Lema, Juan-José Alfaro-Saiz, María-José Verdecho and Raúl Rodríguez-Rodríguez

Abstract Small and medium enterprises (SMEs) are important contributors to the economic growth of a country; however, maintain high performance levels for long periods of time is always a latent challenge. Research activity analysis through a bibliometric analysis helps to understand the development level that has the performance measurements for SMEs. The aim of this study was to analyze this development in the period of (2006–2016), VOS viewer was used as a graphic visualization tool from the extracted records from the Scopus database. Since 388 analyzed records a slight increase in the performance measurement of SMEs in the last decade was identified. Around of the performance measurement terms and supply chain management it was identified a concentration of investigative relationships; countries and leading research groups in the area were recognized, countries and leading research groups in the area were determined.

Keywords Performance measurement · SMEs · Bibliometric analysis VOSviewer

X. Rojas-Lema (✉)
Departamento de Estudios Organizacionales y Desarrollo Humano,
Escuela Politécnica Nacional, Ladrón de Guevara E11-253, Quito, Ecuador
e-mail: ximena.rojas@epn.edu.ec

J.-J. Alfaro-Saiz · M.-J. Verdecho · R. Rodríguez-Rodríguez
Departamento de Organización de Empresas, Universitat Politècnica de València,
Camino de Vera S/N, Valencia, Spain
e-mail: jalfaro@omp.upv.es

M.-J. Verdecho
e-mail: mverdecho@omp.upv.es

R. Rodríguez-Rodríguez
e-mail: raurodro@upvnet.upv.es

Á. Ortiz et al. (eds.), *Engineering Digital Transformation*, Lecture Notes in Management and Industrial Engineering,
https://doi.org/10.1007/978-3-319-96005-0_15

1 Introduction

Currently, there are concern associated to find the best way to measure and report activities that improve performance in an organization or organizations networks. High performance enterprises should be interested in develop and deploying performance measurement systems (PMS) (Keong Choong 2013). Measure and performance management is critical to the effective and efficient management of any business (Melnyk et al. 2014). Research has shown that through proper measurement and performance management organizations are able to benefit in aspects such as: formulation, implementation and strategy communication, communication of results, motivation of employees and partners, increase of productivity, and increase in customer satisfaction (Micheli and Mari 2014; Saunila and Ukko 2013).

In the field of performance measurement, the size of the enterprise is treated in a generic way. However, it is perceived the interest of large, medium and small enterprises in order to consolidate the PMS as the strategy to achieve a competitive advantage in the market. For SMEs, the number of publications that describe characteristics of PMS is small, but strongly demanded (Galdámez et al. 2009).

For a better understanding of the trends in the field of the performance measurement of SMEs, it is necessary to understand the development of the research activity in the area. This understanding can be achieved through bibliometric analysis using well-known databases (Sweileh et al. 2016). This study considers as a relevant source of information to published articles in the last decade, and its results seek to support future researchers, entrepreneurs, investors, public policy makers involved in the sustainable development of SMEs.

2 Methods

This study used synthesized data of Scopus database due to their important features that facilitate bibliometric analysis. Such features include country and author contribution, title, source titles, keywords, abstract, citation index, among others. Scopus is produced by Elsevier and covers more than 20,000 journals, which have 100% of the global coverage, Scopus is the biggest database even more than Web of Science and Google Scholar (Sweileh et al. 2016, p. 2).

The time period analyzed was since 2006 to 2016, and it included the following issues for refinement: Document type—article; Subject areas—Engineering, Business, Management and Accounting, Social Sciences, Economics, Econometrics, Finance, Mathematics, and Decision Sciences. The search strategy used the following key concepts: performance measurement, management, and SME in title-abstract-key, and excluded specific concepts from other study areas. This has generated the following search equation: *(TITLE-ABS-KEY (performance or maturity) AND TITLE-ABS-KEY ((measurement or evaluation or assessment) and management) AND TITLE-ABS-KEY (cluster or network or chain or SME or small or medium)*

Table 1 Most representative languages in published articles

Language	Frequency	(%)
Chinese	6	2
English	366	94
English; Chinese	1	0
English; German	1	0
English; Portuguese	2	1
English; Spanish	2	1
Portuguese	10	3

AND NOT TITLE-ABS-KEY (transaction or circuit or communication or wireless or energy)) AND DOCTYPE (ar) AND PUBYEAR>2005 AND PUBYEAR<2017.

The bibliometric analysis proposed include publication language, published and cited articles per year, research groups, and location of research activity. The search was made on February 1, 2017 and Bibliometric maps and network visualization methods were made using VOSviewer (http://www.vosviewver.com; van Eck and Waltman 2014), which is a software tool specifically designed for constructing and visualizing bibliometric maps (Cobo et al. 2011).

For co-occurrence analysis with all keywords, VOSviewer and thresholds of minimally seven fractionally counted articles for each term; in this map, most frequent terms had dense colored cluster. For co-authorship analysis, the analyses units were authors and countries. Minimum number of documents for authors was three and for countries was five.

For those three types of analyzes was necessary to create a VOSviewer thesaurus files in order to achieve clusters better defined.

3 Results

3.1 General Data

With the implemented search strategy, 3781 records were obtained, this database was refined through the revision of source titles, titles and abstracts, a final database had 388 records. A first analysis regarding the publication language of performance measurement for SMEs (Table 1) indicated that 94% of articles use English as primary language, Portuguese and Chinese represent 5% of published articles.

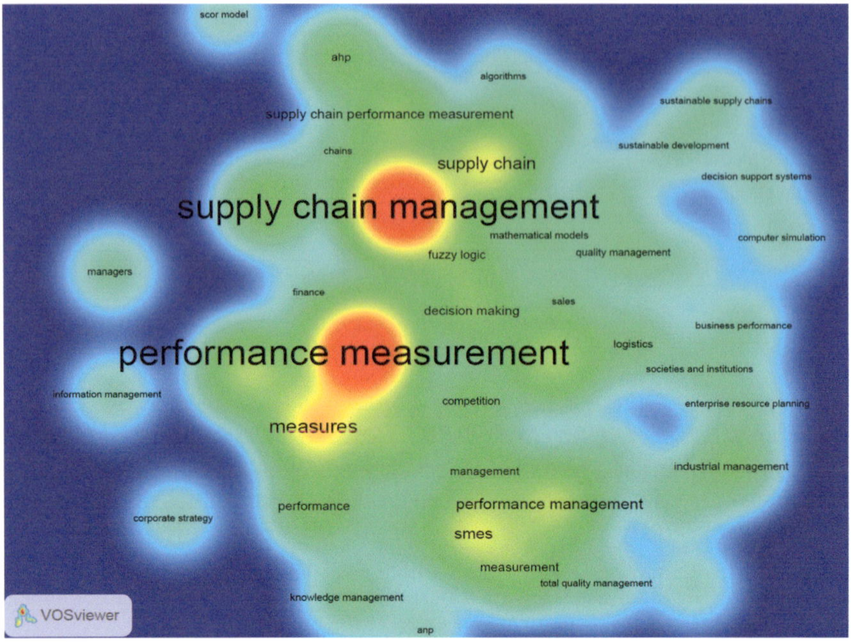

Fig. 1 Most representative terms in the performance measurement for SMEs

3.2 *Keywords Co-occurrence*

The analysis of co-occurrence with all keywords determined 1898 keywords and 48 met the threshold. Figure 1 shows the density visualization map of most frequently encountered terms in all keywords. Five clusters were identified: cluster number one contained 11 terms with supply chain management, performance measurement and measures as terms more representatives; cluster number two contained 10 terms with the SMEs and performance management as most frequent items; cluster number three with 10 items with decision making as the term more representative; cluster number four with 9 items where benchmarking is the most important; and finally, cluster number five had 8 terms, manufacturing industries and business performance are the most frequently encountered.

3.3 *Articles Published in Time*

Table 2 shows the research evolution in the field of performance measurement for SMEs. The annual production of published articles is relatively alike although a slight increase is evident from the year 2013. Figure 2 shows the published articles (principal axis) and the average total citations per article (secondary axis). Articles

Table 2 Evolution of published articles and total citations over time

Year	Published articles	Total citations	Total citations/published articles
2006	25	872	34.9
2007	39	1582	40.6
2008	23	476	20.7
2009	36	699	19.4
2010	35	700	20.0
2011	31	513	16.5
2012	27	341	12.6
2013	33	286	8.7
2014	47	236	5.0
2015	50	103	2.1
2016	42	17	0.4

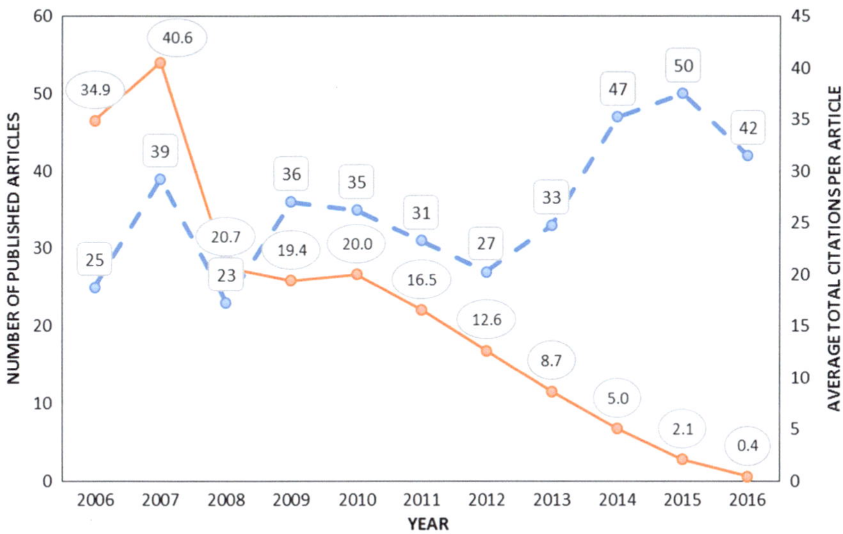

Fig. 2 Evolution of performance measurement research in SMEs over time

published during 2006 and 2007 presented the highest values of citation per article and since 2008 a decreasing attendance for this relation was maintained.

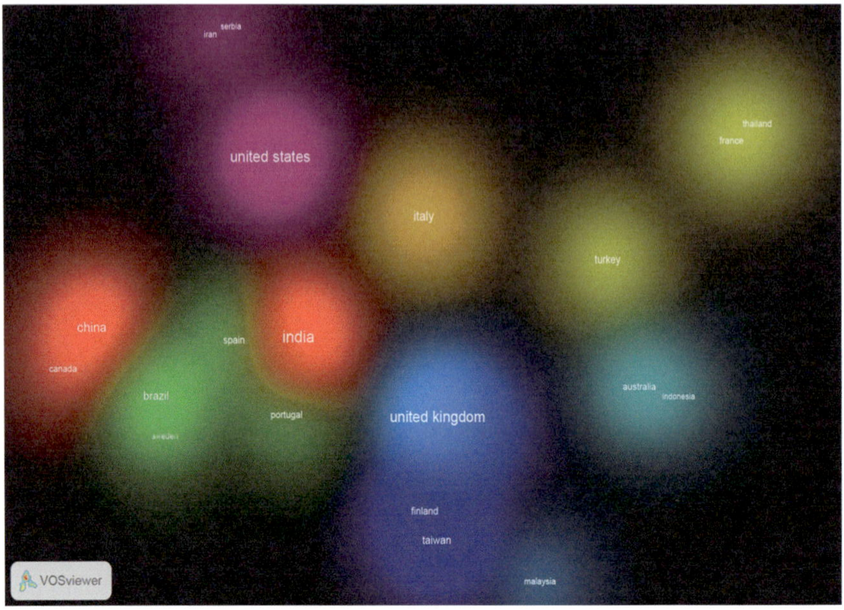

Fig. 3 Most representative countries in studies of performance measurement of SMEs

3.4 Countries

The analysis of co-authorship with countries found that of the 73 countries, 22 met the thresholds. Figure 3 shows the visualization countries density with eight clusters determined: cluster number one contained 4 elements where India was the most important country; cluster number two had 4 elements and Brazil was a representative country; cluster number three contained 3 elements where Taiwan was the most important country; cluster number four contained 3 elements where Turkey was the representative country; cluster number five contained 3 elements where United Stated was the most important country; cluster number six contained 2 elements where Australia was the most relevant; cluster number seven had 2 elements where United Kingdom was the relevant country; and finally cluster number eight with Italy as unique element.

3.5 Authors

Density visualization of co-authorships map with authors showed that of the 903 authors, 20 met the thresholds (Fig. 4). Map showed nine clusters: cluster number one contained 4 authors where Garengo P. was the most important author; cluster number two with 3 members had to Alfaro-Saiz J. J as author more relevant; cluster

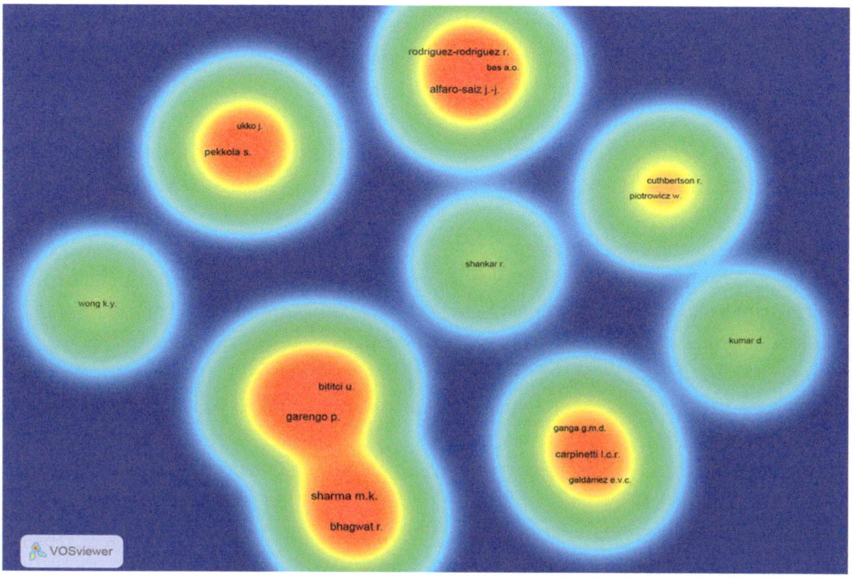

Fig. 4 Density visualization of co-authorships map

number three had 3 members in which Carpinetti L. C. R was the most important author; cluster number four had 3 members where Pekkola S. had highest number of co-authorships; cluster number five with 2 authors had to Sharma M. K as representative author; cluster number six contained 2 members where Cuthbertson R. and Piotrowicz W. had the highest number of co-authorships; clusters number seven, eight, and nine had 1 author, Kumar D, Shankar R, and Wong K. Y respectively.

4 Discussion

English is the research and dissemination preponderant language, through Portuguese and Chinese are also making contributions to the research field.

The analysis of keywords co-occurrence showed zones with greater network density; around to *supply chain management* and *performance measurement* was concentrated an important relations activity; on the other hand, in less measure with a moderate intensity the terms *measures, SMEs, performance management*, and *supply chain* were highlighted. In addition, the map showed terms with an important frequency intensity that although not be found in the search strategy, are associated with the 5 identified clusters, namely: *sustainable supply chain, quality management, knowledge management*, among others. The study pointed out eight countries that has been making significant contributions to performance measurement, and its contribution to the development and competitiveness of SMEs (Yadav and Sagar 2013), studies contemplate specific features of SMEs as factors of inclusion, business environment, innovation criteria, among others (Galdámez et al. 2009; Thakkar et al. 2009). The proximity between countries pointed out a research relationship between them (van Eck and Waltman 2014, p. 288). The study showed the researchers with the greatest scientific production in the last 10 years. It also evidenced the strong internal relation that certain clusters present between their nodes and the global map indicated a weak relation between clusters; due to specific approaches to PMS design, methodology and implementation. The articles analysis published over time showed a growing trend of publication, especially since 2013, which confirm the growing interest of several sectors towards the measurement of performance for SMEs (Garengo and Sharma 2014). On the other hand, the decreasing tendency of the relation between the total of citations and the published articles strengthens the criterion of the increase of the research in this field; however, with a diversity of approaches and applications.

The study identified two important limitations: the first, in relation to the records that might have been left out of the analysis either due to they belong to journals not indexed in Scopus, or due to they correspond to the year 2016, whose records continue to be updated. The second, could have diminished importance to authors and outstanding countries, however recent in the research line.

5 Conclusions

Results of this study have shown a growing but slightly pronounced trend towards research in the field of performance measurement in SMEs from 2013. In this study, the diversity of approaches and own applications of the specificities of SMEs are highlighted.

The density maps showed a concentration of relations around of the terms of performance measurement and supply chain management. The most representative terms around them showed the main research and development trends in the last decade associated with SMEs. The countries analysis identified those with the high distribution to the theme and whose proximity reflects the level of relationship between them. The map of authors highlighted to the main research clusters in the area.

The analysis elements used in this study showed a growing worldwide interest in order to support the measurement of the performance of SMEs and thus contribute to the formulation of competitive business strategies.

References

Cobo MJ, López-Herrera AG, Herrera-Viedma E, Herrera F (2011) Science mapping software tools: review, analysis, and cooperative study among tools. J Am Soc Inf Sci Technol 62:1382–1402. https://doi.org/10.1002/asi.21525

Galdámez EVC, Carpinetti LCR, Gerolamo MC (2009) Proposta de um sistema de avaliação do desempenho para arranjos produtivos locais. Gest Produção 16:133–151

Garengo P, Sharma MK (2014) Performance measurement system contingency factors: a cross analysis of Italian and Indian SMEs. Prod Plan Control 25:220–240. https://doi.org/10.1080/09537287.2012.663104

Keong Choong K (2013) Understanding the features of performance measurement system: a literature review. Meas Bus Excell 17:102–121. https://doi.org/10.1108/MBE-05-2012-0031

Melnyk SA, Bititci U, Platts K, Tobias J, Andersen B (2014) Is performance measurement and management fit for the future? Manag Account Res 25:173–186. https://doi.org/10.1016/j.mar.2013.07.007

Micheli P, Mari L (2014) The theory and practice of performance measurement. Manag Account Res 25:147–156. https://doi.org/10.1016/j.mar.2013.07.005

Saunila M, Ukko J (2013) Facilitating innovation capability through performance measurement: a study of Finnish SMEs. Manag Res Rev 36:991–1010. https://doi.org/10.1108/MRR-11-2011-0252

Sweileh WM, Shraim NY, Al-Jabi SW, Sawalha AF, Rahhal B, Khayyat RA, Zyoud SH (2016) Assessing worldwide research activity on probiotics in pediatrics using scopus database: 1994–2014. World Allergy Organ J 9. https://doi.org/10.1186/s40413-016-0116-1

Thakkar J, Kanda A, Deshmukh SG (2009) Supply chain performance measurement framework for small and medium scale enterprises. Benchmarking Int J 16:702–723. https://doi.org/10.1108/14635770910987878

van Eck NJ, Waltman L (2014) Visualizing bibliometric networks. In: Ding Y, Rousseau R, Wolfram D (eds) Measuring scholarly impact. Springer International Publishing, Cham, pp 285–320. https://doi.org/10.1007/978-3-319-10377-8_13

Yadav N, Sagar M (2013) Performance measurement and management frameworks: research trends of the last two decades. Bus Process Manag J 19:947–971. https://doi.org/10.1108/BPMJ-01-2013-0003

Organizational Structures in Small and Medium-Sized Enterprises (SMEs) and Their Performance Measurement Systems

Ximena Rojas-Lema, Juan-José Alfaro-Saiz, Raúl Rodríguez-Rodríguez and María-José Verdecho

Abstract Several research contributions have shown that SMEs, in their path towards development and competitiveness have incorporated new trends of configuration and organization with the collaborative axis. Which in addition to performance measurement systems play a fundamental role in the management of organizations both individually and agglomerated. The aim of this study was to identify spaces for the research contribution from the review of relevant work in the recent years in the field of performance measurement in SMEs. The evaluated proposals indicated the performance measurement as a key element inside the performance management due to its results support strategic, tactical, and operational decision-making. In addition, the need for validation of the proposed frameworks was also evidenced because they come from specific case studies. Which was achieved reinforced the idea that performance measurements for SMEs is an open space to theoretical and practical contributions.

Keywords SME · Performance measurement · Organizational structures

X. Rojas-Lema (✉)
Departamento de Estudios Organizacionales y Desarrollo Humano,
Escuela Politécnica Nacional, Ladrón de Guevara E11-253, Quito, Ecuador
e-mail: ximena.rojas@epn.edu.ec

J.-J. Alfaro-Saiz · R. Rodríguez-Rodríguez · M.-J. Verdecho
Departamento de Organización de Empresas, Universitat Politècnica de València,
Camino de Vera S/N, Valencia, Spain
e-mail: jalfaro@omp.upv.es

R. Rodríguez-Rodríguez
e-mail: raurodro@upvnet.upv.es

M.-J. Verdecho
e-mail: mverdecho@omp.upv.es

Á. Ortiz et al. (eds.), *Engineering Digital Transformation*, Lecture Notes in Management and Industrial Engineering,
https://doi.org/10.1007/978-3-319-96005-0_16

1 Introduction

There are different structural schemes that SMEs have adopted in order to grow and develop their activities in an increasingly demanding environment, for that high levels of competitiveness and organization are required. The structures formed by SMEs need management strategies that support cooperative work, these strategies allow them to have a continuous improvement. A performance measurement system (PMS) is consolidated as the key tool for performance management and sustained growth of the organization. Several studies of PMS for SMEs are timely and specific with a necessity of the validation of proposals and the far-reaching analysis. From this, it is visualized the space for the investigation towards fields not yet explored in SMEs and their PMS.

The present work is divided into five sections besides the introduction, the first is the methodology used, next the summary of the importance of SMEs in the productive and economic development, and the different organizational structures that have been used in order to potentiate their operations. In the fourth issue highlights the relevance of measuring performance in organizations. Finally, in the last issue an extract of the frameworks and models relevant in the environment of the measurement of performance in SMEs is presented.

2 Methodology

The study was developed through the literature review, using the next keywords: performance, measurement, management, SMEs, organizational structures. In the first place some important issues, models, frameworks, and challenges that have the performance measurement systems in SMEs were identified. Then, the articles that emphasis on SMEs and performance measurement were considered.

3 Small and Medium-Sized Enterprises (SMEs)

New patterns of international and national competitiveness challenge companies to look for new strategies in order to obtain competitive advantages, in this context, the SMEs are subjected to internal and external restrictions normally higher than those of large enterprises, which limits their development (Dolz et al. 2014, p. 162). The SMEs are important for the economic growth of a region and their business activities have become even more prominent over time (Carree and Thurik 1998, p. 630). However, it is a great challenge to them maintain a high sustainable and consistent performance for the long term (Ates et al. 2013, p. 28). The adoption of advanced management practices in the principal business processes is the key to improve the performance and competitiveness for SMEs (Ates et al. 2013, p. 29).

4 Structures and Levels of Organization in SMEs

New ways of relations intra and intercompany give way to the development of enterprises networks, and the industrial agglomerations that gain relevance not only in the industrialized countries economy, but also in those with emerging economies (dos Reis and Amato Neto 2012, p. 346). The potential synergy between enterprises and their networks can generate capacities in learning and increase the skills to innovate and introduce new products and services (Martins et al. 2015, p. 120).

Nowadays, the participation in networks has become in an aspect of high importance for organizations that are looking for increase their competitive advantage in the market. Even more for SMEs that according to Galdámez et al. (2009) the lack of an organizational model limits its industrial development, the competitiveness and decision making.

These new tendencies of configuration and organization have emerged to describe different levels or strategic partnership schemes. Their objective is share resources in order to generate the information interchange, reduce risks, costs, delivery time, and increase knowledge and skills (Pekkola et al. 2013, p. 180); there are some of the concepts developed, for instance, extended enterprise (EE), virtual enterprise (VE), enterprise networks (EN), local production arrangements (LPA) or clusters, supply chain (SC), among others.

"A EE is a conceptual business unit or system that consists of a purchasing company and suppliers which collaborate closely in such a way as to maximize the returns to each partner" (Childe 1998; Bititci et al. 2005, p. 337). Bititci et al. (2005) cite relevant aspects of EE; requirements as: integration of knowledge through improved levels of education, strong critical attitudes and involvement of systems and support tools. Improvement of communication and synchronization between individual enterprises (Martinez et al. 2001). Develop of skills to sharing of learning (Coscia et al. 2002; Preiss 1999).

The VE is represented as a promising form of cooperation. According with Mason et al. (2008) concepts of company's communities and virtual company's communities offer the best description about the performance of a SMEs network that share knowledge to develop regional clusters, through generation of intellectual capital, establishment of innovation culture, creation of value networks, and promotion of social capital (de Oliveira and Martinelli 2015, p. 54). A VE is a cooperation interorganizational on a temporary basis, made up of several legally independent enterprises in an IT-supported network and configured flexibly (Peters et al. 2007).

EN are business structures, where independent signatures are strategically linked horizontally or vertically in order to achieve common objectives (Saiz et al. 2007). EN combined with the performance measurement are attracting a particular interest of managers and researchers. They identify these collaborative relationships as fundamental for the development of business models for managing interorganizational performance in dynamic and global environments.

In the bibliography are collected several concepts around the phenomenon of grouping or agglomeration business (Galdámez et al. 2009, p. 137), such as: industrial

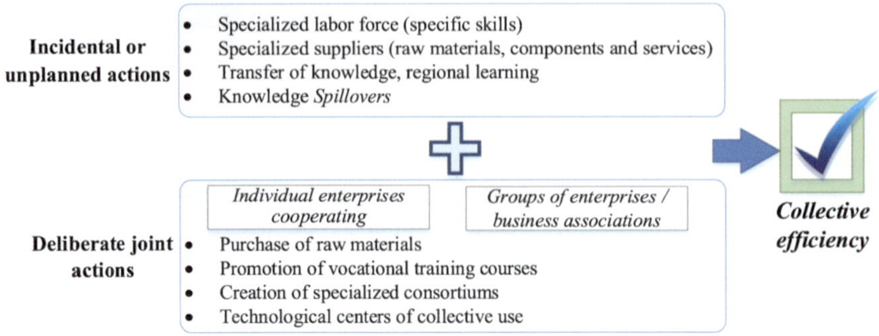

Fig. 1 Collective efficiency, result of LPA (dos Reis and Amato Neto 2012)

clusters, productive systems and innovative locations (Cassiolato and Lastres 2003); local production arrangements (Cassiolato and Lastres 2003; Santos and da Guarneri 2000); industrial districts (Schmitz and Musyck 1994); industrial agglomerations (Naciones Unidas 2011). Figure 1 identifies collective efficiency as a result of the combined development of incidental or unplanned actions derived from the external economies present in the area with deliberate joint actions (dos Reis and Amato Neto 2012).

Awareness about importance of a management strategy to achieve business excellence is shaping a wide range of SMEs. This allows visualization SC and its management as a tool for integrating organizational units and their business processes through the coordination of material flows, information and financial services in order to accomplish the demand of customers with the aim of improving the competitiveness of the chain as a whole (Thakkar et al. 2009, p. 719). The supply chain management is perceived as a tool that ensures the continuous improvement of several enterprises in the competitive market. However, findings in the literature indicate that maximization of the overall benefit of SC through collaborative efforts of members of the chain is little worked (Thakkar et al. 2009).

5 Performance Measurement System (PMS)

Several have been the frameworks with remarkable contribution in the literature that indicate the importance of the Performance Measurement and Management (PMM) in organizations and organizational structures. According to Yadav and Sagar (2013) these can be placed in five categories that attend specific characteristics like: Classical and dominant PMM frameworks, Holistic and integrated PMM frameworks, Frameworks updating balanced scorecard (BSC) approach, Context-specific PMM frameworks and recently developed PMM frameworks (pp. 963–964). These approaches have been basically designed for medium and large enterprises.

According to Ates et al. (2013) most of PMS with an emphasis on SMEs are theoretically valid. However, in practice the results of PMS implementation are poor due to the lack of understanding of the fundamental characteristics of SMEs (p. 35). From this perspective, the authors identify eight main characteristics of influence in the management practices of the SMEs: short-term priorities, internal operational focus and lack of external orientation, tacit knowledge, looking for flexibility, poor managerial skills, entrepreneurial orientation, command and control culture, and limited resources.

A dynamic and integrated PMS should have the following essential characteristics: "identify areas and performance measures associated with the company's strategy and objectives, identify time as a strategic measure, and become a tool for improvement (Ghalayini and Noble 2013)" (Alfaro et al. 2007, p. 54).

The tendency of the countries is to promote the structured performance of SMEs, a PMS is visualized as a tool to improve the performance of SMEs, and to offer a positive impact in the different management areas when it is conducted correctly (Galdámez et al. 2009; Saunila et al. 2014). From the evaluation of the results, subsequent to the measurement of the performance it will be derived specific action plans, incentive mechanisms for the benchmarking between SMEs, coordination actions of improvement, and hence a positive circle of control (Sousa et al. 2006; Galdámez et al. 2009, p. 138).

6 Performance Measurement in SMEs

Table 1 highlights relevant issues of four frameworks or models aligned to the performance measurement focused on SMEs.

The framework developed by Bititci et al. (2005) considers that PSM for EE is structurally and operationally different from a measurement system for a traditional company, because of the structural difference between them. The model is based on the collaborative architecture that forms the EE, where the business units of different enterprises are united and develop a longitudinal structure that integrates processes and activities of the different individual units. This framework proposes to include intra and inter-enterprise coordination measures.

The proposal of Peters et al. (2007) points to VE as a promissory form of cooperation that can be flexibly configure on short notice to process a certain order. The PMS develops information, regulation, explanation, and coordination functions in the network. Besides, it has basic measurements that guarantee its validity and usefulness. The system structure considers the network, project, and process levels with specific performance dimensions in each level. For the network and project levels are considered: financial, process, customer, employees and partners; and for the process level are important measurements of effectiveness and efficiency.

Saiz et al. (2007) propose a PMS-EN methodology that is composed of three phases: definition of strategic framework; definition of process framework; and finally, follow-up and monitoring. In phase 1, it is necessary to distinguish between

Table 1 Performance measurement frameworks for SMEs

Author(s) and year	Structure SMEs	Finding(s) and contribution(s) of paper	Limitation(s)	Methodology
Bititci et al. (2005)	EE	• It recognizes the main characteristics of EE and the importance of specific • Intrinsic and extrinsic coordination measurements are identified • Supports the identification and management of business units processes in EE	• Poor understanding of the tern EE PMS for SMEs • Collaboration between partners, suppliers, customers, designers, and research institutions creates a level of competence very specific	Innovative constructivism (case study)
Peters et al. (2007)	VE	• It needs a IT-support for the implementation and utilization of PMS • It allows the integration of objectives from all levels • Specific process and subprocess measures for each project undertaken • The PMS has a specific functions and measures of validity and utility	The results cannot be generalized	Case study

(continued)

Table 1 (continued)

Author(s) and year	Structure SMEs	Finding(s) and contribution(s) of paper	Limitation(s)	Methodology
Saiz et al. (2007)	EN	• It describes a new PMS to EN • It highlights that both equity and trust will delimit the success or the failure of common objectives and strategies within both inter-enterprise	Study in a specific project	Constructivist approach
Galdámez et al. (2009)	LPA	• It integrates a systematic PMS with a continuous improvement tool • It recognizes the importance of expert managers (SMEs managers, local government agents and the academy) • Influence factors were identified for the performance management process • It helps monitoring the performance of SMEs in a business cooperation network and it addresses collective improvement actions	Generic model without detail of the construction	Investigation—action (case study)

(continued)

Table 1 (continued)

Author(s) and year	Structure SMEs	Finding(s) and contribution(s) of paper	Limitation(s)	Methodology
Thakkar et al. (2009)	SC	• It integrates the best practices of SCOR model and BSC • The model defines metrics and measures for each steps and level and a guide for implementation and use of the PMS • Will help SMEs managers to improve the visibility of their chain • This proposal uses a systematic perspective with procedures, policies and cultural aspects of SMEs	• Study in SMEs manufacturing segment • The *return* activity is not considered inside the model SCOR	Case study

three functional levels: enterprise networks, supply chain and individual enterprise; in phase 2, the performance management elements are defined for vital business process from the network ambit; in the last part the objective is to carry out a follow up of all the performance measurement elements.

Galdámez et al. (2009) propose the integration of the PMS into a systematic process of improvement for the performance management of an LPA where SMEs are closely related by their geographical location, and they have effective participation of local government agents. The system begins with the definition of goals, perspectives of performance management, managers and metrics; below the view of the actors of the cluster is considered, as well as criteria of simplicity and aggregation of value in actions of decision making (Thakkar et al. 2009). This proposal leads more secure and reliable information for decision-making and cooperative work, as well as the allocation of financial resources to critical areas of SMEs; additionally, identifies influence factors in performance management: (a) different visions of agents and managers; (b) limited human resources; (c) deficiencies in the infrastructure and cultural aspects of SMEs; (d) presence of management problems.

The model proposed by Thakkar et al. (2009), integrates a block structure offered by supply chain operation reference (SCOR) model and their decision areas (plan, source, make and deliver) with BSC and its four perspectives (customer service, finance and marketing, internal business and innovation, and learning) in order to obtain the most effectiveness of PMS. The framework starts with de definition of business objectives for SMEs and the identification of specific measures. SCOR model and BSC are linked through the connection of the decision areas with the planning processes to SC (Lockamy and McCormack 2004). The defined processes are categorized in strategic, tactical and operational levels, its elements, metrics and sub processes are also identified. The PMS includes policies, procedures and cultural aspects of SMEs with a systemic vision.

7 Conclusions

This work started with relevance determination of SMEs and its development of the current economy, structures identifications, and organization mechanisms adopted by SMEs. Finally, it highlighted the important points of PMS in SMEs.

- SMEs have been incorporating new trends configuration and organization with collaborative characteristics, which indicate special schemes of grouping and productive ordering.
- Important contributions were identified for SMEs mechanisms and tools for performance measurement. However, the specific application in case studies limited the validity and generalization of the proposed models.
- The work showed that performance measurement systems play a fundamental role in the performance management of organizations both individually and agglomerated. It reinforced the idea that the research towards models or frameworks of performance measurement for SMEs is an open space to theoretical and practical contributions.

References

Alfaro JJ, Ortiz Á, Rodríguez R (2007) Sistemas de medición del rendimiento para la cadena de suministro. Editorial UPV, Valencia

Ates A, Garengo P, Cocca P, Bititci U (2013) The development of SME managerial practice for effective performance management. J Small Bus Enterp Dev 20:28–54. https://doi.org/10.1108/14626001311298402

Bititci US, Mendibil K, Martinez V, Albores P (2005) Measuring and managing performance in extended enterprises. Int J Oper Prod Manag 25:333–353. https://doi.org/10.1108/01443570510585534

Carree MA, Thurik AR (1998) Small firms and economic growth in Europe. Atl Econ J 26:137–146. https://doi.org/10.1007/BF02299356
Cassiolato JE, Lastres HM (2003) O foco em arranjos produtivos e inovativos locais de micro e pequenas empresas. Pequena Empresa
Childe SJ (1998) The extended concept of co-operation. Prod Plan Control 9:320–327. https://doi.org/10.1080/095372898234046
Coscia E, Maiza M, Gauberti C, Arana JM, Zufiaurre M, Meo F, Galassini M, Leveaux JM, Smithers T, Bititci U, Albores P, Martinez V (2002) K-FLOW: knowledge management in the extended manufacturing enterprise. In: Proceedings of the 2002 Ebusiness and eWork conference
de Oliveira MF, Martinelli DP (2015) Desenvolvimento local e arranjos produtivos locais: uma revisão sistemática da literatura. Interações Campo Gd. 15
dos Reis AP, Amato Neto J (2012) Aprendizagem por cooperação em rede: práticas de conhecimento em arranjos produtivos locais de software. Produção 22:345–366. https://doi.org/10.1590/S0103-65132012005000023
Dolz C, Safón V, Iborra M, Dasí A (2014) The antecedents and dynamism of organizational ambidexterity in SMEs. Innovar 24:161–175
Galdámez EVC, Carpinetti LCR, Gerolamo MC (2009) Proposta de um sistema de avaliação do desempenho para arranjos produtivos locais. Gest Produção 16:133–151
Ghalayini AM, Noble JS (2013) The changing basis of performance measurement. Int J Oper Prod Manag. https://doi.org/10.1108/01443579610125787
Lockamy A, McCormack K (2004) Linking SCOR planning practices to supply chain performance: an exploratory study. Int J Oper Prod Manag 24:1192–1218. https://doi.org/10.1108/01443570410569010
Martinez MT, Fouletier P, Park KH, Favrel J (2001) Virtual enterprise–organisation, evolution and control. Int J Prod Econ 74:225–238
Martins I, Rialp A, Rialp J, Aliaga-Isla R (2015) El uso de las redes como propulsor para la orientación emprendedora y el crecimiento de la pequeña y mediana empresa. Innovar 25:117–130. https://doi.org/10.15446/innovar.v25n55.47227
Mason C, Castleman T, Parker C (2008) Communities of enterprise: developing regional SMEs in the knowledge economy. J Enterp Inf Manag 21:571–584. https://doi.org/10.1108/17410390810911186
Naciones Unidas (2011) El desarrollo económico en África: El fomento del desa-rrollo industrial de África en el nuevo entorno mundial [WWW Docu-ment]. http://unctad.org/es/Docs/tdb58d2_sp.pdf. Accessed 7 Feb 2018.
Pekkola S, Niemi P, Ukko J (2013) Building understanding of the development of performance management for collaborative networks with a knowledge maturity model. Int J Netw Virtual Organ 12:179–200
Peters M, Odenthal B, Schlick C (2007) Performance measurement system for virtual enterprises in the aerospace industry. In: Supply chain event management. Springer, pp 85–102
Preiss K (1999) Modelling of knowledge flows and their impact. J Knowl Manag 3:36–46. https://doi.org/10.1108/13673279910259376
Saiz JJA, Bas AO, Rodríguez RR (2007) Performance measurement system for enterprise networks. J Product Perform Manag, Int. https://doi.org/10.1108/17410400710745324
Santos AMMM, da Guarneri LS (2000) Características gerais do apoio a arranjos produtivos locais. BNDES Setorial Rio Jan
Saunila M, Pekkola S, Ukko J (2014) The relationship between innovation capability and performance: the moderating effect of measurement. Int J Product Perform Manag 63:234–249. https://doi.org/10.1108/IJPPM-04-2013-0065
Schmitz H, Musyck B (1994) Industrial districts in Europe: policy lessons for developing countries? World Dev 22:889–910. https://doi.org/10.1016/0305-750X(94)90060-4
Sousa SD, Aspinwall EM, Guimarães Rodrigues A (2006) Performance measures in English small and medium enterprises: survey results. Benchmarking Int J 13:120–134. https://doi.org/10.1108/14635770610644628

Thakkar J, Kanda A, Deshmukh SG (2009) Supply chain performance measurement framework for small and medium scale enterprises. Benchmarking Int J 16:702–723. https://doi.org/10.1108/14635770910987878

Yadav N, Sagar M (2013) Performance measurement and management frameworks: research trends of the last two decades. Bus Process Manag J 19:947–971. https://doi.org/10.1108/BPMJ-01-2013-0003

Resilience Engineering: Concepts of the New Paradigm

Juan Carlos Rubio-Romero, María del Carmen Pardo-Ferreira, María Martínez Rojas, Antonio López-Arquillos and Manuel Suarez-Cebador

Abstract Resilience Engineering is a new safety paradigm that considers that the way duties are normally performed is something that is subject to variability, and since change is necessary to achieve success it should not be restricted. People in this complex context are continuously making adjustments to the original design, enabling them to achieve success, but occasionally accidents also arise due to an incomplete analysis of the current conditions. Thus, linear, simple or complex causation models do not reflect the ongoing reality, which is essentially non-linear, besides being complex. The accident "emerges" from normality, due to concurrent events that "resonate", and are not "caused" by a simple chain of errors. Therefore, new tools are required to analyse accidents and indicators to monitor processes, even though most simple incidents continue to be dealt with in the usual manner.

Keywords Resilience engineering · Health and safety · Normal complex models Causation · Resonance

J. C. Rubio-Romero (✉) · M. del C. Pardo-Ferreira · M. M. Rojas · A. López-Arquillos · M. Suarez-Cebador
Dpto. de Economía y Administración de Empresas, Escuela de Ingenierías Industriales, Universidad de Málaga, C/Dr,
Ortiz Ramos S/N, Málaga 29071, Spain
e-mail: juro@uma.es

M. del C. Pardo-Ferreira
e-mail: carmenpf@uma.es

M. M. Rojas
e-mail: mmrojas@uma.es

A. López-Arquillos
e-mail: investigacioncatedra@gmail.com

M. Suarez-Cebador
e-mail: suarez_c@uma.es

Á. Ortiz et al. (eds.), *Engineering Digital Transformation*, Lecture Notes in Management and Industrial Engineering,
https://doi.org/10.1007/978-3-319-96005-0_17

1 Resilience Engineering

The word "resilience" comes from the Latin "resilio", which means to turn back, jump back, rebound. The term was adopted in social science to describe individuals who, despite suffering hardships, are psychologically sound and successful (Montero Martínez 2011). At the end of 2014, the Spanish Royal Academy included a definition for resilience:

1. f. Psychology. Capacity of humans to recover quickly from difficulties and overcome them.
2. f. Mechanics. Capacity of an elastic material to absorb and store energy when deformed.

A general definition of Resilience Engineering (henceforth RE) could be that of the safety management paradigm that focusses on "how to help people deal with complexity under pressure to achieve success" (Woods 2005).

Resilience Engineering first appeared in 2006 as a collective movement when the minutes of a conference in Sweden were published (Hollnagel et al. 2006), led by Hollnagel, Woods and Levenson. It is also undoubtedly an individual movement led by Erik Hollnagel and David Woods, who started from Cognitive Systems Engineering (henceforth CSE) and restructured it. Other authors, such as Sidney Dekker, have also contributed to developing RE. In the field of health and safety in the construction industry, one of the most prolific authors has been Saurin (Costella et al. 2009; Saurin et al. 2008; Saurin and Carim Junior 2011).

Le Coze (2013) stated that the main ideas in RE could be summarized as follows:

(1) The understanding of variability is more appropriate than a study of errors.
(2) It is more important to study normal operation than to study incidents or accidents.
(3) Control and contextual models are better than normative models.
(4) Engineering and risk assessment are fundamental.

Despite the varying definitions, the concept of RE has been evolving, so that it can be said that there are currently 4 types of RE (Woods 2015).

(1) Resilience as bouncing back: This refers to the capacity to recover and resume normal performance, regain balance, go back to the situation prior to a disruption, to deal with surprise and return to the initial state. This skill relies heavily on structures implemented before the disruption arises, in order to adequately cope with surprise. This is a case of responding to unexpected circumstances, specific disruptions that are not envisaged within normal operation and which, therefore, the system will be able to manage. Surprise represents a challenge and brings a process of learning and review with it.

(2) Resilience as robustness: This refers to the skill of absorbing disturbances, so that many people confuse robustness with resilience. Logically, an increase in robustness increases the capacity to absorb disturbance. However, the issue is that robust control only works in cases where the disturbances are modelled correctly. Again, it is impossible to overcome a disturbance if it exceeds the designed management capacity of the system. Thus, the system will collapse if it cannot withstand demands greater than its capacity.

(3) Resilience as an opposing concept to fragility: Or how to extend adaptive capacity to deal with surprise. Systems in changing situations and with finite resources are always striving to fit events in with challenges. With no capacity to continue working to manage extremes, the system is more fragile than tough. Obviously, a difficulty normally signifies that the limits of the situation are not certain. Woods's "graceful extensibility" is based on the dynamism required to confront a wave of disturbances

(4) Resilience as sustained adaptability: This refers to the capacity for sustained management of adaptability, not to the simple ability to adapt. Thus, for example, some systems can adapt to specific changes, but collapse in the face of new types of change.

The principles of RE were set out by Wreathall (2006): (1) Commitment from senior management. Resilient safety is an important objective for an organization's managers and must be on the same level or above other objectives. (2) Culture of justice or equality. An atmosphere of trust in the organization that encourages workers to raise issues on health and safety at work without fear or rejection. (3) Culture of learning. A company culture of not only learning from problems and incidents, but also from normal performance. (4) Awareness and clarity. Awareness and a lack of clarity are critical when a judgement has to be made on an issue, and something has to be sacrificed, and also to keep ahead of future changes in the environment, as these may affect the capacity of the system to function. (5) Flexibility. The capacity to actively anticipate threats and be prepared to handle these. (6) Preparation. The capacity of an organization to restructure itself in response to changes and modifications, being able to support human error, and with employees who can make critical decisions without having to wait for managers to decide.

2 Methodology

The main aim of current study is to analyze this new paradigm of Resilience Engineering in relation to other already existing causation models, both simple and complex. This study was developed using the main academic search engines and data base available online as: Web of Science (WOS), Science Direct, Scopus, Wiley Online, PubMed, Mendeley and Google Scholar.

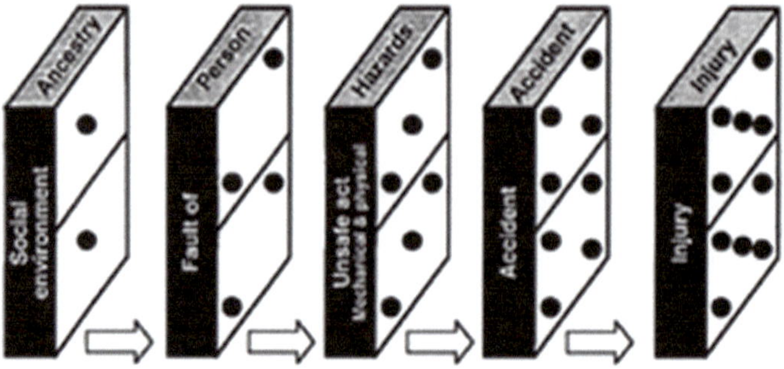

Fig. 1 Heinrich's simple linear causation model (Domino) (Hollnagel 2004)

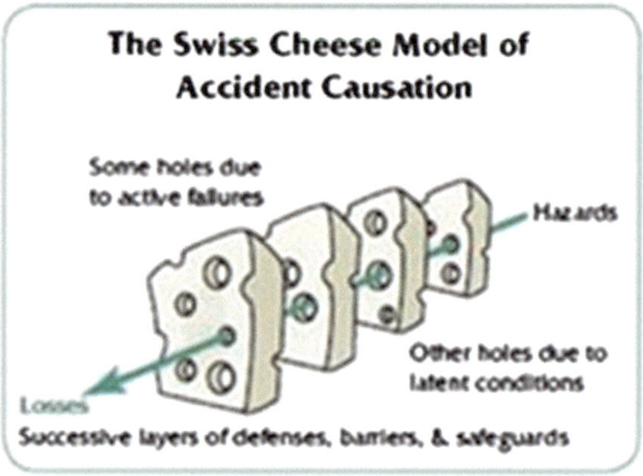

Fig. 2 Reason's complex linear causation model (Swiss cheese) (Reason 2008)

3 Linear, Simple and Complex Causation Models

On delving further into the ideas, Resilience Engineering finds that the accident models used are not accurate, or lack sufficient veracity in many situations in the current environment. Basically, the previous models, from Heinrich's simple linear model (Domino) (Fig. 1) to Reason's complex, linear models (Swiss cheese) (Fig. 2) and later ones, assume that an accident is merely a disruption in a naturally stable system. Thus, safe systems can be obtained either by finding the weak domino and removing it, or in the complex case, blocking the holes in the cheese, meaning interrelationships between unsafe actions and the weakness of buffers or defences. Both models focus on the "components" and "structures" rather than the functions linked to such components and structures. These models see accidents as the result of a linear combination of events (Hollnagel et al. 2006).

According to this approach, accidents happen because people are careless, inexperienced or have the wrong attitude. In addition, technology and materials are not perfect, so failure is inevitable unless variability is restricted and reduced. Correct functioning comes from thorough, well-designed procedures, as the designers can act in advance and plan for even the smallest contingencies, and because people behave as they should and do as expected of them, and even more important, as how they have been shown and trained. The approach is for accomplishment within the designed "mode". The current model accepts that teams are highly reliable, that workers and managers are vigilant and well-trained, and the system is friendly. In this context, humans as machines that can break down are a responsibility, and variability is a threat. Restrictions are the answer (buffers, procedures, standardization, regulation, etc.) (Steen and Aven 2011; Hollnagel 2007).

Effort is made to ensure that the number of adverse results (either from something going wrong or because a risk that may go wrong has been identified) is kept as low as possible. Causes are sought and attempts made to control risk. It is a reactive approach to a negative finding. At high frequency, it may become impossible or difficult to take care of basic activities in the company, the primary ones that constitute the lifeblood of the business. This happens in several situations, such as overburdened emergency services, disasters, etc., but also in fairly complex companies in normal situations, such as an industrial plant, which nevertheless expends a great deal of energy on keeping up with the reports on minor accidents required by law. When something completely unknown occurs, the system is very slow to respond: analysis, finding a solution, what does or does not work, etc. The negative side to this is that inefficient solutions tend to be found in a hurry (Hollnagel 2009).

4 The Non-linear Normal Systemic Model of Causation

However, the work environment has changed drastically, and these cases are no longer valid for many situations. Some accidents do not conform to the models (Perrow 1984). This leads to an explanation of the unexpected combination or accumulation of events, also known as "concurrence" or "resonance". Thus, a view of the accident emerges as a non-linear phenomenon arising from a complex system, what is called "systemic accident models". This model recognises that systems are always variable, due as much to change in the surroundings (exogenous) as to alterations in the component sub-systems (endogenous). Endogenous variability is mainly attributable to people, either individually or in groups. However, this does not in any way imply that human performance is in error or flawed, but the opposite variable performance is necessary to achieve success by confronting the complexity of the real world (Hollnagel 2012).

Therefore, in this new paradigm, "normal performance" is differentiated from "normative performance". The rules and procedures are not what is performed, but the result of adjustments and variations of these, which are necessary in a changing, unpredictable environment. The results of actions taken can sometimes differ from what was intended, what was expected or required, and when this occurs, it is more often due to variability in the context and conditions than to faults in the actions, components or functions. With individual human performance, local optimization or adjustments are the rule more than the exception. In fact, adaptability and flexibility of work done by humans is exactly the reason for efficiency. "Normal" actions are successful because people adjust their behaviour to local conditions. They rapidly learn to keep ahead by using variations, enabling them to be proactive, in addition to saving time, which is needed to assess the situation and success.

Equally, adaptability and flexibility in work done by humans is also the reason for failures, although it is rarely the "cause" of failing (Rasmussen 1997). Actions and failures are almost always based on limited analysis of current conditions, rather than a full analysis. Since this is the normal mode of operation, "normal" actions cannot be erroneous, by definition. Failures occur due to adjustments going wrong, but both the actions and principles of adjustment are technically correct (Hollnagel 2009).

In this context, except for extremely simple systems, adjustments are a sine qua non, and procedures and instructions are incomplete due to the complexity. Failures and undesired results cannot be avoided by removing variability, as that would also affect the desirable results. Therefore, efforts are required to support the performance adjustments needed. Variability must be managed via recognition, monitoring and control. This leads to the assumption that failures and success happen in the same way. So, by ensuring that things go right, those that go wrong can be reduced. We seek more than the causes of what goes wrong in single events, patterns and the relationship among events.

According to this paradigm, following procedures to the letter is inefficient and could be unsafe. In order to complete them, people and organizations habitually make adjustments, so that resources can be paired or harmonized with demands and buffers. The skill in achieving this is what helps people successfully resolve such situations. Whatever the resources, information and time are finite, so adjustments will always be approximate. In any case, variability is inevitable and must be recognized as the key to success as well as failure, and is the way to ensure that things are going smoothly (Levenson 2003). A system with flexible buffers that tolerate failures, as well as the actors in the system being aware of its limits, are features of a resilient organization (Hollnagel 2009). Reason said that error and success are two sides of the same coin, and that a good error theory necessarily requires better understanding of success. Hollnagel takes the opposite view and says we need to understand success in order to understand error (Le Coze 2013).

No one is unaware of the difficulty in thinking about things that are going right. This is because people normally think in terms of cause and effect, so it is very difficult to pay attention to things that go well. In addition, there is the process of "habituation". Habit reduces conscious attention paid to what we are doing. This makes sense as it is important to pay attention to the unusual and unexpected. It may be a waste of time to expend effort on the commonplace and expected. Regulations understandably focus on things that go wrong. There is much more information on what could go wrong and what to do to avoid it. Find and fix is the basic principle. Look for faults, find the causes and remove them, or improve buffers and defences. Obviously, when things go right:

(1) There is no difference between expectations and events, nothing attracts attention.
(2) There is no motivation to learn since things are going right and this is evidently because the system is working, things are working and no setbacks occur.

Unfortunately, the latter is not true on many occasions, setbacks could have occurred and success was usually achieved. Who stops to analyse these situations when success is achieved in a context of incremental pressure to produce and improve profits? Who would compensate or trade off production against safety in order to stop and analyse situations leading to success due to natural and normal adjustments? (Hollnagel et al. 2006).

5 Conclusions

To sum up, we must study what goes right in addition to what goes wrong, since people achieve continuing success by making adjustments to their original intentions. Thus, things go right for the same reasons as they go wrong, but it is much less incriminating and much easier to study things that go well, besides being a much larger database.

In this respect, it is better to analyse frequent occurrences and focus on how often, rather than how severe the events are. It is much easier to be proactive with frequent occurrences than with occasional ones. A small improvement in daily events can count for more than a large improvement in exceptional cases. It is clearly necessary to be aware that failures occur and know the limits of the system.

In addition to the above, most incidents are relatively simple and can be handled as they always have, but the number of cases where that will not work will increase. In short, traditional methods can and must still be applied, but at the same time there is a clear need to develop new measurement, monitoring and assessment methods consistent with the complex, non-linear nature of the current context.

Obviously, new methods of risk assessment and investigation into accidents must be designed to agree with the concepts put forward here.

Acknowledgements We wish to thank the Spanish Ministry of Economy and Competitiveness for financing project BIA2016-79270-P, of which this paper forms a part, and also to the University of Malaga for its support.

References

Costella MF, Saurin TA, de Macedo Guimarães LB (2009) A method for assessing health and safety management systems from the resilience engineering perspective. Saf Sci 47(8):1056–1067
Hollnagel E (2004) Barriers and accident prevention. Ashgate, Aldershot, UK
Hollnagel E (2007) http://www.sites.google.com/site/erikhollnagel2/whatisresilienceengineering%3F

Hollnagel E (2009) The ETTO principle. In: Efficiency-thoroughness trade-off. Why things that go right sometimes go wrong. Ashgate, Paris
Hollnagel E (2012) FRAM: the functional resonance analysis method. In: Modelling complex socio technical systems. Ashagate, Paris
Hollnagel E, Woods D, Leveson N (2006) Resilience engineering: concepts and precepts. Ashgate
Le Coze JC (2013) New models for new times. An anti-dualist move. Saf Sci 59:200–218
Levenson N (2003) White paper on approaches to safety engineering
Montero Martínez R (2011) Ingeniería de la Resiliencia. Nueva tendencia en la gestión de la seguridad y salud laboral, Seguridad y Salud en el Trabajo 63:13–19
Perrow C (1984) Normal accidents. In: Living with high risk technologies. Basico Books
Rasmussen J (1997) Risk management in a dynamic society: a modelling problem. Saf Sci 27(2/3):183–213
Reason J (2008) The human contribution. Accidents, unsafe acts and heroic recoveries. Ash-gate
Saurin TA, Carim Junior GC (2011) Evaluation and improvement of a method for assessing HSMS from the resilience engineering perspective: a case study of an electricity distributor. Saf Sci 49:355–368
Saurin TA, Formoso CT, Cambraia FB (2008) An analysis of construction safety best practices from a cognitive systems engineering perspective. Saf Sci 46:1169–1183
Steen R, Aven T (2011) A risk perspective suitable for resilience engineering. Saf Sci 49:292–297
Woods DD (2005) Creating foresight: lessons for resilience from Columbia. In: Farjoun M, Satarbuck WH (eds) Organization at the limit. NASA and the Columbia Disaster, Oxford, Backwell
Woods DD (2015) Four concepts for resilience and the implications for the future of resilience engineering. Reliab Eng Syst Saf 141:5–9
Wreathall J (2006) Properties of resilient organizations: an initial view. In: Hollnagel E, Woods D, Leveson N (eds) Resilience engineering: concepts and precepts. Ashgate, Aldershot

The Importance of Intangible Liabilities to Business Management

R. Alcalde Delgado, L. Sáiz-Bárcena, M. A. Manzanedo del Campo and R. Del Olmo

Abstract The intangible assets of the company are those non-tangible aspects that have been identified as internal strengths and external opportunities, while intangible liabilities are internal weaknesses and external threats of a tacit and non-explicit nature. Both, in most cases, do not become visible outside the company. However, these intangible assets have a specific accounting item, an issue that does not occur for intangible liabilities, which prevents their recognition and effects for the company. Consequently, the objective of this paper is to identify and analyze the existence of intangible liabilities in the company as factors that, as opposed to intangible assets, produce losses that are difficult to quantify, but absolutely necessary to know the generation of strategic value And accounting, of the company.

Keywords Intangible liabilities · Intangible assets · Knowledge management Business strategy

R. Alcalde Delgado
Departamento de Economía y Administración de Empresas, Facultad de Ciencias Económicas y Empresariales, Universidad de Burgos, Pza. de la Infanta Dª. Elena, s/n, 09001 Burgos, Spain
e-mail: roberto.alcalde.delgado@gmail.com

L. Sáiz-Bárcena (✉) · M. A. Manzanedo del Campo · R. Del Olmo
Departamento de Ingeniería Civil, Escuela Politécnica Superior, Área de Organización de Empresas, Universidad de Burgos, Avda. Cantabria, s/n, 09006 Burgos, Spain
e-mail: lsaiz@ubu.es

M. A. Manzanedo del Campo
e-mail: mmanz@ubu.es

R. Del Olmo
e-mail: rdelolmo@ubu.es

Á. Ortiz et al. (eds.), *Engineering Digital Transformation*, Lecture Notes in Management and Industrial Engineering,
https://doi.org/10.1007/978-3-319-96005-0_18

1 Introduction

For some authors, the concept of Intellectual Capital is broader than the term intangible assets, although it has been used as a synonym for many other terms, such as: human capital, intangible assets, hidden intangible assets that do not appear on balance sheets, goodwill, or acquired or internally generated intangible assets. (Rojo Ramírez and Sierra Fernández 2000).

However, other authors have suggested that intellectual capital would itself be one of the elements of intangible assets, although it is not the only one, as can be seen in Fig. 1. Other elements of intangible assets are: the company's reputation, organizational capital, relational capital, patents, etc. They also establish that intangible liabilities are those obligations of neither a monetary nor a physical nature with the stakeholders or third parties of the organization whose non-satisfaction causes a depreciation of the intangible assets of the organization. Therefore, intangible capital is presented as the difference between intangible assets and intangible liabilities, whereas intellectual capital is an element of intangible assets (Simó Guzmán 2008).

Moreover, the plural term "intangibles" is used as a synonym for the intangible set of capacities, agreements, skills, resources, leadership, organization, knowledge, relationships and obligations of a given organization (Simó Guzmán 2008). In this connection the term "intangibles" refers to a broader and more general concept, which allows it to be associated with intangible assets and intangible liabilities (García-Parra et al. 2004; Pastor et al. 2017).

A new definition is suggested for intellectual capital that is equivalent to intangible capital, whereby it is considered to be the difference between the intangible assets held by the company and the intangible liabilities that have a negative effect on the company. As examples, there are: weaknesses in strategic plans or inadequate actions, loss of company reputation or unsafe working conditions, among others, which represent a must and an undervaluation of the worth of the company (Rodriguez Antón 2005; Lozano Gutiérrez and Fuentes Martín 2005).

Therefore, the objective of this paper is to identify and analyze the existence of intangible liabilities in the company as factors that, as opposition to intangible assets, provoke losses that are difficult to quantify and necessary to value the company. To this end, the article is structured in four sections, including this introduction.

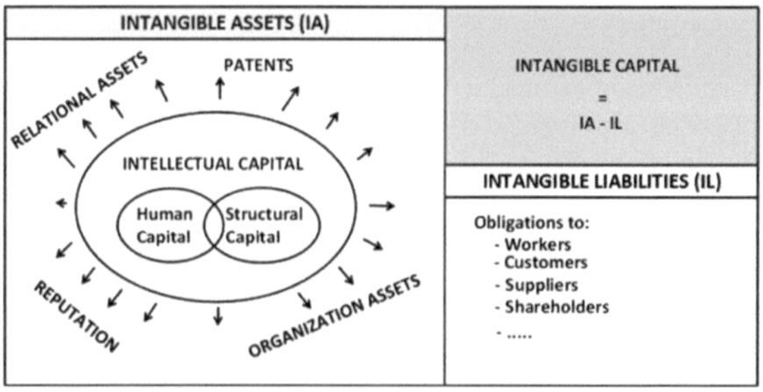

Fig. 1 Proposed conceptual model on intangibles, intangible assets, intangible liabilities, intangible capital and intellectual capital. *Source* Simó Guzmán (2008)

The second chapter identifies intangible liabilities and approaches their study from a dual perspective of business management and knowledge management. The third section contains the conclusions and limitations of the study and finally lists the references consulted for its preparation.

2 Intangible Liabilities

The concept of liability represents what the company owes to another person or entity, designated as a creditor, or the event giving rise to the obligation that has already been incurred with a creditor (Ortiz 2006). It can be defined as:

- Those obligations incurred by intangible assets (García-Parra et al. 2004)
- The company's responsibility or obligation to transfer economic resources or provide service to other entities in the future that may reflect lower profitability (Harvey and Lusch 1999)
- The obligations to the organization's stakeholders, of a non-monetary or physical nature, whose dissatisfaction causes a depreciation of the organization's intangible assets (Simó Guzmán 2008).

Consequently, some authors propose the existence of an intangible liability, just as there is a tangible liability, which counterbalances the effect of the asset. For this reason, there are two mainstreams of intangible liabilities, according to the following authors:

- They are a reduction in the value of intangible assets (Caddy 2000; García Ayuso and Larrinaga 2004; Lozano Gutiérrez and Fuentes Martín 2005).
- They are an obligation associated with intangibles that is closer to the accounting definition of liabilities (García-Parra et al. 2004; Konar and Cohen 2001).

The two definitions suggests that the origin of intangible liabilities is associated with the tacit knowledge of business agents, in which only present non-monetary debts, obligations and contingencies are considered. Intangible liabilities could be considered as weaknesses due to actions to strengthen intangible assets (Nevado et al. 2012).

A first approximation to the existence of these intangible liabilities in accounting, can be found in the appearance of negative goodwill, which arises in those acquisitions of companies in which the purchase cost is lower than the book value (García-Parra et al. 2004).

Accounting standards make some differences in the face of impairment of intangible assets. Whereas the IASB (International Accounting Standards Board) and the PGC (General Accounting Plan) indicate that this impairment occurs when the carrying value exceeds the asset's recoverable amount (the higher of fair value and value in use). The FASB (Financial Accounting Standards Board) considers that this loss occurs when the book value of the asset is greater than fair value. All of them consider that this impairment loss, if any, should be classified as an expense in the income statement. The IASB further clarifies that such a loss should also be reflected in the revaluation of the asset if it had been carried out previously (Álvarez Villanueva 2010).

Consequently, within the accounting standards there is no express reference to intangible liabilities, although it does include some nuances of these that can help to measure them (Nevado et al. 2012).

Therefore, the appearance of intangible liabilities is due to these three aspects (Nevado et al. 2012):

- Impairment of the value of intangible assets and wealth, which would be covered by the same model that measures intellectual capital through a decrease in the value of intellectual capital.
- Debts, obligations and contingencies of the company due to tacit knowledge with all the economic agents of the sector and the organizational structure, that is, its intellectual capital. Its cancellation would entail the elimination of resources that would include economic benefits.
- Actions to strengthen or improve intangible assets.

2.1 Business Management Perspective

From a business management perspective, intangible liabilities would be all the actions that need to be taken to achieve intangible assets. On this basis, intangible liabilities can be classified using the same criteria as intangible assets. Thus, intangible liabilities are the sum of three elements (Nevado et al. 2012):

- Human liabilities that correspond to the debts, obligations, contingencies and actions that can be derived from the tacit knowledge of people, their skills, training and motivation.
- Structural liabilities that are due to debts, obligations, contingencies and actions derived from tacit knowledge that materializes in the structure of any organization. This can be disaggregated into: process liabilities (in relation to the quality of processes, products and services), relational or commercial liabilities (in relations with customers and suppliers), communication liabilities (in marketing, advertising, promotion, image) and R&D&I liabilities (to achieve innovation and development potential)
- Liabilities not explicit or not included in the foregoing liabilities.

Consequently, intangible liabilities are derived from the existence of intellectual capital in its different aspects; human and structural (processes, relational, communicational, R&D&I), as is considered in the intellectual capital model (Alfaro Navarro et al. 2011).

In the scope of management and strategy, there are no studies that propose the management of intangible liabilities; moreover, they only focus on improving intangible assets (Gu and Lev 2011), without addressing the causes of their strengthening, which are often motivated by intangible liabilities (Nevado et al. 2012).

Other authors also acknowledge that there is insufficient literature and studies on intangible liabilities, and urge that research be continued (García-Parra et al. 2004). Even the latest published definitions of intellectual and intangible capital do not include references to intangible liabilities (Simó Guzmán 2008).

Moreover, it highlights the need to consider intangible liabilities in the management, measurement and reporting practices of intellectual capital, as it allows a more complete and balanced view of the concept of intellectual capital itself as a lever to create value (Giuliani 2013).

Fragouli (2015) also expresses a growing interest in understanding intangible liabilities. Thus, the value of the change in intangible assets is best explained by an appreciation or depreciation due to the context (market forces, speculation, etc.) and the efficient or ineffective use and management of it. For this author, liabilities and intangible assets do not work analogously to accounting, i.e. they do not need to have a double entry. Furthermore, it presents intangible liabilities as a promise of intangible payment, which decreases the value of intangible assets by reporting them to a creditor.

2.2 *Knowledge Management Perspective*

The perspective of knowledge management has a significant human dimension, since it transforms individual capacities and knowledge through codification processes, and makes individual expertise a collective asset that enables the organization to resort to people with appropriate skills or knowledge repertoires (Stewart 1997). Therefore, the knowledge of individuals is the impulse for the development of the rest of the assets that constitute the company's capital (Brooking 2010).

But also, employee knowledge can be the brake on asset value. As indicated in the accounting perspective, intangible liabilities are the company's debts, obligations and contingencies due to tacit knowledge with all economic agents in the sector and the organizational structure (Nevado et al. 2012).

In this work, the intangible liabilities related to knowledge are considered to be mainly associated with tacit knowledge, but also with explicit knowledge that is secret or hidden within the organization and, therefore, also outside of it.

It is evident that in the case of tacit knowledge, it is not shared because of the difficulties of making it explicit. However, in explicit secret or hidden knowledge, there is a willingness on the part of the organization to be selectively shared within it, but in no circumstances should it be transmitted outside of the organization. Most likely because of the consequences it could have for the company, as it is contrary to existing legislation and/or social responsibility disclosed by the company.

A case of the secret or hidden explicit knowledge liability is what has been called planned or provoked obsolescence, which is defined as the deliberate introduction of alterations by companies in their products to shorten their useful life and thus oblige them to buy the item in question again (Sempere 2009). One of the most popular cases is an Epson printer model that was scheduled to automatically lock at 18,000 copies, forcing the user to buy another printer (Pinto 2015).

Another case of secret or hidden explicit knowledge liability is the recent scandal involving the automobile manufacturer Volkswagen (United States Environmental Protection Agency Epa 2015), which deliberately and with the knowledge of several employees, made a modification in the operation of its diesel engines to ensure that their emissions comply with the legislation, when the car detected that it was being tested; but not in their normal operation, achieving better performance that its.

Probably, within the source code of the engine control software, were hidden and at the same time explicit, the conditions for activating the engine's operating mode that would allow it to comply with environmental emission regulations, at the expense of reducing the diesel engine's performance. The group of employees who worked on engine development, or at least a small group of them, knew the software's manoeuvre to achieve the challenges posed in development of engines (high power, low cost and low emissions). It is possible that Volkswagen may have had to choose between complying with engine performance or standards at the time and decided to modify the software to comply with both conditions.

The public visibility of this deliberate modification of the software by the Volkswagen company has caused possibly one of the biggest scandals in the automotive industry of all time. The company has suffered significant economic losses, as well as in its stock market valuation, and a great damage to its reputation and image, which led to the resignation of its chairman.

3 Conclusions and Limitations

As there is an intangible asset accounting item, it is appropriate that the accounting plans include an item on the liability side of the balance sheet to reflect the provision for undetermined losses due to intangible assets recognised in the balance sheet.

As discussed, knowledge management may be at odds with the company's values, mission, vision or strategy. This suggests the existence of a dark side in knowledge management, which is what we have called knowledge liabilities, which are part of the intangible liability, and whose disclosure may harm the company.

Secrets or confidential documentation for the strategic management of the company can be classified into two categories: those aligned with ethics and corporate social responsibility (formal secret knowledge of the company), and those that are contrary to the values and principles promulgated by ethics and corporate social responsibility (explicit, secret or hidden knowledge liabilities).

In our opinion, knowledge liabilities are all the tacit and individual knowledge that is necessary for the company, but that has not been transferred to the organization. Also, the explicit knowledge that if it were disclosed, the company would not be proud, since it contradicts what it promulgates with its business ethics, its image and its corporate social responsibility.

The liabilities of knowledge cannot be considered in the accounting balance sheet, since it is a knowledge that the company has not been able to make explicit, or that it wishes to keep hidden and secret. Nevertheless, knowledge liabilities should be included in the balance sheet of the company as accounting provisions to compensate future rights or obligations for the loss of tacit knowledge or the disclosure of explicit knowledge, which is secret or hidden.

References

Alfaro Navarro JL, Lopez Ruiz VR, Nevado Peña D (2011) An alternative to measure national intellectual capital adapted from business level. Afr J Bus Manage 5(16):6707–6716

Álvarez Villanueva C (2010) Hacia un nuevo modelo de valoración de intangibles. Thesis Doctoral. Centre for Reputation Leadership

Brooking A (2010) Intellectual capital: core asset for the third millennium. CENGAGE Learning EMEA, Hampshire

Caddy I (2000) Intellectual capital: recognizing both assets and liabilities. J Intell Capital 1(2):129–146

García-Ayuso M, Larrinaga C (2004) El lado oculto de los intangibles: activos y pasivos ligados a la sostenibilidad. Harvard-Deusto Finanzas & Contabilidad (57):34–41. Enero-Febrero

García-Parra M, Simo P, Mundet J, Guzman J (2004) Intangibles: activos y pasivos. Intangible Capital (37):1–17. Junio

Giuliani M (2013) Not all sunshine and roses: intellectual liabilities in action. J Intell Capital 14(1):127–144

Gu F, Lev B (2011) Intangible assets: measurement, drivers, and usefulness. In: Schiuma G (ed) Managing knowledge assets and business value creation in organizations: measures and dynamics, IGI Global

Fragouli E (2015) Intellectual capital & organizational advantage: an economic approach to its valuation and measurement. Int J Info Bus Manage 7(1):36–57

Harvey MG, Lusch RF (1999) Balancing the intellectual capital books: intangible liabilities. Eur Manag J 17(1):85–92

Konar S, Cohen MA (2001) Does the market value environmental performance? Rev Econ Stat 83(2):281–289 (May 2001)

Lozano Gutiérrez MC, Fuentes Martín F (2005) La importancia del intangible en la empresa de internet: una propuesta de medición contable, Febrero (Online). Available: http://repositorio.bib.upct.es/dspace/bitstream/10317/815/1/iie.pdf

Nevado D, López VR, Alfaro JL (2012) Intangible liabilities: a new key in SCM. Valahian J Econ Stud 3(3):17–22

Ortiz MAA (2006) Intellectual capital (intangible assets) valuation considering the context. J Bus Econ Res 4(9):35–42 (Sep, 2006)

Pastor D, Glova J, Lipták F, Kováč V (2017) Intangibles and methods for their valuation in financial terms: literature review. Intangible Capital 13(2):387–410

Pinto RF (2015) Obsolescência Programada, 15 Junio, (Online). Available: http://comum.rcaap.pt/bitstream/123456789/9635/1/T-EE_15_11_%20Rui%20Pinto.pdf

Rodriguez Antón JM (2005) El Capital Intelectual como la diferencia entre los activos y los pasivos intangibles. http://www.madrimasd.org/revista/revista28/aula/aula2.asp (Online)

Rojo Ramírez AA, Sierra Fernández M (2000) La revelación de información sobre intangibles en empresas españolas: un estudio exploratorio. VI Jornadas de Trabajo sobre Análisis Contable, Almería, pp 61–78

Sempere J (2009) Mejor con menos. Necesidades, explosión consumista y crisis ecológica, 2nd edn. Critica, Barcelona
Simó Guzmán PSLJM (2008) Capital intangible y capital intelectual: Revisión, definiciones y líneas de investigación. Estudios de Economía Aplicada 26–2:65–78
Stewart TA (1997) Intellectual capital: the new wealth of nations. New York
United States Environmental Protection Agency EPA (2015) EPA, California notify Volkswagen of clean air act violations. In: Carmaker allegedly used software that circumvents emissions testing for certain air pollutants, 18 Sep, (Online). Available: http://yosemite.epa.gov/

Part II
Production

Bucket Brigades: Simulating a Production Line in the Classroom

Pablo Aparicio-Ruiz, Jesús Muñuzuri-Sanz, Alejandro Escudero-Santana and María Rodríguez-Palero

Abstract The experience is described in this paper, it is a small game for drawing a little house. The experience tries to innovate in the traditional methodologies of learning of theoretical subjects and is shown a small game that can be used to introduce the concept of production assembly orders or in order picking systems by relays or "Bucket Brigades". This concept is explained in the subject of "Logistical decision systems" developed in the field of "Master's Degree in Industrial Engineering" at the University of Seville.

Keywords Bucket brigades · Game · University teaching
Teaching innovation · Production line

1 Introduction

The experience is described in this paper is a small game for drawing a little house. The experience tries to innovate in the traditional methodologies of learning of theoretical subjects and is shown a small game that can be used to introduce the concept of production assembly orders or in order picking systems by relays or "Bucket Brigades" (BB). This concept is explained in the subject of "Logistical decision systems" developed in the field of "Master's Degree in Industrial Engineering" at the University of Seville.

P. Aparicio-Ruiz (✉) · J. Muñuzuri-Sanz · A. Escudero-Santana · M. Rodríguez-Palero
Dpto. de Organización Industrial y Gestión de Empresas II, Escuela Técnica Superior de Ingeniería, Universidad de Sevilla, Camino de los Descubrimientos S/N, 41092 Seville, Spain
e-mail: pabloaparicio@us.es

J. Muñuzuri-Sanz
e-mail: munuzuri@us.es

A. Escudero-Santana
e-mail: alejandroescudero@us.es

M. Rodríguez-Palero
e-mail: mariarodriguez@us.es

Á. Ortiz et al. (eds.), *Engineering Digital Transformation*, Lecture Notes in Management and Industrial Engineering,
https://doi.org/10.1007/978-3-319-96005-0_19

The main objective of this paper is to present a simple example of a game that allows us to explain the concept of line production or assembling lines, the game can be incorporated into the training of engineers, as a pre-activity explanation of the theoretical concept, so that the students observe the importance in the organizations of this type of decisions.

The objective is to instil in the students the need to study every problem or organizational structure of a company, the engineer must have an open mind to improve production processes. It was decided to make a playful proposal, to experimental strategies aimed at improving the commitment and attention of students in the development of the course.

The analysis of the results of the activity reveals that the great majority of the students consider that this type of methodologies facilitates the understanding of the main concepts of the subject, and the implication in the described problem.

That allows us to conclude that the application of play activities also allows the development of other skills such as: tolerance, respect, teamwork and speak in public, attitudes that should be fostered in the training of an engineer, among other aspects.

The structure of the document is as follows. Section 2 introduces the context in which the BB concept emerges. Section 3 presents the theoretical framework. Section 4 describes the experience. Section 5 introduces the concepts added to the case. Section 6 presents the results. Finally, Sect. 7 contain the conclusions.

2 The Context of Bucket Brigades

The term Bucket Brigades arises from the classic assembly lines, in which workers were assigned fixed workstations and the station with the highest working content determined the production rate. With this situation, there were only two ways of changing production costs, or changing the number of shifts or redistributing tasks, tools, and parts over different stations.

A flexible manufacturing system is a manufacturing system in which there is some amount of flexibility that allows the system to react in case of changes, this system arose when products have a low seasonality and short life cycles. The most used example was associated with the clothing industry, to increase production flexibility in this industry, a variation of the assembly line was introduced in which there were fewer workers than stations and workers walked to adjacent stations to continue working on a product.

The idea of a decentralized line control where each worker independently follows a simple rule that determines what to do next was called TSS (Toyota Sewn Products Management System) and is used in the manufacturing of many types of sewn products, including clothing, furniture, shoes, bags, suitcases, etc.

3 Theoretical Framework

The term Bucket Brigades was created by (Bartholdi and Eisenstein 1996). The authors provided the first comprehensive analysis of the dynamics of these systems and pointed out that if workers can be sequenced from the slowest to the fastest along the production line, then there is a stable fixed point at which the system will converge independently of the initial positions of the workers, providing optimum performance.

The concept of BB is represented as in Fig. 1, where the instantaneous position of the worker on the line is expressed as the cumulative fraction of the work completed in his product at a given moment of his working speed. Each station is associated with a certain partition of the total work-content executed. In the structure of the line, workers are obliged to keep their order on the line, since the worker with higher index always has priority. Thus, the index worker i can eventually be blocked (waiting) at the beginning of a station occupied by worker i + 1, and leaves this state if worker i + 1 advances, finishes working at that station, or if the worker i + 1, moves back, and takes his product. When this happens, the worker must give up the current product and begin his or her way back. The only one ever blocked is the last worker (index n).

The application of BB in production assembly orders or in order picking is a system applied in the industry, against automated systems that are sometimes too expensive to be implanted.

The resolution of work orders based on BB is a way to organize the workers in an assembly line or order preparation so that the line is balanced.

In the case of an assembly line, the products are progressively assembled as they move down the line from worker to worker until completion. In the case of order picking in distribution centres, the BB application refers to the process of retrieving warehouse products to meet customer orders. In the case of order picking systems (OPS), an average of 55% of the retail cost of a distribution centre is consumed (Tompkins et al. 2010).

In the classroom, the teacher explains to the students that a production line is, in general, a set of “workstations” in a set of sequential operations established in a factory where it carries out a series of works that result in the final product. In the

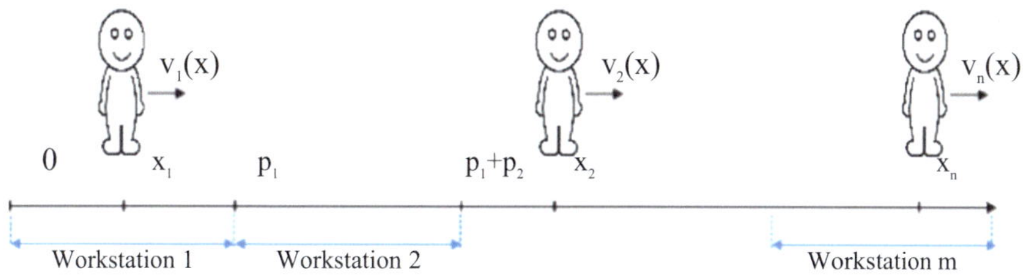

Fig. 1 Example of BB

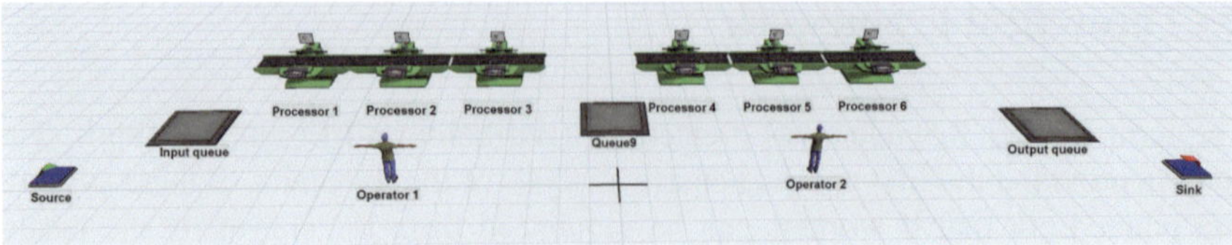

Fig. 2 Example of elements of a production line

line, each workstation will be composed in turn, of one or more identical machine, attended by one or more operators and one or more queues (Fig. 2).

The situation in Fig. 2 reflects a way of coordinating workers who are progressively assembling product along a flow line in which there are fewer workers than the number of stations or working places.

At BB, each worker follow this simple rule: "go ahead, from station to station, until someone takes over your work, and then back for more". When the last worker has completed a product (or customer's order), the worker walks back upstream and takes over the task of his predecessor, who then walks back and takes over the work of his predecessor, and so on, until the first worker starts with a new product (customer order) at the beginning of the line.

In this system, no working process inside the chain is left unattended, since the workflow would be broken. Likewise, it is emphasized from this system that workers are not restricted to a subset of stations or workstations; but are used to bring the product as far as possible towards the end of the line or the termination of the product. Keep in mind also that a worker could reach his successor and be blocked in the procedure; the rule requires the locked worker to remain inactive until the station is available.

The final requirement is that workers can sequence tasks from the slowest to the fastest along the direction in which the flow of material development. These protocols, on the whole, make the chain assembly line a perfect traction system.

The ideal situation is that all the stations work in a synchronized way, so that production occurs in a fluid way, without waiting, jams or unproductive times. However, in reality, this is not possible due to a multitude of factors: scarcity of resources, variable processing times, breakdowns, supplier failures, poor programming, etc. All this leads to the appearance of bottlenecks in the installation, which will be those processes that are determining the final productivity, i.e. Any improvement that occurs in them, will be directly reflected in the improvement of productivity, and the conversely, when you improve a bottleneck, you may move to another process.

4 Description of the Experience

The experience proposed is to simulate a line of houses, in our case, it is a house painted on a paper with coloured pens (Fig. 3). It defines 7 work processes, which

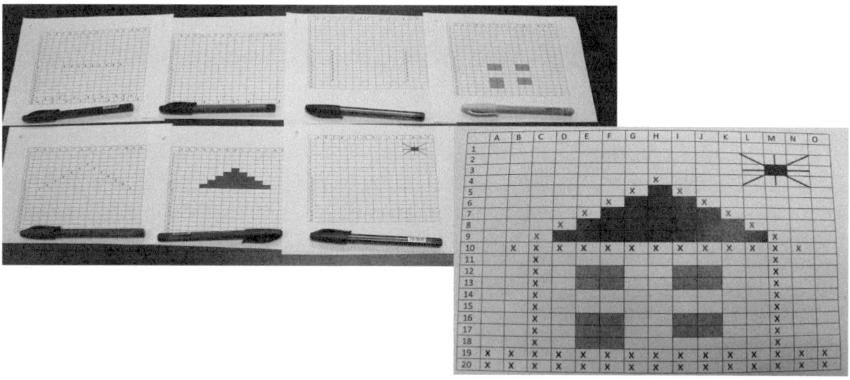

Fig. 3 (Left) Example of jobs in a production line. (Right) Example of house (product) of a production line

will be associated to 7 jobs, with their corresponding drawing resources. The work will be carried out by 4 workers (students). They are told that it must be assumed that all workers have the same capabilities to perform any of the processes (because it is a matter of painting, or an "X" or a "blot").

Figure 3 shows the 7 jobs, each one has a template, which reflects the work to be done, and the pen, or tool with which the process will be performed. In our case it was 6 colours. The template, in the form of an image, was made to make the process fast and easy, although a gradient could be made of the rest of the house in the template, to facilitate the task and to link the concepts of Lean Manufacturing, even so, this experience is mentioned in the course as an example of an idea for the process of "standardization of tasks".

The procedure of experience can be guided or free. In the case of having few students, it could be done in a freeway, and students would be asked to explain the strategy decided or applied in the game. For after the experience, make a discussion.

In the case of making a guided explanation, we would propose a set of strategies to be analysed:

Case 1: The first experience would be for one player to perform all the operations that involve the manufacture of the house (product) until finished, so that the total production will be the sum of the individual production of each player. Also, it could be done with 7 workers and 7 positions, to know the time, and the rigidity that the system has when each one has an activity, and the cost that this would suppose, since as can be observed in Fig. 3, in the manufacture of the cottage not all processes require the same time.

Case 2: Each player is assigned a set of activities in the production process so that when a player completes all the operations that correspond with him on the product, he passes to the next player and so on until the product is finished. The process would lead to a reduction in personnel costs, but bottlenecks would not be solved, unless the line's activities were balanced.

Case 3. Development of the activity, under the rules of BB. For this purpose, the players are placed sequentially, so that one after another, begin the manufacturing, so that the first begin the processes and continue until the end, performing the operations necessary to make a finished product. When it finishes its work it turns back in the line to the player that precedes it, takes its product, and it finishes, and the predecessor player in turn moves towards the player who is before him and so on until the player located first in the line goes to the beginning of this one and takes another order of production. On this last experience, it is only necessary to emphasize to the students that if the partners were sequenced from the slowest to the fastest along the production line, then a stable fixed point is established in which the system will converge independently of the positions workers, providing an optimal performance (the latter concept will be shown in theoretical classes).

5 Related Concepts

After describing the experience and the environment in which the experience is developed, it is important to emphasize that as it is a teaching activity, around the explanation of the game, references are incorporated to other concepts and to how they are defined in the activity.

The players, it is remarked that they are the workers, and that these must have the same capacities to develop all the tasks of the line.

The system does not require a person to organize the work, if it is not the operators who, with their own activity, manage the chain.

On the other hand, students who do not participate in the action, they will be in charge of performing another activity. The analysis and the study of the time or the analysis of the quality of the final product. Since these two elements are necessary for the study and measurement of productivity and efficiency. Essential elements in the decision making of the teams responsible for the management and control of the operations of a plant. In addition, with respect to the times, the analysis of the time consumed in the activity or process regarding the time of transport is emphasized.

Likewise, during the discussion time, the possibility of analysing the ergonomic is discussed (although it has not been represented in the game, nor taken into account), while the design techniques of the systems manufacturing or assembly, since both are closely related (Battini et al. 2011).

Other examples of reference, which can be associated are the concepts of production management, such as Pull and Push systems, the line configuration or the way to make orders are executed, and so on.

In the development, reference is also made examples of activities in which BB is used, for example in manual assembly lines, reference is made in the case of fashion bag assembly lines, in which there are a number of products with low production volumes, while the line organization is often under the responsibility of the foreman, who balances workloads in an empirical way, in the case of (De Carlo et al. 2013), the results were excellent, showing the advantages of the BBs in terms of flexibility, the reduction of work in the process and the ability to manage small anomalies.

6 Results

Among the results of the activity carried out, it should be noted that the main motivation of the activity is the conceptual introduction of the term BB, and introduce this form of work as a good solution, which can be clearly understood without much previous knowledge. Of the activity, we can observe the capacity for analysis, observation, criticism and common sense that are fundamental elements in professional activities and especially in engineering.

Even so, there are situations where learning arises, which will surely mark the students and in the case of the teacher are useful for linking with their explications, such as:

- Strategies in which students skip part of the chain, to other workstations making large trips.
- Students who stand without product by a bottleneck, in front of this system, in which it is not allowed that any work process of the chain is unattended, so as not to break the workflow.
- Students taking the resource (pen) to the next processing point.
- Workers waiting for the task to arrive, in front of workers who return in the chain to look for homework, etc.
- Orders that accumulate at a point in the process, and accumulate in the form of a pile, so that the arrival of the products arrives disorderly, and therefore orders that are made before, they arrive later.

This is an example set of situations arising from the game, as you can see in Fig. 3, the exercise is easy and fast. The activity is very suitable for classes where there are wide corridors in which students can move from one job to another. In addition, if the group of students is not small, you can do the activity with several groups at a time. Competitive activity is also stimulating when students are allowed to experiment with a free organization.

7 Conclusions

The experience developed in the subject was positive. Play activities are highly accepted by the students, this is reflected in the surveys carried out in the teacher analysis, where the students rated the interest in this block as very interesting (4/5) and this type of activities allows them to qualify with the question "The teacher tries to generate a good atmosphere in the classes" with a value of 4.4/5. In addition, it is observed that this type of activities improves the attention and the involvement of the students in the subject.

References

Bartholdi JJ, Eisenstein DD (1996) A production line that balances itself. Oper Res 44(1):21–34

Battini D et al (2011) New methodological framework to improve productivity and ergonomics in assembly system design. Int J Ind Ergon 41(1):30–42. Available at: http://dx.doi.org/10.1016/j.ergon.2010.12.001

De Carlo F, Borgia O, Tucci M (2013) Bucket brigades to increase productivity in a luxury assembly line. Int J Eng Bus Manag 5(Spl. Issue):1–9

Tompkins JA et al (2010). Facilities planning. Wiley

Impact of Lean Manufacturing on Productivity and Quality in a Food Company

M. Victoria de la Fuente Aragón, Lorenzo Ros-McDonnell and Manuel Cardós Carboneras

Abstract This paper describes a project to increase a food manufacturer's productivity by applying several Lean Manufacturing techniques to the company's manufacturing organization and production quality. As a conclusion, the paper highlights the company's developed performance as a result of addressing such issues as clean and order at the workplace, machinery breakdowns, and quick tools changeover.

Keywords Lean manufacturing · Productivity · 5S · SMED · TPM

1 Introduction

The company Ice Cream Factory Comaker (ICFC) is an enterprise that combines the artisan tradition of ice cream and adapting to the new needs of the market, betting on innovation, collaborations with other brands, MDD (distributor's label) and internationalization, factors that during the last years have made ICFC to grow facing up its competitors.

Furthermore, Ice Cream Factory Comaker is considered one of the great Spanish companies of this sector. The company currently owns factories in Spain, Italy and France, and ICFC manufactures for nineteen countries on five continents. It is on the fourth place in the ranking of the 10 Spanish companies leaders in the market according to sales volume in year 2014.

M. V. de la Fuente Aragón (✉) · L. Ros-McDonnell
Research Group "Industrial Engineering", ETSII, Technical University of Cartagena, C/Dr. Fleming, s/n, 30202 Cartagena, Spain
e-mail: marivi.fuente@upct.es

L. Ros-McDonnell
e-mail: lorenzo.ros@upct.es

M. C. Carboneras
Dpto. Organización de Empresas, Universidad Politécnica de Valencia, Camí de Vera, s/n, 46021 Valencia, Spain
e-mail: mcardos@doe.upv.es

Á. Ortiz et al. (eds.), *Engineering Digital Transformation*, Lecture Notes in Management and Industrial Engineering,
https://doi.org/10.1007/978-3-319-96005-0_20

These successes are due to the fact that in recent years ICFC has been promoting the culture of Lean Manufacturing, continuous improvement and teamwork to achieve a global efficiency objective, placing the company in adequate levels of quality, service cost and management.

This paper presents the initial diagnosis of the company, the improvement plan proposed and the analysis of the results obtained during the initial phase of the Lean Project on year 2015. The results obtained are one more example of how the application of Lean techniques during the last decade in the company is a key point to increase their competitiveness. In this sense, the best way for a company to become more competitive is full customers' satisfaction, by offering them better products and services, together with a lower costs and delivery times.

2 Lean Manufacturing: Techniques Used

The Lean Manufacturing implementation strategy at ICFC is oriented towards techniques whose characteristics (clarity, real possibility of implementation and practical approach) (Shah and Ward 2003; Hobbs 2004; Rajadell and Sanchez 2010) make them required at the first stage of the implementation of Lean Thinking, which will cause a total change in the company and also high investments in its personnel (Lathin and Mitchell, 2001; Fawaz et al. 2006).

The Lean Manufacturing techniques selected by the company and the research group (5S, OPL-TPM and SMED), were tools that produce tangible and quantifiable results without large financial investments, with a large visual component and high impact in the short term (Davis 2003; Cerda 2013; Garbie 2010; Lizarralde and Ferro 2013). Likewise, these techniques directly involve staff, who perceive that their environment depends on themselves, and that quality begins with very immediate things, so that a positive attitude towards the workplace is achieved.

1. The implementation of the 5S technique has ensured a precise order and cleanliness in the selected areas initially, and it has allowed its progressive extension, reaching all manufacturing lines and storage areas. The stages of this technique were progressively implemented (Seiri, Seiton, Seiso, Seiketsu and Shitsuke), in order to a later control of their compliance (internal audits).
2. The second tool applied is SMED. It was initially implemented in four production lines, due to their low efficiency (losses due to breakdowns, setup times, settings, defective products, cleaning, etc.). After verifying the higher efficiency of the initial system, the SMED was implemented in two new production lines: GRAM-3 and EXTRUSION-4.
3. Finally, the company implemented the OPL tool (One Point Lesson) in order to standardize autonomous maintenance. A group of OPLs on basic knowledge and problem solving were developed for the production line EXTRUSION-4, to facilitate the work of the operators assigned to that manufacturing line.

3 Improvement Actions at ICFC

All different improvement actions developed in ICFC are related to the constant detection and elimination of main waste and every activity that does not add value, from the point of view of the internal and external customer of the company.

3.1 Implementation of 5S

The development and implementation of the 5S tool, which will ensure a proper order and cleaning throughout the plant, compels to divide the production plant into small areas in which each stage of the 5S will be progressively applied, thereby ensuring Control of compliance (Levonson and Rerick 2002; Cerda 2013).

The divisions (areas A, B, C,…, N) were made according to criteria of use and separation between wet areas (areas of product manufacture, with high risk of product' contamination) and dry areas (packaging of the product, with no risk of contamination) (Fig. 1).

The development of the different stages (Seiri, Seiton, Seiso, Seiketsu and Shitsuke) has allowed the workers and staff to get a correct order and cleaning throughout the plant:

- To list all the necessary auxiliary elements in each area.
- To identify unnecessary items (using color cards) and take decisions about disposal or new location.
- Visual control and location marking: quickly detecting if an item is out of place.

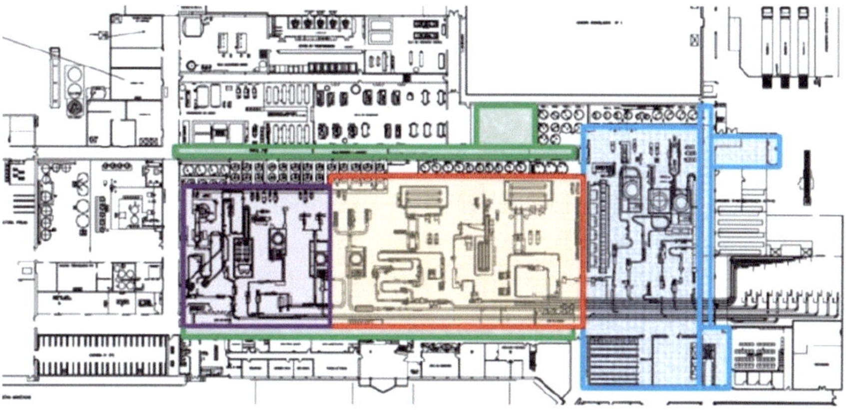

Fig. 1 Areas of implementation of lean manufacturing tools: Line GRAM-3 (blue zone), line EXTRUSION-4 (red zone), palletizing area (purple zone)

- To define a 5S map, showing the situation of usual and special elements (tools, stairs, fire extinguishers, eyewash, emergency corridors, warehouses, etc.) and registering their location and order in the plant.
- To create guides, manuals and cleaning routines.
- To monitor with periodical audits (5S), in order to comply with the actions developed until this moment.

3.2 Implementation of SMED

The SMED tool has been developed by blocks in several production lines of the plant. As proof, the company implemented SMED in lines VENTURA, CONOS, EXTRUSION-3, and POLO-14. Once the efficiency of this system was checked, SMED was added to the production lines GRAM-3 and EXTRUSION-4.

The detailed development of SMED in GRAM-3 and EXTRUSION-4 has led to the detection of several inefficiencies in the lines, some of them are:

- The need to incorporate a 2nd CLM in GRAM-3, to avoid waiting during cleaning.
- The pack changes in EXTRUSION-4 were made with the production line stopped, without being necessary.

The following conclusions were drawn:

- Definition of the main families of change in each production line (example shown in Fig. 2).
- Training has been given to the three shifts of workers in each line.
- A SMED panel with indicators has been installed to monitor the three families of products by collecting data and incidences in each shift (see Fig. 3).
- Several actions were implemented to convert internal to external operations and to reduce the time of nozzle change in the line GRAM-3.

However, several actions are still pending:

- To implement meetings before and after each change.
- To perform analysis of deviations with SCRA's.
- To follow the evolution of the shrinkage and follow-up of the shrinkage rates.

Another type of data, which is displayed in SMED panels (depending on the area): garbage count, shrinkage, breakdowns, list of format changes, OEEs, wood splinter control, etc.

3.3 Implementation of OPL—TPM

OPL (one point lesson) is a tool initially developed within the Total Productive Maintenance (TPM) technique to standardize autonomous maintenance (Levonson

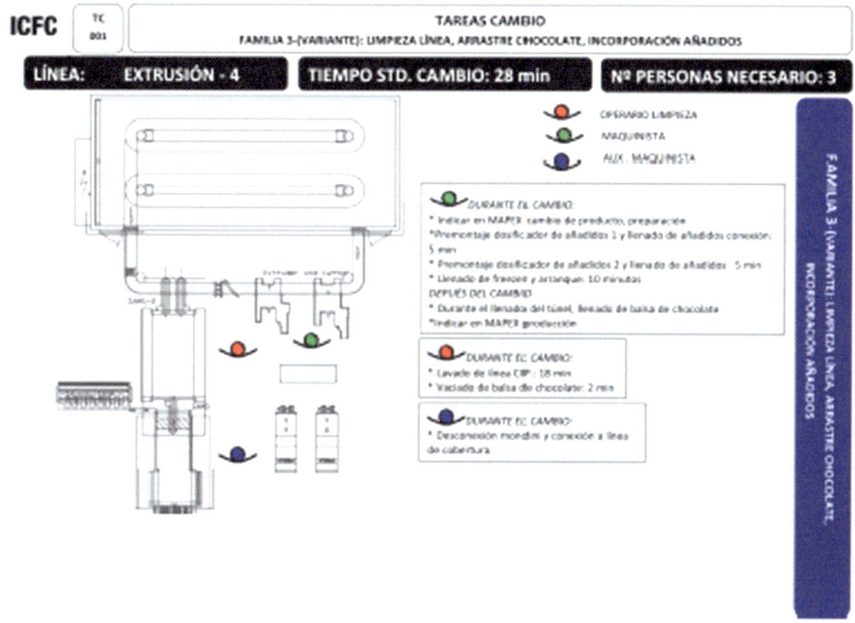

Fig. 2 Division of tasks for several changes in family 3 (production line EXTRUSION-4)

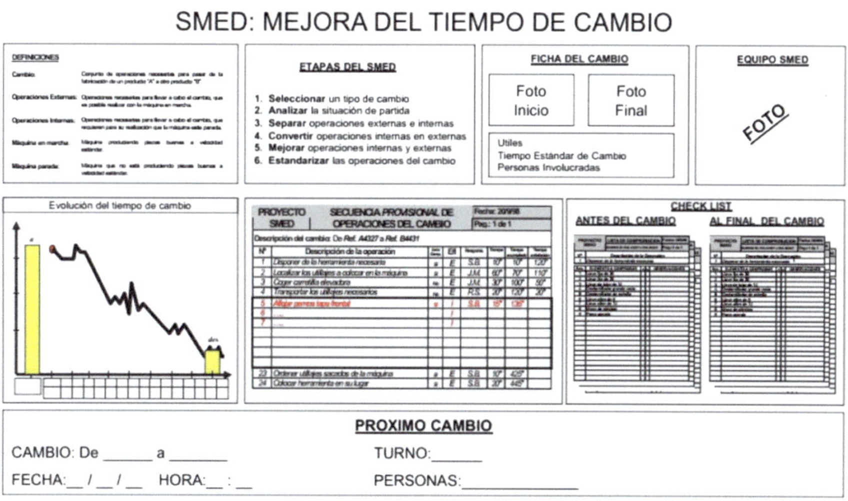

Fig. 3 SMED panel design, which will be located in each factory zone

and Rerick 2002; Garbie 2010). At ICFC, OPLs have been designed in all those areas and departments where it is necessary to standardize a task, share a modification or improvement, report problems, or develop workers' skills.

There are three groups of OPLs:

1. OPL of Basic Knowledge: to cover knowledge gaps in some operations along the production process, with the objective of ensuring efficiency and safety in the development of the work. Actions developed:
   a. OPL for adjustment and control of the parameters at the cream dispenser, the cutter and the squeegee machine.
   b. OPL for the start-up of the labelling machine, change of ink roll and labels.
2. OPL about problem examples: show real situations of breakdowns, defects or errors, allowing to define the guidelines for their identification, analysis of causes and avoiding their reappearance. Actions developed:
   a. OPL on how to solve a jam in the punch press machine.
   b. OPL on how to act before a machine that closes boxes of products and it is not working.
3. OPL about improvement examples: show the different steps that a workgroup has made to implement an improvement action. ICFC has not developed any OPL in that stage of the project, remaining for later stages (Fig. 4).

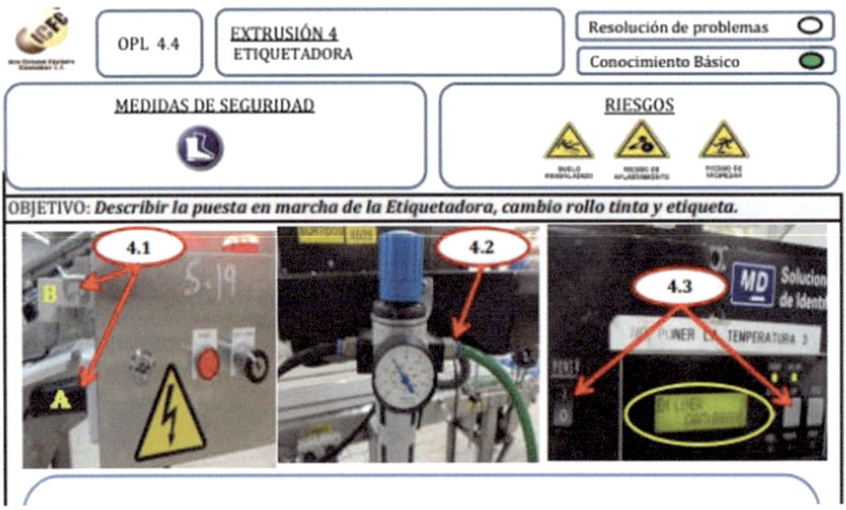

Fig. 4 Partial example of OPL about labeler machine operation (OPL of basic knowledge)

4 Results

After implementing the improvement actions previously described, several Lean audits were performed in the areas selected to apply the selected tools. The improvements for the objectives pursued are shown in the following Tables 1 and 2:

5S tool has been successfully applied in each of the areas selected for this purpose, raising the awareness of staff and workers about the improvements obtained from its application, and keeping collaboration after their project's participation.

The benefits obtained are reflected in the attitude of the people who have been able to see their workplace clean, organized and orderly, facilitating the development of their daily tasks.

Improvements in productivity have been achieved by controlling wear and tear on machinery, detection of imperfections, reduction of defective products, management of waste from production lines, reduction of risk of cross-contamination with allergens, reduction of repetitive problem solving times, etc.

The order has facilitated the reduction of time lost by workers in searching tools, or fixing places where to leave the tools and accessories to avoid accidents.

In addition, the 5S tool has led to improvements in product quality due to the clarity of work instructions as well as greater safety and health conditions for employees as well as for final customers.

Regarding the SMED tool, its application to the ICFC company is the technique that has obtained the best results. Its application in two manufacturing lines has

Table 1 Comparison between the initial situation and the improvements performed on ICFC

Indicators	Before the project (%)	After the project (%)
Quality	65	78
Cleanliness	46	72
Visual management	56	75
Machinery reliability	60	85

Table 2 SMED technique results in production lines GRAM-3 and EXTRUSION-4

Production line	Product families	Actions	Time improvement (%)
EXTRUSION-4	1	Line cleaning	31
	2	Nozzle change	41
	3B	Can change	78
GRAM-3	1	Pack change	79
	2	Cleaning of chocolate pond	45
	3	Drag-mix, Chocolate drag	42

achieved a time reduction greater than 40% in the pack change operations, which is currently achieved with the line in progress.

5 Conclusions

This paper presents the application of several Lean Manufacturing techniques in one of the main Spanish ice cream companies. In the framework of change and productivity improvement of the company, the development of the 5S, SMED and TPM techniques, have involved all the staff in its application, getting cleaner and orderly workplaces, defective product reduction, early flaw detection, better waste management, faster problem solving with machinery (wear and tear, breakdowns, setup time, downtime, etc.), which has resulted in better working conditions at the plant, as well as higher product quality.

The results obtained are one more example of how the application of Lean techniques during the last decade in the company is a key point to increase their competitiveness. In this sense, the best way for a company to become more competitive is full customers' satisfaction by means of through by offering them better products and services, together with lower costs and delivery times.

References

Cerda JA (2013) Manual de las 5's en las industrias. Editorial TPG

Davis JW (2003) Fabricación libre de despilfarros. Editorial TPG

Fawaz A, Rajgopal J, LaScola K (2006) A classification scheme for the process industry to guide the implementation of lean. Eng Manage J 18:15–25

Garbie I (2010) Enhancing the performance of industrial firms through implementation of lean techniques. In: IIE Annual Conference, Proceedings, Norcross

Hobbs DP (2004) Lean manufacturing implementation. J Ross Publishing

Lathin D, Mitchell R (2001) Lean manufacturing: techniques, people and culture. In: ASQ's annual quality congress proceedings, Milwaukee

Levonson WA, Rerick RA (2002) Lean Enterprise: a synergistic approach to minimizing waste. ASQ Press

Lizarralde E, Ferro E (2013) Lean manufacturing. Conceptos, Técnicas e implantación. Fundación EOI

Rajadell M, Sanchez JL (2010) Lean Manufacturing: La evidencia de una necesidad. Ed. Diaz de Santos

Shah R, Ward PT (2003) Lean manufacturing: context, practice bundles, and performance. J Oper Manage 21:129–149

Analysis of the Customer Service Rate and Reliability of Suppliers in a Production and Assembly Multinational Company

Alexander Iñurrieta Martín, Josefa Mula Bru and Francisco Campuzano-Bolarín

Abstract This study arises from the need to improve the goals or objectives set for the supply chain management department of a multinational company in the electric sector given the successive poor results achieved in service rates, and also from the need to merge two different management models: the traditional company model and the multinational company model. To fulfill this objective, a simulation model was built by the system dynamics method and using the Vensim® simulation software. After simulating the different processes that encompass the activity of the supply chain under study, the obtained results were analyzed and different scenarios were simulated according to the proposed objectives. Then whether these objectives were actually fulfilled was verified, the best economic results were selected and possible future lines of action were set out.

Keywords Service rate · Supplier reliability · Simulation · System dynamics Electricity supply chain

1 Introduction

One of the essential aspects of company management is what cannot be measured, cannot be managed (Sink and Tuttle 1989). Measuring supply chain performance

---

A. I. Martín · J. M. Bru (✉)
CIGIP (Research Centre on Production Management and Engineering),
Universitat Politècnica de València, Plaza Ferrándiz y Carbonell,
2, 03801 Alcoy, Alicante, Spain
e-mail: fmula@cigip.upv.es

A. I. Martín
e-mail: alexinurrieta@gmail.com

F. Campuzano-Bolarín
Business Management Department, Universidad Politécnica de Cartagena, Campus Muralla del Mar, s/n, 30202 Cartagena, Murcia, Spain
e-mail: Francisco.campuzano@upct.es

Á. Ortiz et al. (eds.), *Engineering Digital Transformation*, Lecture Notes in Management and Industrial Engineering,
https://doi.org/10.1007/978-3-319-96005-0_21

is extremely important for industry. Measuring and managing performance cannot be done only in the main company, but must be done in all the elements that make it up. Thus performance measuring has become an important factor to control the degree of fulfilled objectives. According to Moullin (2002), performance measurement involves evaluating how good managed organizations are, and the value they give to their customers and shareholders. Other authors like Neely et al. (2002) state that performance measurement is the process by which the efficiency and efficacy of actions are quantified. For De Meyer and Wittenberg-Cox (1994), today's performance measurement focal point, which underlies most manufacturing companies, is based on cost accounting systems, even for assessing operational performance. These systems center on costs and ignore other measures which could better reflect performance in terms of quality, flexibility or manufacturing opportunities. Other authors like Kaplan (1990) and Bititci et al. (2000) coincide in that cost accounting systems are not sufficient to measure the performance of manufacturing operations because today's competition is not based only on prices or costs, but it is necessary to be capable to forecast other production priorities, which are intangibles in some cases, to be able to attract attention and customer orders. KPIs (*key performance indicators*) are the financial or non-financial metrics employed to quantify the objectives that reflect an organization's performance. KPIs tend to be linked to the organization strategy, and their main objectives include: measuring level of service, diagnosing the situation, communicating and informing about the situation and objectives, motivating the teams responsible for fulfilling the objectives reflected in the KPI, making constant progress, etc.

Customer service can be broadly defined as critically measuring the logistic's system's action to provide a product or service on time and in a given place. In this context, the customer service level is directly related with the supply chain's management and effectiveness: information flows, materials, products, etc. The more efficient supply chain management is, the higher the added value included in the given customer service. Customer service plays a key role in developing and maintaining customer fidelity and customer satisfaction. The appearance of large business groups, whose success is based on controlling the costs of processes and the image they portray to their customers, means that traditional companies have to adapt to new work and operational models.

The other alternatives that large multinational companies tend to take is to outsource production activities or processes. Outsourcing, or externalizing, is the contracted use of resources, assets and capacities from a third party with guaranteed levels of quality, performance and value, as opposed to cost, in order to provide services that were previously given within the company (Bravard and Morgan 2007). Outsourcing can offer many benefits, of which the main ones are: cost reduction, flexibility and specialization.

This article has developed a simulation model to study the customer service rate and supplier reliability in a supply chain of the electric sector. The systems dynamics methodology was used (Forrester 1961; Sterman 2000), along with its applications in the supply chain management and the production planning and control context (Campuzano et al. 2010; Campuzano and Mula 2011; Campuzano-Bolarín et al.

2013; Jeon and Kim 2016) as a modeling approach using the Vensim® simulation software. Simulation allows us to analyze whether objectives are fulfilled as far as the supplier's response is concerned to future changes in demand or other variables in the model. For this work, a specific article was considered as it is the most relevant one, it is the main product made at the plant, its production requires a coating process to avoid the article from corroding and rusting, and this process is outsourced. The fact that more suppliers in the area are lacking means having to completely depend on the only supplier of this service.

The rest of the article is arranged as follows: Section 2 presents the supply chain under study; Sect. 3 presents the simulation proposal; Sect. 4 validates the simulation model and assesses the obtained results. Finally, Sect. 5 presents the conclusions and some identified future research lines.

2 Description of the Supply Chain

A multinational company group that acquires a traditional or family company which specializes in the design, production and commercialization of low-voltage electric wiring systems, and offers all kinds of products and accessories for electric fittings. This implies that all the processes and the way management is done have to adapt to the business models of large multinational companies. This transformation involves incorporating new metrics, objectives or changes in the way management is done which, to date, are not done or analyzed due to lack of resources or because these analyses are not necessary. The present work models a supply chain whose main actor is one of the plants of an all-round specialist company in power management, with operations in more than 100 countries, which offers integral solutions for different market segments, and occupies leading positions in energy and infrastructures, industry, buildings and data centers, and is present throughout the residential sector. It provides solutions to manage energy, which range from the generation, transport and distribution of, to the use of, electric power. This multinational company in Spain has: nine production centers from which it designs, develops and produces its catalog with a wide range of products, services and solutions that it presents the electric market with; one logistics center, which distributes all over the Iberian Peninsula and to over 50 countries; six regional management centers and 49 commercial delegations, which cover any point in the Spanish geography in order to always be close to customers. The wide range of products that its offers means having to have between 100 and 200 different suppliers for each production plant that actively operate nationally and internationally, and which supply the plant with raw materials and components on a daily basis.

The logistics centers that form part of the supply chain arise from the need to achieve a more efficient, flexible and dynamic distribution; i.e., ensure a rapid customer response capacity as demand becomes increasingly more specialized. The logistics centers are a key point in the supply chain as from them products are supplied to different regions worldwide. The logistics partners network, PAC (Personalized

Attention Center) agents, the people in charge of managing orders, and the commercial network work together to optimize and meet customers' logistic requirements as far as delivery dates, packaging and following up orders are concerned. Finished products are sent every day via land transport, for which a 4-axle trailer is used, and a delivery time is set from the time the material is informatively dispatched to the time it arrives at the logistics center, which is a maximum 2-working day period.

The present case study centers on the assembly activities of some components, and also on the galvanizing process (coating iron and steel by immersing them in a molten zinc bath) and the electrolytic zinc-coating process (adding the layer of zinc electrolytically at room temperature), applied to some materials. The main aim of this procedure is to avoid the rusting and corrosion these materials may suffer due to environmental humidity and pollution.

The Sales Department is in charge of selling products, and of promoting and offering the multinational energy company's service. In parallel, it must know and follow up the market in its area, and obtain information for its databases. PAC receives customer demands as it receives these demands from its appointed customers and, depending on these demands, it deals with them, it makes inquiries, and then answers or resends information to different departments if it does not possess information to answer directly. According to the demand of customers or the organization, promotional actions of the multinational company's products and services are taken. When a delegation confirms a customer order, it will appear in the plant's MRP (*material requirement planning*) tool. Orders can be sales-order-related, can be special in that they request a given feature not included in the catalog of offered products, e.g., different thicknesses or accessories, or can also be made-to-stock orders.

Next the indicators that are the study object are defined, whose results will help to make decisions and in future corrective actions. The ESSR (*external supplier service rate*) provides a measure of a supplier's capacity to fulfill its commitments in terms of delivering the placed purchase orders on time. This is obtained from all those orders whose delivery has been recorded by the customer. The ESSR is calculated in relation to the last replanned delivery date. The ESSR is expressed as a percentage, can refer to a given period, and is normally measured monthly, and in an accumulated manner yearly. A lot is considered to be delivered "on time" if it arrives on, or before, the date set out in the purchase order. The position of an order will be considered "complete" if this implies at least 98% of the quantity indicated on the customer's purchase order. Some points of the organization's internal policies to bear in mind include: not accepting products arriving that exceed the quantity set out in the order by 10%, unless this has been agreed on beforehand with the supplier; products should not be delivered by more than 5 working days (the equivalent to 1 calendar week) before the dates set out in the purchasing documents (orders), unless explicitly requested by the customer, as this involves costs to store material. The set objective is for the service rate (ESSR) for each supplier to equal or exceed 95% during the period to be measured.

The OTDS (*on time delivery service*) calculation measures the order lines sent to customers within the set delivery dates from among the number of complete lines to be sent to the customer. This indicator completely measures the existing capacity to meet the logistic demand; i.e., the degree of commitment for the lead time or the set delivery date. This indicator is measured by leaving the material in the distribution center, but not when it is at the customer's door; in other words, it acts as the service rate to the logistics center. It also includes all the lines sent to customers from both the distribution center and the factory (for special reasons). Only blocked order lines are excluded for the reasons customer's have. This rate is expressed as a percentage and the fulfillment objective is 96%.

The OTDC (*on time delivery at the customer's door*) calculation measures the complete and delivered orders whose delivery equals or is less than that referred to by customers between placing an order and it being delivered to the customer. It is used to assess the degree of commitment when the order is delivered at the customer's door; i.e., it measures the capacity to fulfill the set delivery time. This calculation includes either the complete orders delivered to the customers of the group located beyond the distribution center, or direct deliveries from the factory, but excludes any orders blocked by customers. If customers initially deny partial deliveries of their order, the complete order lines are considered to make the calculation, although scarcity leads customers to accept partial deliveries as a commitment. If the delivery date that the customer requires is longer than the delivery period, the order will be supplied on the set date. Conversely, if the order requires a shorter time than the set lead time, this lead time will determine the date the order is delivered to the customer's door. This rate is expressed as a percentage and the fulfillment objective is 98%.

3 Model Formulation

To draw the causal diagram of the studied supply chain, the APIOBPCS (*automatic pipeline, inventory and order-based production control system*) model is taken as the reference model (John et al. 1994).

This study considers a small-structured supply chain to which to add some modifications. A small supply chain is characterized by the elimination of one member or more compared to a traditional supply chain (Disney et al. 2004). In this case, immediate customers, and those to which it sells directly, tend to be mainly large infrastructures or works, which means that the retailer disappears. Moreover, it is a small supply chain as the analysis is done from the supplier to the distribution center, which is where the data of the indicators are obtained. For the simulation model, it is necessary to consider the supplier's causal diagram in a small supply chain.

Unlike the suppliers with which it places purchase orders, and which deliver the ordered material on the agreed date, the company under study manufactures the material in order to provide it to the supplier and to request the corresponding electrolytic zinc-coating process of the delivered products so that it can then deliver them to its installations. Hence it is a network-like supply chain. Once the material has been delivered and accepted, it is still not ready to be supplied to the distribution center or wholesaler. The coated product moves to another section known as "film packing". This section performs the process of covering every two product units of the material with plastic film, and then places them on pallets to be sent to the distribution center. The distribution center stores the delivered product and serves the end customer if the corresponding order has been placed and the desired quantity is available. If this quantity does not totally complete the order, whatever quantity is available will be delivered. The rates or indicators that are the object of this study are determined by two clearly defined stages: on the one hand, checking that the received order fulfills the delivery date agreed on by the manufacturer and supplier; on the other hand, measuring the degree of order fulfillment; i.e., the requested quantity when the order is delivered.

For the suppliers analysis, all those that provide the galvanizing service for the semi-manufactured products which, as previously mentioned, they are supplied with to receive this service, and with the respective protection to avoid products from rusting and corroding. The Vensim® simulation software is used to perform simulation.

Although the objective of the simulation model is to analyze supplier reliability and the service rates at the various supply chain levels, it is necessary to complete the study with some of the most significant costs, which are easy to detect, to be able to deal with them if any changes need to be made. Such changes will result from using the merchandise transport and the product's storage cost. The transport cost will be measured at all the model's different levels. For this purpose, it is necessary to add counters to quantify the number of shipments issued. As the total cost per level is the number of shipments during the period multiplied by the transport cost, the transport cost differs at each level as it depends on the respective rates of these levels.

4 Evaluation of the Results

According to Sterman (2000), the validation of dynamic models can be done by the test run to check the evolution of the used parameters, among other tests. This test verifies the equivalence between the parameters used in the simulation model and their value with those that actually exist. After starting the model for an initial scenario, the model indicates the quantity of the supplied product to render the corresponding service of electrolytically zinc-coating the material, as well as the quantity of material coated and delivered to the manufacture before the set deadline delivery date (Fig. 1).

With this scenario, the service rates of the respective levels of the supply chain do not meet the objectives set. The indicator measuring the provider is the ESRR and the degree of commitment is 95%. However, the lack of supply of the supplier causes many orders to fail in the delivery period between the manufacturer and wholesaler, obtaining values close to 65%, far from the target of 96%. Regarding the level of response of the distribution center to customer demand (OTDC), the values are close to the target. This indicates that the level of response to customer demand is around 93%, although it does not meet the target of 98%.

	MONTH	ORDERS	RECEPTION	%
RELIABILITY	January	4,200	2385	57%
	February	6,000	5382	90%
	March	4,800	4173	87%
	April	3,600	2985	83%
	May	4,800	2985	62%
	June	5,400	4667	86%
	July	3,000	2364	79%
	August	4,200	2367	56%
	September	5,400	3000	56%
	October	3,600	3000	83%
	November	5,400	2367	44%
	December	4,200	3582	85%
			Average	**72%**

Fig. 1 Supplier reliability

5 Conclusions

The proposed model proves useful to analyze both supplier reliability and service rates among the different downstream supply chain levels with distinct scenarios or when modifying parameters. It also provides further information about aspects like costs, lack of supplies, response capacity, etc. The model also compares traditional methods to measure service rates with measures by order posts or lines. Traditional methods are seen to penalize the supplier more and do not bear in mind the partial quantity delivered, which is received, but not computerized as a partial order. Data about suppliers' reliability and response capacities can also be collected when posts are measured.

Some future research lines are proposed: on the one hand, complete the model with consumer forecasts to reinforce communication with the supplier; on the other hand, the inclusion in scenarios of the effects of losing customers from not having enough material available in the wholesaler's warehouse.

References

Bititci US, Turner T, Begemann C (2000) Dynamics of performance measurement systems. Int J Oper Prod Manage 20(6):692–704

Bravard JL, Morgan R (2007) La externalización inteligente, Deusto, 24–26. In: Campuzano F, Mula J Supply chain simulation. A system dynamics approach for improving performance. Ed. Springer, London

Campuzano F, Mula J (2011) Supply chain simulation. A system dynamics approach for improving performance. Springer

Campuzano F, Mula J, Peidro D (2010) Fuzzy estimations and system dynamics for improving supply chains. Fuzzy Sets Syst 161(11):1530–1542

Campuzano-Bolarín F, Mula J, Peidro D (2013) An extension to fuzzy estimations and system dynamics for improving supply chains. Int J Prod Res 51:3156–3166

De Meyer A, Wittenberg-Cox A (1994) Nuevo enfoque de la función de producción. Folio

Disney SM, Naim MM, Potter A (2004) Assessing the impact of e-business on supply chain dynamics. Int J Prod Econ 89(2):109–118

Forrester JW (1961) Dinámica industrial. Editorial El Ateneo, Buenos Aires, Argentina

Jeon SM, Kim G (2016) A survey of simulation modeling techniques in production planning and control (PPC). Prod Plann Control 27(5):360–377

John S, Naim MM, Towill DR (1994) Dynamic analysis of a WIP compensated decision support system. Int J Manuf Syst Des 1(4):283–297

Kaplan RS (1990) Measures for manufacturing excellence. Harvard Business School Press, Boston

Moullin M (2002) Delivering excellence in health and social care. Open Univiversity Press, Buckingham

Neely A, Adams C, Kennerley M (2002) The performance prism: the scorecard for measuring and managing business success: the scorecard for measuring and managing stakeholder relationships. Prentice-Hall

Sink S, Tuttle T (1989) Planning and measurement in your organization of the future. Industrial Engineering & Management Press, Norcross, GA

Sterman JD (2000) Business dynamics: systems thinking and modelling for a complex world. McGraw-Hill, Boston

Quantitative Techniques to Effectively Link Enterprise's Performance and Lean Production

José Luis Perez-Ruano, Raúl Rodríguez-Rodríguez, Juan-José Alfaro-Saiz and María-José Verdecho

Abstract Lean Production has been widely applied in the last decades and it is generally accepted as a tool to improve the operations of organisations. However, how to properly quantify such an improvement is still not clear from a global and integrated point of view. The fact that the data gathered from applying LP production techniques—at the operative level—should be somehow correlated with the enterprise strategic objectives one—at the strategic level—presents important challenges to decision makers. This papers sets up this research topic, covering some relevant scientific literature, moving then to define a classification of important factors to take into account when deciding what quantitative technique to apply to this topic. Then, following such a classification, it assesses some quantitative techniques that could be used, mainly broken down into either objective or subjective. From this evaluation, it concludes that, from the evaluated, the most appropriate technique to be used should be the Analytic Network Process.

Keywords Enterprise performance · Lean production · Statistical techniques ANP

J. L. Perez-Ruano
Escuela de Doctorado, Universitat Politécnica de Valéncia,
Camino de Vera sn, 46022 Valencia, Spain
e-mail: joruape@etsid.upv.es

R. Rodríguez-Rodríguez (✉) · J.-J. Alfaro-Saiz · M.-J. Verdecho
CIGIP—Universitat Politécnica de Valéncia,
Camino de Vera sn, 46022 Valencia, Spain
e-mail: raurodro@cigip.upv.es

J.-J. Alfaro-Saiz
e-mail: jalfaro@cigip.upv.es

M.-J. Verdecho
e-mail: mverdecho@cigip.upv.es

Á. Ortiz et al. (eds.), *Engineering Digital Transformation*, Lecture Notes in Management and Industrial Engineering,
https://doi.org/10.1007/978-3-319-96005-0_22

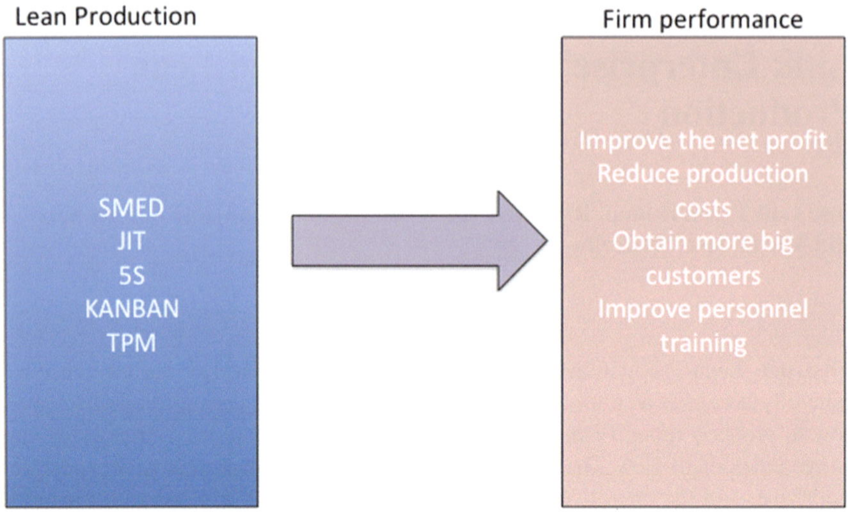

Fig. 1 Impact of LP on firm performance

1 Introduction

In the last decades, Lean Production (LP) has attracted great interest from both researchers and practitioners. The application of LP techniques will, theoretically, bring tangible advantages to organisations, which will mainly take place at the lowest organisational level: The operative one. Then, although it is widely accepted that a firm applying LP techniques will, generally speaking, improve its performance at several levels—i.e. cost reduction, increasing of customer service, etc. —there are not many works that directly link LP and enterprise's performance in a solid and coherent way (Hofer et al. 2012; Maga et al. 2010). This may be due to the difficulty of reconciling the decision levels inherent to both LP (operative) and to the organisational strategic decision-making (strategic). In other words, there are lot of data related to the improvement at the operative level as a consequence of implementing LP techniques but it is difficult to correlate this data and to establish how it has affected to the organisation's improvement level.

Figure 1 illustrates this thinking, where it is possible to look at the problem from two different perspectives:

- To establish how the implementation of LP techniques affects to the achievement of firm performance from an integrated approach. In this case the LP techniques are shown as an aggregation and firm performance is seen individually; in other words, it is assessed the impact of all the LP techniques as a whole over each of the strategic objectives defined for the organization.
- To consider individually the different LP techniques and what is their impact on each of the strategic objectives defined. This a more difficult approach from a

practical point of view, as it should be consider and quantify the effect of every single technique, i.e. SMED, over the variation of every strategic objective, i.e. to improve the net profit for the current year.

Looking at the scientific literature it is possible to find some works that point to this topic. Then, Alsyouf et al. (2011) carried out a process map where they measure productivity, quality and strategy performance altogether. On the other hand, Bhasin (2012) proposed specific LP performance measures integrated within a Balanced Scorecard for large organizations. Further, some other studies have found that LP has got a relevant effect on business performance (Hofer et al. 2012; Agus and Hajinoor 2012). The main problem is that these works reduce business performance to financial performance, lacking of a holistic organizational approach. Additionally, a recent study (Abreu-Ledón et al. 2018) has pointed out the difficulty of correlating the impact of LP on business performance. To this end, they carried out a meta-analytic study, finding a low positive relationship but at the aggregate level between LP and business performance.

In general terms, there are other authors who have also pointed out the mentioned difficulties when aiming to measure and quantify the impact of implementing LP techniques over firm performance (Kannan and Tan 2005; Klingenberg et al. 2013; Camacho-Miñano et al. 2013). These authors have got reservations about the positive of LP on firm performance and how to apply different methods and approaches to assess it.

Then, it is possible to affirm that it is certainly difficult to establish objective (data based) relationships between the application of LP and business performance. The question now is: are there other possibilities to better establish such a relationship? If the objective techniques (statistical techniques) are difficult to be applied and their results are, as presented above, limited, would make sense to apply subjective techniques? To what extent and under what conditions?

These questions are answered in the next points of this research. Next section presents the objectives and the methods followed; then, the main results obtained are highlighted, establishing finally some conclusions.

2 Objectives and Methods

The main objective of this research is to explore and identify what techniques should be applied, under different conditions and contexts regarding specially data availability and decision-making stages, to establish a proper link between the application of LP and the impact that this fact provokes over firm performance (measured this with the defined strategic objectives).

To this end, the authors have mainly explored statistical techniques and Multi-Criteria Decision Aid (MCDA) ones—the Analytic Hierarchical Process and the Analytic Network Process—and their feasible application in situations of existence/absence of performance historical data sets and considering other important characteristics of the techniques.

3 Results

3.1 Factors for Classifying the Techniques

From the above stated, it is possible to affirm that, when aiming to assess the impact that LP techniques has got on enterprise performance, decision-makers should carefully choose the quantitative technique they are going to use to this end. Then, there are different factors to take into account, being some of the most important ones the following:

- Objective/Subjective. This means that the technique to be used is either based on historical data or rather based on experience and subjective judgment. In this particular research, it should be desirable to be able to apply historical data to measure the effect of LP on firm performance. However, this would make the analysis pretty difficult, as this would bring together cause-effect relationships between data coming from LP indicators (operative level) and data gathered through strategic objectives indicators (strategic level). Such a mixture should be carefully dealt with. Additionally, the variation of the values of the strategic objectives will be due not only to the effect of LP techniques but also to many other factors and activities of the organization. Then, when following the approach of using objective techniques, the decision-makers should somehow establish to what extent the implementation of LP techniques have contributed to the global variation of the strategic objectives. This latter approach is not easy from an objective point of view.
- Data. The type of data needed to apply the different quantitative techniques. In the case of objective techniques, it would be the data that has been gathered overtime regarding both the LP techniques and the strategic objectives. On the other hand, when using subjective techniques no historical data is needed, as the analysis is based on experience of organizational experts.
- Relationships. It will be important to choose a quantitative technique that is able to identify relationships between more than two variables. In other words, able to identify how a group of variables (i.e. several LP techniques) are affecting over other group of variables (i.e. several strategic objectives) at the same time. This would also allow identifying inter-relationships among the different variables.
- Complementary. This factor contemplates the possible need of some techniques of applying complementary techniques in order to be able to properly set up the effect of LP techniques on firm performance.

Table 1 shows the results of classifying different quantitative techniques following the above describe factors.

Table 1 Quantitative techniques classification

Technique	Type	Data	Relationships	Complementary
Correlation analysis	O	Historical	Pairs	N
Regression analysis	O	Historical	Multiple	N
Principal component analysis	O	Historical	Multiple	Y
Partial least squares	O	Historical	Multiple	Y
Analytic hierarchical process	S	Experience	Pairs	N
Analytic network process	S	Experience	Multiple	N

Table 1 includes different techniques, which might be broken down into objective and subjective techniques.

3.2 Explanation of the Classification

Table 1 highlights, four objective techniques, which are described next in the context of this research:

- Correlation analysis. This is the simplest technique and it provides pair-wise comparisons between variables, not being able to identify interrelationships between variables. It has been widely used and, in our context, some recent studies such as Abreu-Ledón et al. (2018) applied obtaining some aggregate results, as commented previously.
- Regression analysis. This technique is more complex than the correlation analysis but is able to carry out global analysis where the independent variable could be a specific strategic objective and the dependent variables could be all the variables representing the LP techniques (their performance indicators).
- Principal Component Analysis (PCA). This is a more complex technique, being also its results more complete and adequate than from the previous techniques. The idea is that PCA will reduce the variability of the initial data and will allow, focusing on a low number of principal components, identify what variables maintain relationships over time. Its main drawback is that it is necessary keep a relationships of at least 3:1 in terms of number of variables analysed. In this research, it would be necessary that the number of variables of LP techniques is not more than three time the number of variables of strategic objectives. Additionally, its capa-

bility of quantifying the relationships identified is limited and it woks better, in this context, as an exploratory technique that could be complemented with another more powerful one such as Partial Least Squares to establish the quantification of the relationships.

- Partial Least Squares (PLS). As mentioned, this technique could be implemented as complementary to the PCA and it will allow to design and implement both PLS and PLS2 models. These models will outcome, in a very accurate way, the quantification of the variables under study.
- Analytic Hierarchical Process (Saaty 1980). This is a subjective technique, as it is based on experts' judgment and experience. In this research could easily used to quantify, according to these opinions, to which extent the implementation of LP techniques is affecting to firm performance. Its main drawbacks are that it carries out pairwise comparisons between variables and its rigidity, as changes on the AHP network would lead to construct and calculate again everything.
- Analytic Network Process (Saaty 1996). ANP can be seen as the extension of AHP and it allows not only multiple comparisons at the same time between variables but also add a high degree of flexibility, as its matrices can be easily modified when new variables are introduced in the network. It have widely used and applied in the literature and, more concretively, it is possible, focusing on this research, mention the work of Verdecho et al. (2012) and Boj et al. (2014). The former applies ANP to quantify the effect of collaborative practices on performance of a network of organizations whereas that the latter applies ANP to quantify the effect of intellectual capital on the performance of a public research center.

At the light of the problematic treated in this research, the antecedents from the literature and the analysis carried out of some representative techniques, it is possible to affirm that the most adequate technique to quantify the effects of implementing LP techniques on firm performance is the ANP.

4 Conclusions

The application of Lean Production implies to invest resources and enterprises need to be aware of to what extent such an investment is paying back. Then, this research identifies what quantitative techniques could be applied, under some important conditions, in order to effectively link the application of LP techniques and firm performance. These techniques have been broken down into objective and subjective ones.

The objective ones are based on dealing with historical data, which should lead to more accurate results, as the main input is real data from the organisation. However, to correlate data coming from the operative level (from LP techniques) with data of the strategic level (firm performance is represented with the strategic objectives) may lead to decision-making problems. Additionally, it should necessary to identify, in an accurate matter, to what extent the implementation of LP techniques have contributed to the global variation of the strategic objectives, as such a variation will have other sources from the organization.

On the other hand, the subjective ones overcome these problems, being their main drawback that they are based on subjective judgment and expert opinions rather than on historic data. However, from the classification and analysis of the techniques carried out, it is possible to state that the most adequate technique to quantify the effects of implementing LPO techniques on firm performance is the ANP.

References

Abreu-Ledón R, Luján-García DE, Garrido-Vega P, Escobar-Pérez B (2018) A meta-analytic study of the impact of Lean production on business performance. Int J Prod Econ 200:83–102

Agus A, Hajinoor M (2012) Lean production supply chain management as driver towards enhancing product quality and business performance: case study of manufacturing companies in Malasya. Int J Qual Reliab Manag 29(1):92–121

Alsyouf I, Al-Aomar R, Al-Hamed H, Qiu X (2011) A framework for assessing the cost effectiveness of lean tolos. Eur J Ind Eng 2(5):170–198

Bhasin S (2012) Performance of Lean in large organisations. J Manuf Syst 31:349–357

Boj JJ, Rodriguez-Rodriguez R, Alfaro-Saiz J-J (2014) An ANP-multi-criteria-based methodology to link intangible assets and organizational performance in a balanced scorecard context. Decis Support Syst 68:98–110

Camacho-Miñano MM, Moyano-Fuentes J, Sacristán-Díaz M (2013) What can we learn from the evolution of research on lean management assessment? Int J Prod Res 51(4):1098–1116

Hofer C, Eroglu C, Hofer AR (2012) The effect of lean production on financial performance: the mediating role of inventory leanness. Int J Prod Econ 138:245–253

Kannan VR, Tan KC (2005) Just-in-time, total quality management, and supply chain management: understanding their linkages and impact on business performance. Omega 33:153–162

Klingenberg B, Timberlake R, Geurts TG, Brown RJ (2013) The relationship of operational innovation and financial performance—a critical perspective. Int J Prod Econ 142(2):317–323

MaGa Y, Hong P, Modi SB (2010) Impact of lean manufacturing and environment al management on business performance: an empirical study of manufacturing firms. Int J Prod Econ 129:251–261

Saaty T (1980) Multicriteria decision making: the analytic hierarchy process. McGraw Hill, New York

Saaty TL (1996) The analytic network process: decision making with dependence and feedback. RWS Publications, Pittsburgh, PA

Verdecho M-J, Alfaro-Saiz J-J, Rodriguez-Rodriguez R, Ortiz-Bas A (2012) A multi-criteria approach for managing inter-enterprise collaborative relationships. Omega 40(3):249–263

Laser Additive Manufacturing: A Patent Overview

Enara Zarrabeitia, Iñaki Bildosola, Rosa María Río Belver, Izaskun Alvarez and Ernesto Cilleruelo-Carrasco

Abstract Additive manufacturing technologies have the potential to change the paradigm of manufacturing. In this scenario, the main objective of this research work is to gain, through a patent study, an overall view of the trends of laser-based additive manufacturing technologies. The database used in order to retrieve patent information is *Patseer* and the data have been analyzed through the text mining and analytics package called *VantagePoint*. From the data obtained it can be concluded that, Laser-based additive manufacturing is an emerging technology with huge market potential. Undoubtedly, China is the prime mover of basic research in this technology development. However, Germany and the United States are the ones who are developing the market applications.

Keywords Additive manufacturing · Laser additive manufacturing · Patent overview

E. Zarrabeitia (✉) · I. Alvarez · E. Cilleruelo-Carrasco
Industrial Organization and Management Engineering Department,
Faculty of Engineering, University of the Basque Country,
Alameda Urquijo s/n, 48013 Bilbao, Spain
e-mail: enara.zarrabeitia@ehu.eus

I. Alvarez
e-mail: izaskun.alvarez@ehu.eus

E. Cilleruelo-Carrasco
e-mail: ernesto.cilleruelo@ehu.eus

I. Bildosola
System Engineering and Automatics Department,
Faculty of Engineering, University of the Basque Country,
Alameda Urquijo s/n, 48013 Bilbao, Spain
e-mail: inaki.bildosola@ehu.eus

R. M. Río Belver
Industrial Organization and Management Engineering Department,
Faculty of Engineering, University of the Basque Country,
C/Nieves Cano, 12, 01006 Vitoria, Spain
e-mail: rosamaria.rio@ehu.eus

Á. Ortiz et al. (eds.), *Engineering Digital Transformation*, Lecture Notes in Management and Industrial Engineering,
https://doi.org/10.1007/978-3-319-96005-0_23

1 Introduction

The ASTM international F42 Technical Committee defines Additive Manufacturing (AM) as the "process of joining materials to make objects from 3D model data, usually layer upon layer, as opposed to subtractive manufacturing methodologies". It is also known as additive fabrication, additive processes, additive techniques, additive layer manufacturing, layer manufacturing, and freeform fabrication (ASTM 2013).

AM technologies have the potential to change the paradigm of manufacturing. Not only is it a disruptive technology that has the potential to replace many conventional manufacturing processes, but it also symbolizes an enabling technology allowing new business models, new products and new supply chains to flourish (Quarshie et al. 2012).

The additive manufacturing market in 2013, consisting of all AM products and services worldwide, grew 34.9% (Compound Annual Growth Rate, CAGR) to $3.07 billion. This compares to growth in 2012 of 32.7% to $2.275 billion. Industry wide growth in 2011 and 2010 was a healthy 29.4 and 24.1%, respectively (Wohlers Report 2014).

In terms of industry specifics, AM has thrived considerably in particular areas over the last few years (AM Platform 2014). The technology cuts across a large number of industries and applications, and that is part of what makes its potential so compelling. Aerospace, automotive, medical and consumer products will drive AM into the future (Bourell et al. 2009).

Although various AM processes have been introduced to the commercial market by industrial companies, this paper only analyzes laser-based processes.

Laser-based processes have generated considerable interest in the past few years. The main reason for this interest is that the quality of the laser additive manufactured parts is on such a high level that the parts can be used in different industrial fields as functional parts. The possibility of creating geometrically complex parts and working prototypes has also generated increased enthusiasm (Matilainen 2012).

Laser-based AM implies layer-by-layer shaping and consolidation of feedstock, typically powder materials, to arbitrary configurations, using a computer controlled laser as an energy source (Gu 2015).

Laser sintering (LS), laser melting (LM), and laser metal deposition (LMD) are presently regarded as the three most versatile laser-based AM processes. However, different institutions and companies throughout the world use different phrases to denominate these three most prevalent variants of AM technology, as shown in Table 1 (Gu 2015).

Also, patent information is a useful indication of strategies for technological development. The details of some technologies at the state of the art and those technologies that might possess a great potential are disclosed within patents (Tsai et al. 2016).

Consequently, in the context of this rapidly developing technology area, this study has been performed looking at a dataset comprising a large number of individual

Table 1 Synonyms from different institutions/companies for laser-based AM processes

Laser-based AM processes	Synonyms
Laser sintering (LS)	Selective laser sintering (SLS)
	Direct metal laser sintering (DMLS)
Laser melting (LM)	Direct metal laser sintering (DMLS)
	Selective laser melting (SLM)
	Direct metal laser re-melting (DMLR)
	Lasercusing
Laser metal deposition (LMD)	Laser material deposition
	Direct metal deposition (DMD)
	Direct metal laser deposition (DMLD)
	Laser engineered net shaping (LENS)
	Direct light fabrication (DLF)
	Direct laser deposition (DLD)
	Direct laser fabrication
	Laser rapid forming (LRF)
	Laser melting deposition
	Laser cladding

Source Gu (2015), Guo and Leu (2013), Cotec (2011)

patents and analyzed to provide a macroscopic view of this technology "space" over time (Intellectual Property Office 2013).

2 Objectives

The main objective of this research work is to achieve an overall view of the trends of laser-based additive manufacturing technologies. Specifically, through patent study, with the intention being to analyze the evolution in time of these technologies, the leader countries, inventors and entities behind them, collaboration between entities, and the markets where this area is being investigated and patented the most, with special emphasis on medical applications (Adimen Lehiakorra 2015).

3 Methods

In this study, the database used in order to retrieve patent information has been *Patseer*. *Patseer* is an inclusive online global patent database and research platform containing the world's most comprehensive full-text patent collection along with inte-

Table 2 Variables analyzed to get an overview of Laser additive manufacturing processes

Objectives to achieve	Variables analyzed
Understand the technological evolution	Publication date
Identify technological leader countries, inventors and entities	Priority country code
	Inventors
	Assignee normalized
Study the technology transfer	Collaboration (Cluster map)
Identify markets and applications	International patent classification (IPC)
Medical applications	IPC: A61

Source Own elaboration

grated analytics, project workflows, and collaboration capabilities. *PatSeer* includes global patent data coverage of 42 Full Text Authorities (Sinha and Pandurangi 2016).

Terms were searched by title or/and abstract field, as they describe the main characteristics of the patents. The query used was developed as follows:

WORD = "laser sintering" or "selective laser sintering" or "direct metal laser sintering " or "laser melting" or "selective laser melting" or "direct metal laser re-melting" or "lasercusing" or "laser metal deposition" or "laser material deposition" or "direct metal deposition" or "direct metal laser deposition" or "laser engineered net shaping" or "direct light fabrication" or "direct laser deposition" or "direct laser fabrication" or "laser rapid forming" or "laser melting deposition" or "laser cladding"

In order to improve the accuracy of the research, patents were grouped by patent families, resulting in 3459 families.[1] A patent family is a set of either patent applications or publications taken out in multiple countries to protect a single invention by a common inventor(s) and then patented in more than one country (European Patent Office 2016). Hence, analysis by patent family gives more accurate results regarding the level of innovation taking place (Intellectual Property Office 2013).

The overall view of the trend in laser-based additive manufacturing technologies is analyzed through the following variables (Table 2).

The main computer software used for this work is a text mining and analytics package called *VantagePoint* produced by *Search Technology* in the USA. The patent documents exported from *Patseer* database are imported into *VantagePoint*, where the data is cleaned and analyzed.

[1]To be more precise, there are 2589 applications, 537 granted utility models, 317 patents, 14 others, 1 granted innovation patent and 1 utility model application.

4 Results

4.1 Technological Evolution of Laser Additive Manufacturing

Although the first patent is dated 1969, the greatest activity has been in recent years. Figure 1 shows that growth is exponential and that the instant growth rate is 0.2.

It is observed that, in recent years, laser-based AM has become a technology of growing interest and with very intense patent publication.

Figure 1 shows that the patent publication growth follows the bibliometric law of exponential growth (Derek j. de Solla Price's law) (Ardanuy 2012). According to that law the publication of patents in this field is in the phase of exponential growth, i.e., it is an emerging technology.

4.2 Leader Countries, Inventors and Entities in Laser Additive Manufacturing

Following this are the countries and entities that have most patented in development of technologies related to laser additive manufacturing.

As shown in Fig. 2, for the set of patent families analyzed, China stands out as the leader in the generation of innovations (2282), followed by the United States (371) and Germany (274).

From the first 20 researchers in this area, 19 are also Chinese (see Fig. 3), with only 1 of a different nationality (German).

As for the entities, in this case too, a simple glance shows that the bulk of patents are from Chinese entities, followed by entities from Germany and the United States (see Fig. 4). However, whereas in China all entities are universities or centers (basic

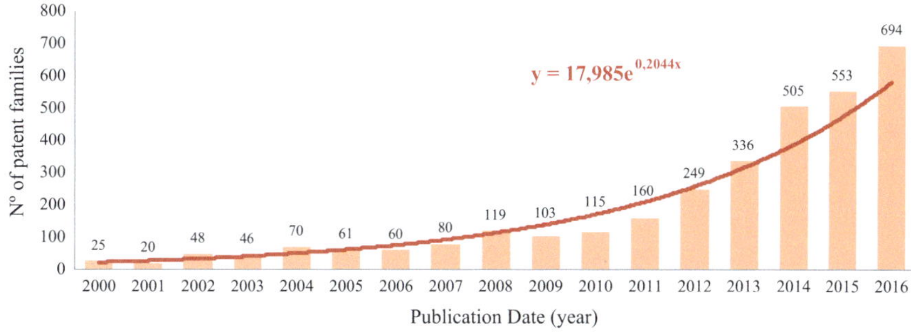

Fig. 1 Evolution of the number of patent families published (2000–2016)

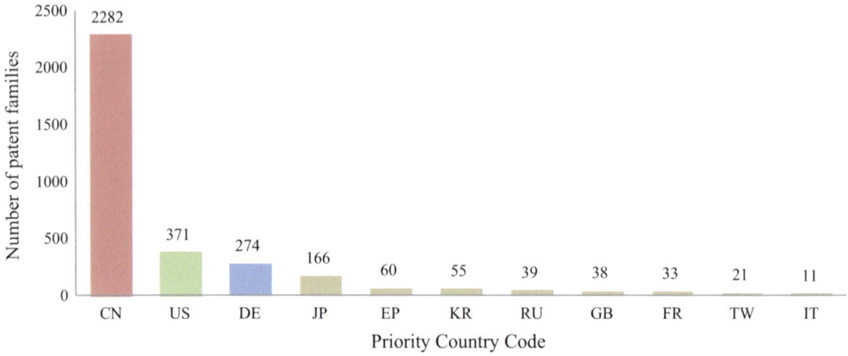

Fig. 2 Priority country codes with more than ten patent families

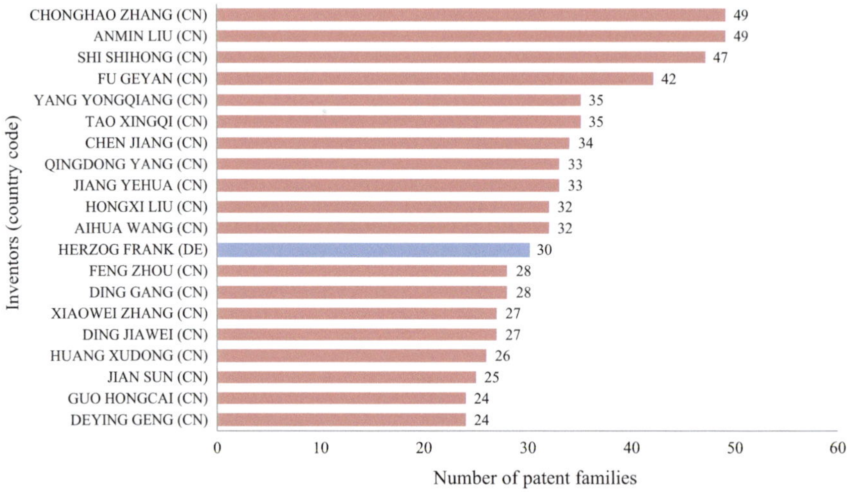

Fig. 3 The top twenty inventors

research), in the case of Germany and the United States entities are companies (applied research).

Finally, most of the top entities in this dataset have not collaborated with other entities, and they are the sole applicants on all of their patents in this field of technology. Only two Chinese universities have collaborated in one patent (see Fig. 5).

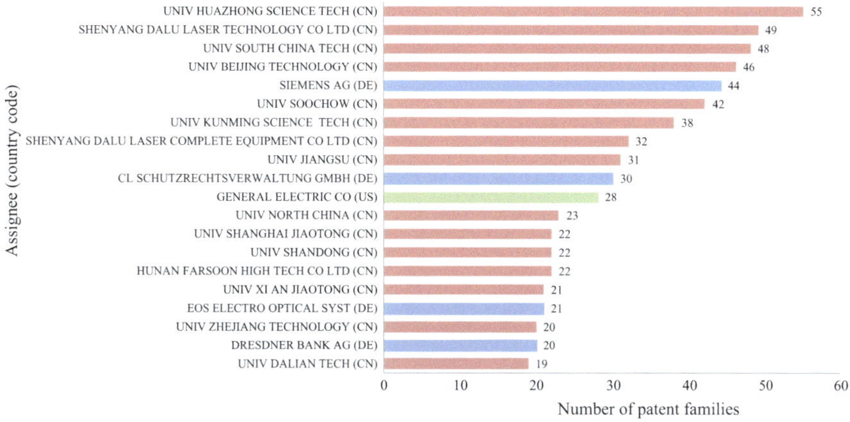

Fig. 4 The top twenty assignees

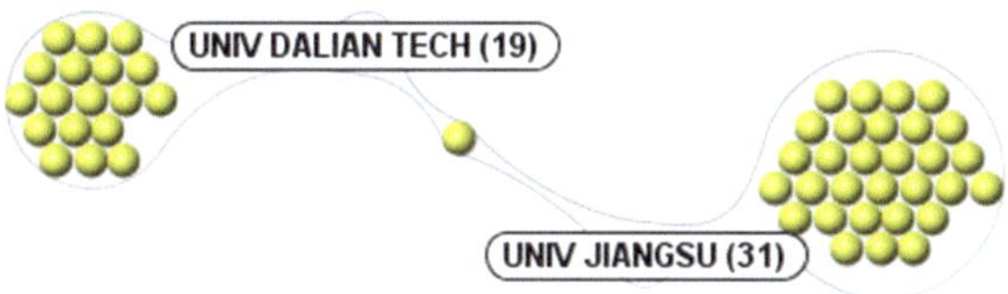

Fig. 5 Collaborations between entities

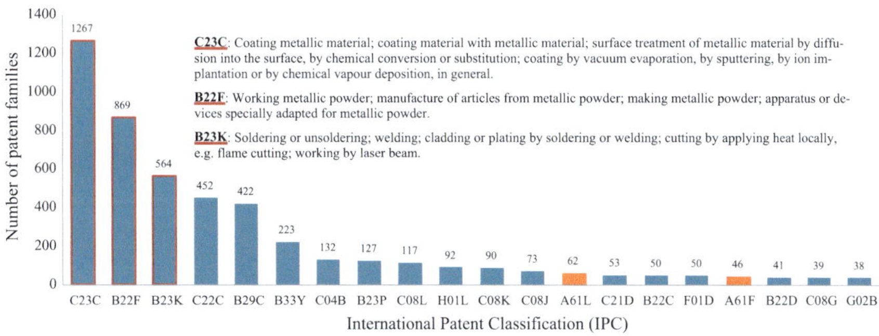

Fig. 6 Top international patent classifications

4.3 Markets and Applications of Laser Additive Manufacturing

Figure 6 shows the top International Patent Classification sub-groups. As might be expected, most of them pertain to Chemistry and Metallurgy (C) and Performing Operations and Transporting (B) groups. However, the presence of patents in Human Necessities (A), among others, is also observed.

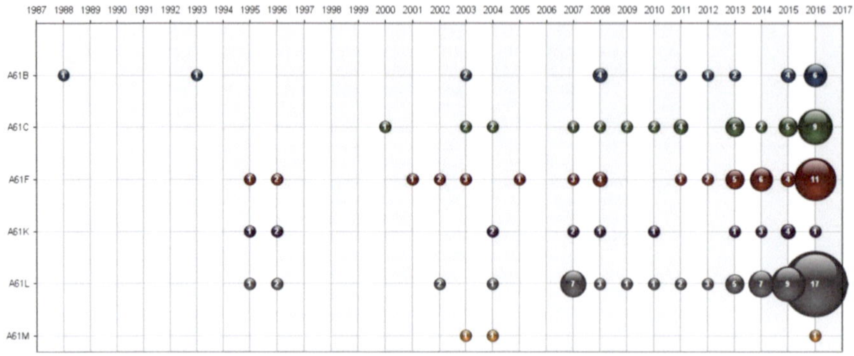

Fig. 7 Evolution of the number of main patent families published in A61 group

4.3.1 Medical Applications (IPC: A61)

In the IPC analysis, it is curious to observe how in recent years an important group of patent families in class A61, directed toward medical applications, have been published (see Fig. 7).

This corroborates the highly applicable nature that this technology has been enjoying recently in the medical industry generally, and in subgroups A61L and A61F specifically.

5 Conclusions and Future Lines of Research

Laser-based additive manufacturing is an emerging technology, i.e. it is an early stage of development.

Within this scenario, China is undoubtedly the central character in basic research of this technology development. However, Germany and the United States are the ones who are developing the market applications.

Nonetheless, most of the top entities in this dataset have not collaborated with other entities. This may be due in part to the relatively new area of such technology, which encourages most companies to do their own research in-house (Intellectual Property Office 2013).

Despite the fact that most patents are grouped into “Chemistry and Metallurgy” and “Performing Operations and Transporting”, areas such as medicine are also benefiting from these technological advances. It can be said that it is a technology with huge market potential.

Finally, future works related with this study could take many forms given the diversity of the current situation for additive manufacturing. It would also be interesting to expand this study to include other processes and technologies; and to analyze the maturity of the technological market of the laser-based AM.

References

Adimen Lehiakorra (2015) Análisis de patentes de fabricación aditiva de fusión de lecho de polvo. Diputación Foral de Guipúzcoa, Departamento de Innovación, Desarrollo Rural y Turismo
AM Platform (2014) Additive manufacturing: strategic research agenda. AM SRA Final Document
Ardanuy (2012) Breve introducción a la bibliometría. Universitat de Barcelona
ASTM (2013) ASTM F2792–12a, standard terminology for additive manufacturing technologies
Bourell DL, Leu MG, Rose DW (2009) Roadmap for additive manufacturing identifying the future of freeform processing. The University of Texas at Austin Laboratory for Freeform Fabrication Advanced Manufacturing Center
Cotec (2011) Fabricación Aditiva. Documentos Cotec sobre oportunidades tecnológicas. Fundación Cotec para la innovación tecnológica, Madrid
European Patent Office (2016) Patent families. Retrieved from http://www.epo.org/searching-for-patents/helpful-resources/first-time-here/patent-families.html
Gu D (2015) Laser additive manufacturing of high-performance materials. College of Materials Science and Technology, Nanjing University of Aeronautics and Astronautics, China, Springer
Guo N, Leu MC (2013) Additive manufacturing: technology, applications and research needs. Front Mech Eng 8(3):215–243
Intellectual Property Office (2013) 3D printing. A patent overview. UK Intellectual Property Office Patent Informatics Team, United Kingdom
Matilainen V (2012) Benchmarking of laser additive manufacturing process. BK10A0401 Bachelor's thesis and seminar, Lappeenranta University of Technology
Quarshie R, MacLachlan S, Reeves P, Whittaker D, Blake R (2012) Shaping our national competency in additive manufacturing. Additive Manufacturing Special Interest Group, UK
Sinha M, Pandurangi A (2016) Guide to practical patent searching and how to use PatSeer for patent search and analysis. Retrieved from http://patseer.com/book-on-patent-searching/
Tsai YC, Huang YF, Yang JT (2016) Strategies for the development of offshore wind technology for far-east countries—a point of view from patent analysis. Renew Sustain Energy Rev 60:182–194
Wohlers report (2014) Additive Manufacturing and 3D printing state of the industry, Wohlers Associates 2014. Available from: http://wohlersassociates.com/2014report.htm

Part III
Methods and Applications

Short-Term Forecasting in Office Consumption with Identification of Patterns by Clustering

Pablo Aparicio-Ruiz, Elena Barbadilla Martín, José Guadix Martín and Pablo Cortés Achedad

Abstract Fuzzy clustering algorithm has demonstrated advantages in data mining, especially in problems with large collections of imprecise or fuzzy data. This methodology is widely used in the context of pattern recognition. It is proposed this methodology to use in the forecast of the comfort area of building or in the forecasting of consumption. In this paper the concept of fuzzy clustering, which is widely used in the context of pattern recognition. Based on the study of the fuzzy algorithm, we propose a method to forecast the comfort and consumption.

Keywords Pattern recognition · Forecasting · Clustering · Comfort · Energy

1 Introduction

The problem of modelling and forecasting the energy use in a building can be very complex due to the number and diversity of factors influencing it. In the case of the HVAC system, the system depends on a conjunction of factors like: thermal properties of the building envelope, weather conditions, the number of occupants and their behaviour, etc.

In (Zhao et al. 2014) are shown numerous studies have developed various control systems and modelling methods to better assist occupants to play their "active" roles in building. In the literature 10 out of 15 studies demonstrate an energy impact of

P. Aparicio-Ruiz (✉) · E. B. Martín · J. G. Martín · P. C. Achedad
Dpto. de Organización Industrial y Gestión de Empresas II. Escuela Técnica Superior de Ingeniería, Universidad de Sevilla, Camino de los Descubrimientos s/n, 41092 Seville, Spain
e-mail: pabloaparicio@us.es

E. B. Martín
e-mail: ebarbadilla@us.es

J. G. Martín
e-mail: guadix@us.es

P. C. Achedad
e-mail: pca@us.es

Á. Ortiz et al. (eds.), *Engineering Digital Transformation*, Lecture Notes in Management and Industrial Engineering,
https://doi.org/10.1007/978-3-319-96005-0_24

occupant behaviour by using various individualized control systems and dashboards for building heating, ventilation and air conditioning (HVAC), lighting and plug load systems. In Zhao et al., it is highlighted the impact of the occupant feedback can be useful for building diagnostics. On the one hand, the providing occupants with individualized controls can influence their thermal comfort, and on the other hand, if it is provided the occupants an individual control can improve their subjective satisfaction towards their working environment.

The impact of the occupant feedback can be used in the forecasting, thus helping to meet the consumption to make important decisions, including decisions on purchasing and generating electric power and load switching. Load forecasts can be divided into three categories: short-term forecasts which are usually from one hour to one week, medium forecasts which are usually from a week to a year, and long-term forecasts which are longer than a year (Feinberg and Genethliou 2006). Short-term forecasts are important in day-to-day operation of the equipment, they are useful for building energy managers to schedule the operation of HVAC systems. Our interest focuses on short-term forecasting for demand response application.

There are many methods for short-term forecasting: statistical techniques, artificial intelligence algorithms, neural networks, fuzzy logic, expert systems, etc.

Energy efficiency in the HVAC systems is a critical issue for buildings. Energy presents a high percentage of the running cost while it defines at a large the thermal and optical comfort of the building users. In general, the system defines the forecasting energy consumption in the base on the hour or the weather prevision. This paper proposes an energy rating using clustering techniques.

Clustering techniques are based on numerical data. The cluster analysis is a mathematical procedure to identify natural groupings of objects, ‘clusters’, in such a way that the characteristics of objects belonging to the same cluster are very similar while the characteristics of objects in different clusters are quite distinct, producing thus a concise representation of the dataset behaviour.

Clustering analysis treats each input in the data set as an object having a similar characteristic, the analysis is based on the calculation of a distance metric that defines the similarities and dissimilarities between the data used in the set.

Each data cluster is defined by the member objects belonging to the cluster and by its centroid, or centre. The centre of each cluster is the point where the sum of the distances of all data in that cluster is minimized. Different parameters or criteria for clustering can be defined like the total energy consumption per unit surface, or the specific energy consumption of the buildings.

The scope of this research is the investigation of applying clustering techniques on a database in an office for the forecast of consumption. It is analysed the classification schemes for each case: heating demand and cooling demand.

The establishment of these classifications will permit the analysis of the consumption in a short-term. The analysis is based on the use of the system. The aim of this research is to propose a methodology for classification and improvement of energy performance of the office building sector.

The paper is structured in five more sections. Section 2 describes the adopted classification methodology in detail and the description of the selected clustering techniques that are investigated. The building dataset that these methods have been applied in the criteria for clustering algorithm evaluation or pattern identification. The Sect. 3 is a short definition of the case study. In Sect. 4 the numerical results and discussion about the proposed classification schemes for energy consumption (heating and cooling mode) and for the indoor thermal comfort in an office. In Sect. 5 presents the conclusion.

2 Methodology

The clustering technique is the task of assigning a set of objects into groups (clusters), where the objects in the same cluster are more similar (in some sense or another) to each other than to those in other clusters.

In (Bezdek 1974) the cluster is defined as a set, $[x_i] = \{x_j | (x_i, x_j) \in \text{R}\}$, as the equivalent class of x_i on a universe of data points, X. This class is contained in a special relation, R, known as an equivalence relation. This class is a set of all elements related to x_i that have the following properties:

1. $x_i \in [x_i]$ therefore $(x_i, x_i) \in R$
2. $[x_i] \neq [x_j] \Rightarrow [x_i] \cap [x_j] = \emptyset$
3. $\mathbf{U}x \in \text{X}[x] = \text{X}$

The clustering techniques, in this area, different approaches to clustering data can be described, as in (Jain et al. 1999). In our case, the problem has aspect relates to algorithmic structure and operation.

Our problem is defined by a group of categorized situations. In the Fig. 1, is shown the real consumption in an office for heating demand and cooling demand. The X-axis shows the mean outdoor temperature (the average external temperature in the day), and the Y-axis shows the internal temperature. The consumption data are presented in a classification in kWh every 15 min. The data have been selected because it is known the user comfort through surveys. The data presented, have a diffuse pattern, for this reason, it has been decided to take a range to define a classification: <0.2, 0.2–0.5, 0.5–1, 1–1.5, 1.5–2 and >2 kWh.

This data defines an agglomerative approach begins with each pattern or range with a distinct (singleton) cluster. In our case, this is shown in the Fig. 2. The cluster defines ranges of consumption, the number of ranges defines the number of centre of gravity.

This algorithm works by assigning membership to each data point corresponding to each cluster centre based on the distance between the cluster centre (v_j) and the data point (x_i). '$||x_i - v_j||$'is the Euclidean distance between *i*th data and *j*th cluster centre. The membership and cluster centres are updated per the minimization of the:

$$\sum ||x_i - v_j||^2 \tag{1}$$

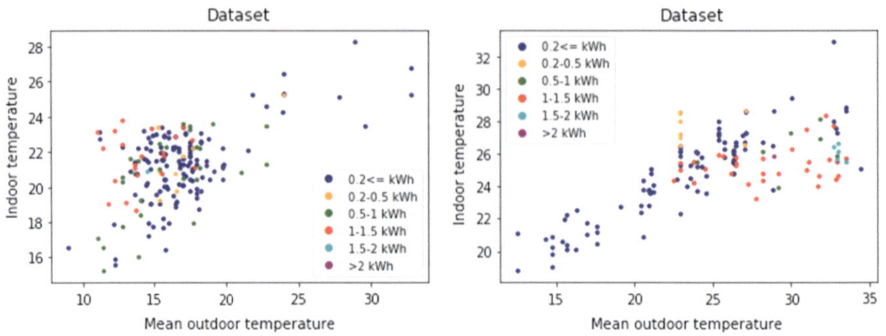

Fig. 1 Example of dataset of this problem. On the left, heating consumption, and on the right, consumption in cooling mode

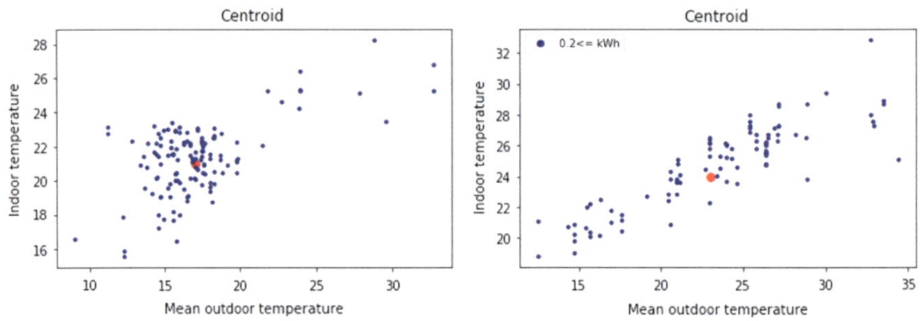

Fig. 2 Example of centroid for less than 0.2 Kwh. On the left, heating consumption, and on the right, consumption in cooling mode

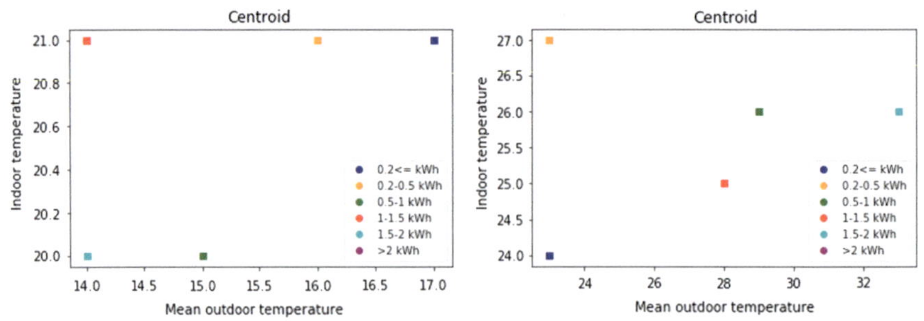

Fig. 3 Example of centroids. On the left, heating consumption, and on the right, consumption in cooling mode

Our classification in clusters intends to partition n objects into a cluster in which the centroid is the nearest mean distance. This method produces exactly 6 different clusters of greatest possible distinction, the centroid is shown in Fig. 3. The solution of this centroid is the same as a Voronoi diagram indicates the areas that are closer to one cluster centre than any other.

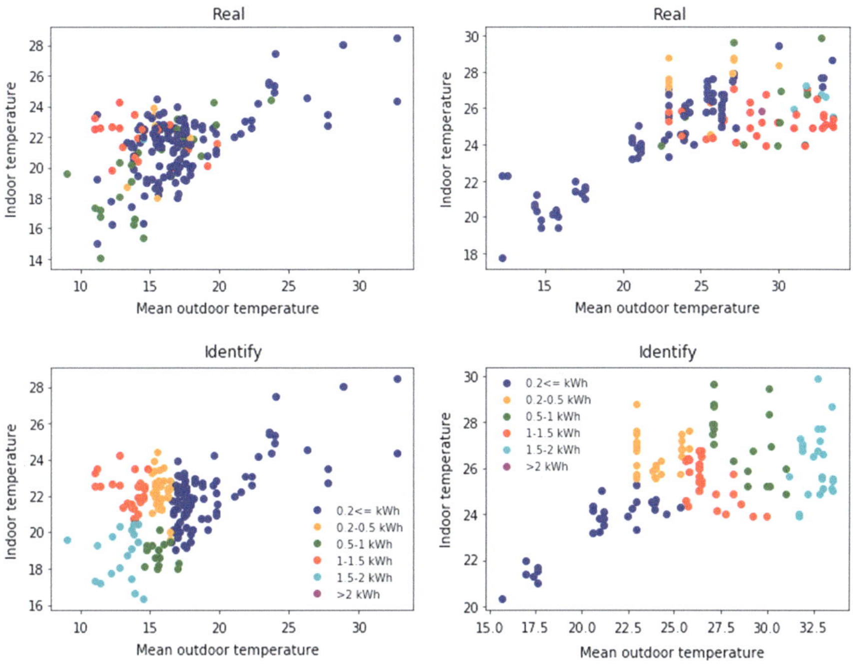

Fig. 4 On the left, heating consumption, and on the right, consumption in cooling mode

In Fig. 3, it can be observed that the centroids appear from lower to higher consumption depending on the situation in which the users are in respect to a state of comfort, and the variation of the mean outside temperature of one day with respect to the interior. For example, if the room is very cold, more consumption is required if the average temperature of the day is lower. Likewise, in the summer, more indoor temperatures require more consumption.

Figure 4 shows the actual situation in an office (top) compared to the situation identified by the methodology (bottom). The figure shows the consumption needed in the office in a day whose average temperature is X and the interior temperature is Y. Therefore, known a weather forecast of the day, or based on the running mean in previous days (Nicol and Humphreys 2002), and in the interior temperature. The methodology can be useful for the forecast of the consumption.

The same methodology can be used to forecast the comfortable status of users in that situation, using the same technique. This analysis explores the pattern of variation of the desired thermal sensation on the ASHRAE scale (ASHRAE 2013), applying the method of direct enquiry. For this, the information about a set of surveys related to an office was processed. The users answered the question: How do you feel just now? The possible answers were: Cold, cool, slightly cool, neutral, slightly warm, warm, hot.

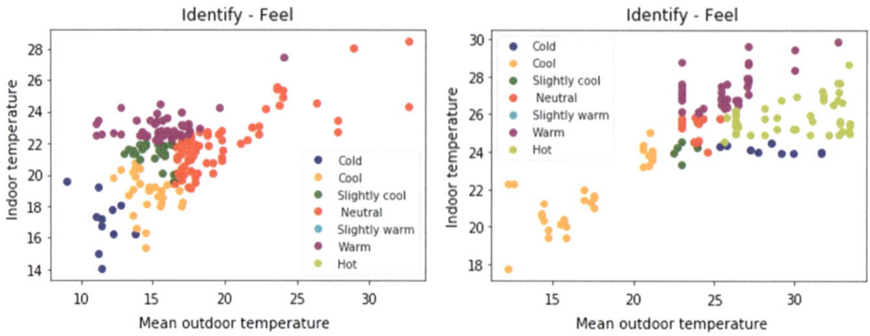

Fig. 5 Example of thermal sensation identification. On the left in heating mode and on the right in cooling mode

Data from the survey were analysed. A set of 7 clusters was defined, each cluster corresponding to one of the 7 responses. Figure 5 shows the identification of the thermal sensation of the consumption situations that can be seen in Fig. 4. From these clusters, decisions could be made in a system based on the needs of consumption and sensation thermal of the users. A set of logical rules associated with energy consumption and user comfort could be defined.

3 Case Study

A study of the application of the technique has been made based on a sample of 1000 surveys, and the physical data of the instant where those surveys were done. 400 surveys were discarded, because they represent situations without the use of HVAC system, since it was wanted to make forecasts with the system in heating or cooling mode.

The figures in the previous section are based on those real data, these data contain situations in a period of one year.

The study was developed into an office in which four users sit (2 women and 2 men). The office is not conditioned by external radiation. The study used half of the information for learning, and the other half to verify the ability to identify the pattern.

4 Results and Discussion

Suppose we developed a system, whose input correspond to the question of comfort indicated, instead of a temperature to select. It could be developed an intelligent system that decides the actions to be taken with low consumption and high comfort.

Table 1 Results of identification applying clustering

Number of error levels	Heating mode			Cooling mode		
	Comfort	Comfort*	Consumption	Comfort	Comfort*	Consumption
0	0.29	0.85	0.46	0.40	0.94	0.51
1	0.34	0.06	0.18	0.20	0.04	0.27
2	0.27	0.07	0.18	0.14	0.01	0.06
3	0.09	0.01	0.10	0.13	0.00	0.11
4	0.01	0.01	0.06	0.13	0.00	0.02
5	0.00	0.00	0.00	0.01	0.01	0.00
6	0.00	0.00	0.01	0.00	0.00	0.02

Due to the problem of thermal comfort and consumption in HVAC system in offices depends on many factors, and considering that the limits of comfort are diffuse. It seems logical to use a model based on clustering, to define the state of the office.

The indoor temperature that is set in a building in the heating or cooling season is key to the energy used in the building. The heat loss of, or gain in, the building depends on the indoor-outdoor temperature difference (Nicol et al. 2012). In our case, the clustering is based on the indoor temperature, and the mean outdoor temperature. The selection of this data is based on the adaptive nature of the comfort of the users, since the external climatology affects the interior comfort.

The experiment was performed using real data from a project was being developed. The results of the application of this methodology can be observed in Table 1. In Table 1 the following data are observed:

Column 1: In this column is shown the number of degrees in the classification in which the analysis or identification of the pattern has failed or has been successful (if the error is 0 levels).

Column 2 and 5: In this column is shown the percentage of error or success (in the case of 0 in column 1). The percentage is shown to compare the number of situations where a value is identified on the scale, whose difference from the actual situation appears in column 1. In this column, the reference is the ASHRAE scale of 7 responses.

Column 3 and 6: In this column is shown the same result than column 2, but in this case, it is assumed no error between slightly cool to slightly warm levels. This is because it is an acceptable in the comfort range, the idea is based on the (Humphreys and Hancock 2007) analysis, they found that people often were where they would like to be on the scale, even if this differed from 'neutral'. So, you cannot assume that people always like to feel neutral.

Column 4 and 7: In this column is shown the percentage of error or success (in the case of 0 in column 1) in the classification of energy consumption.

In Table 1, the error rate is observed, but besides, an analysis of the deviation in kWh between the actual situation and the maximum value corresponding to the consumption range indicated in the classification. In the analysis, the mean deviation is 0.56 kWh in heating mode and 0.51 kWh in cooling mode. In the study, 300 situations were used for learning, and 300 other situations to analyse the results. The 87% in heating mode and in the 88% in cooling mode, the consumption analysed was below the upper limit of the classification range identified.

5 Conclusion

An analysis methodology has been applied with actual survey data made during a year of study. The methodology, applied to a set of diffuse data, to detect patterns of comfort and consumption that are presented in an office. The technique can be useful for the development of a control system based on asking the user about their thermal sensation. Instead of allowing the user select a temperature of the office, this technique can be joined with other techniques in forecasting and in artificial intelligence.

References

ASHRAE (2013) Standard 55-2013: thermal environmental conditions for human occupancy R. and A. conditioning E. American Society of Heating, ed

Bezdek JC (1974) Numerical taxonomy with fuzzy sets. J Math Biol 1:57–71

Feinberg EA, Genethliou D (2006) Load forecasting. Appl Math Restruct Electr Power Syst:269–285

Humphreys MA, Hancock M (2007) Do people like to feel "neutral"?: exploring the variation of the desired thermal sensation on the ASHRAE scale. Energ Build 39(7):867–874. Available at: http://www.sciencedirect.com/science/article/pii/S037877880700059X

Jain AK, Murthy MN, Flynn PJ (1999) Data clustering: a review. ACM Comput Surv 31(3): 264–323

Nicol F, Humphreys MA, Roaf S (2012) Adaptive thermal comfort: principles and practice. Routledge, New York. Available at: http://dx.doi.org/10.4324/9780203123010

Nicol JF, Humphreys MA (2002) Adaptive thermal comfort and sustainable thermal standards for buildings. Energ Build 34(6):563–572. Available at: http://www.sciencedirect.com/science/article/pii/S0378778802000063

Zhao J et al (2014) Occupant behavior and schedule modeling for building energy simulation through office appliance power consumption data mining. Energ Build 82:341–355

Competence Acquisition, Encouragement and Development in Smart Organizations: Some Examples

Lourdes Canós-Darós, Pilar I. Vidal-Carreras, Julio Garcia-Sabater and Julien Maheut

Abstract A good competence and knowledge management is required in smart companies to achieve and maintain competitive advantages. In this context, competence acquisition, encouragement and development are needed to hire and retain talented employees in these companies. In this paper, we describe briefly some examples according to these three dimensions.

Keywords Competence management · Knowledge management · Acquisition Encouragement · Development

1 Introduction

Competence assessment has evolved over time, together with knowledge management. In traditional human resources management, economic and social aspects of the company are incompatible and employees are considered a cost rather than an

PIME A23 "Análisis de metodologías y su uso en la evaluación de las competencias transversales", Universitat Politècnica de València (Spain).

L. Canós-Darós (✉)
ROGLE. Dpto. de Organización de Empresas, Escuela Politécnica Superior de Gandia, Universitat Politècnica de València, C/Paranimf, 1, 46730, Grau de Gandia, Valencia, Spain
e-mail: loucada@omp.upv.es

P. I. Vidal-Carreras
ROGLE. Dpto. de Organización de Empresas, Escuela Técnica Superior de Ingeniería del Diseño, Universitat Politècnica de València, Camino de Vera S/N, 46022 Valencia, Spain
e-mail: pivicar@omp.upv.es

J. Garcia-Sabater · J. Maheut
ROGLE. Dpto. de Organización de Empresas, Escuela Técnica Superior de Ingeniería Industrial, Universitat Politècnica de València, Camino de Vera S/N, 46022 Valencia, Spain
e-mail: jugarsa@omp.upv.es

J. Maheut
e-mail: juma2@upv.es

Á. Ortiz et al. (eds.), *Engineering Digital Transformation*, Lecture Notes in Management and Industrial Engineering,
https://doi.org/10.1007/978-3-319-96005-0_25

investment that must be optimized. It is an error to assess functions, responsibilities and tasks that a worker must perform without taking into account the potential that can be developed. Today, competences and knowledge are valued in order to make good decisions.

In any industry, personnel selection, motivation and team work have been usually considered as a determining factor of performance, productivity and quality of work. Moreover, in smart companies, human resources practices are a key point to achieve and maintain competitive advantages. Notice that highly skilled employees are responsible for performing complex, multidimensional, and interdependent tasks. In concrete, knowledge workers commonly follow good practices as to share their knowledge with teammates or to be proactive, flexible and adaptable (Badoo et al. 2006).

In this paper, we describe some key concepts related with human resources management and show some examples according to acquisition, encouragement and development of competences in smart companies.

2 Competence Management and Knowledge Management

Competences are characterized by the individuality of each person, who has his or her own competences. These competences cannot be copied or imitated, constituting a valuable resource for the company. In addition, the correct development of the appropriate competences for a job allows the employee to achieve success in the performed work. Consequently, we can affirm, first, that not all the people will be able to succeed in performing their work, and second, that the competences required by the company will allow to distinguish between workers with excellent performance from others with normal performance.

When developing a list of competences, the company must take into account a precise and specific definition, and if they are useful to describe the job requirements. Therefore, although it is possible to establish a list of commonly accepted generic competences, it is increasingly important to develop individual lists corresponding to the activities, strategies, structure and culture of the company, because all of them contribute to value creation. We can find different classifications for the competences, which usually take as a classification criterion a cognitive approach (intellectual abilities, verbal information, cognitive strategies, motor skills, and attitudes), or job requirements (cognitive, psychomotor, physical and perceptual requirements).

One of the fundamental characteristics of competence management is that it allows the integration of different human resources management processes, taking as a unit the concept of competence. Competence management set up on three basic dimensions (Ordóñez 1995) that are summarized in Table 1.

This is a not unique belonging relationship. Then, there are no pure systems, policies or procedures within the dimensions of competence management, and each of them is able to gather typical aspects of the three dimensions.

Table 1 Basics dimensions in competence management and associated human resources systems

Acquisition	Encouragement	Development
Recruitment, personnel selection and hiring	Reward and incentives	Training
Short, medium and long-term forecasting management	Job valuation	Communication
Career plan	Performance assessment	Promotion
Succession plan	Motivation	Potential assessment

A few decades ago, to know the value of a company, it was enough to look at its balance sheet and analyze some ratios. The work was simple, mainly because the company presented itself as a set of tangible assets that, adequately combined, allowed achieving the proposed goals. Now the process is more complex and the value of the companies is determined not only by their material assets but also by the set of its intangible assets. They are responsible for the creation and maintainance of competitive advantages. Within this set, we make special mention of a fundamental strategic asset that allows companies not only to survive, but also to overcome competitors: knowledge (Canós-Darós et al. 2002).

We distinguish between tacit and explicit knowledge. Explicit knowledge can be expressed in words, in numbers, in symbols. It is easily transferable, so it can be shared mainly in the form of databases and universal principles, taking the consideration of public good. It is composed of technical know-how, some skills or abilities, and few activities. Tacit knowledge is acquired and retained by individuals, being difficult to articulate or codify, and consequently, to share. It is closer to the talent, and it consists basically of attitudes, capacities and certain abstract and complex knowledge. These characteristics make tacit knowledge a strategic asset, a source of sustainable competitive advantage (Nonaka and Takeuchi 1995; Osterloh and Frey 2000; SubbaNarasimha 2001). Consequently, managing the knowledge in the company is an important task, in order to take advantage of the workers' intellect. Following Ulrich (1998), we define intellectual capital simply as "capacity × commitment".

The intellectual capital component we are interested in is the one related to the individual-centered assets defined as collective expertise, creative ability, problem solving, leadership, and entrepreneurial and managerial skills. These elements do not belong to the company but to the individuals, so it is necessary to carry out formulas to retain them if they are considered as fundamental for success. It is required that the individual had a certain level of knowledge and skills to perform the job corresponding activities, but what really allows to generate knowledge is his or her entrepreneurial attitude, that is, eagerness to set goals and do anything to achieve them, a process that generates know-how. Commitment is defined as the identification of an individual with the company, and requires a strong desire to remain in it, as well as the acceptance of its goals, values and willingness to strive for a collective benefit. Therefore, to reach a balance between commitment and competence is necessary, because employees with a high level of competence but low commitment will not be willing to contribute the most in their task performance, while those with a low level of competences and high commitment will be unable to achieve the desired result. Employees will be committed if they have the resources to achieve the optimal results (Ulrich 1998). If resources and requirements are not in harmony, it is necessary to foster commitment, either through a reduction of demands (eliminating less relevant ones and automating or simplifying tasks), or through an increase in resources (greater control, collaboration and teamwork, communication and training). Once a talented professional is captured, the company must maintain and reinforce his or her commitment through continuous development, maintaining his or her competitiveness all the time by giving an appropriate value. A study of talent management practices and their reflection in Spanish medium-sized companies can be seen in Valverde et al. (2013).

Differentiation via innovation has become a key factor to compete and survive in a turbulent environment. Differentiation is directly related with the talent of professionals, so it is imperative to detect those whose knowledge, together with certain emotional competences such as communication, leadership, teamwork, motivation and commitment, can become fundamental to the success of the company. This individual talent can be strengthened through teamwork, that is, interaction with other professionals to participate in and share their knowledge, competences, skills, and join efforts to achieve higher goals by building capacities.

Organizations that follow these patterns of behavior are called smart organizations. We can define them as those that know how to learn and that, thanks to their organizational learning processes, surpassing an individual character, are capable of permanently transform themselves, giving an answer to the present and potential threats, challenges and opportunities. Senge (1992) defines them as those where people continually expand their aptitude to create the results they want, where new and expansive thought patterns are cultivated, where collective aspiration is released, and where people continually learn to learn together. Smart organizations are interested in professionals who, in addition to their knowledge, contribute with their attitude to leadership and to create talent for the group, taking into account the distribution of the responsibilities of human resources management between internal and external agents to the organization (Valverde et al. 2006).

3 Competence Acquisition, Encouragement and Development in Smart Organizations

3.1 Competence Acquisition in Smart Organizations

Traditionally, the personnel selection process has been based on threshold or basic skills. Today, not only the IQ and the skills required for a job are evaluated, but also emotional competencies. If certain emotional factors are not considered, the generation of organizational talent would be jeopardized and, therefore, innovation. The challenge is to attract professionals with concerns, with innovative projects, professionals who question the established way of work and reflect about procedures. Then, the personnel selection process has to identify the competences of the candidates and also their willingness to share knowledge. In this context, a transformation in traditional techniques has happened in order to provide new formulas for reflection.

Personnel selection is the process in which one person or more are chosen to develop the tasks and responsibilities of a vacancy (Canós-Darós et al. 2014). A good management of the selection policy that takes into account the circumstances of the company allows it to optimize costs and achieve corporate goals. It is usual that in most management problems this process is complicated and implies focusing on concepts such as validation, trust and setting criteria (Canós-Darós and Liern 2008).

In the personnel selection an inflexible treatment of candidates' evaluations can obstruct and fail the ordering process if not all the requirements are considered (Casasús et al. 2012). In addition, the global assessment can neutralize the positive assessment of the competences with the negative, and this is unfair. It is interesting to present different complementary and flexible personnel selection methods to order candidates to a job according to their fitting. Among them, it is worth mentioning the use of intervals that allow more flexibility and better reflect the usual ways of assessment in companies and the incorporation of useful efficiency to find optimal solutions (Canós-Darós et al. 2011).

In order to carry out a personnel selection process, we can use some techniques based on a comparison with an ideal candidate, in the assessment of a group of experts, in a pairwise order, or in efficient parametric techniques. These techniques have been applied in a multinational company in a global industry. The responsible for this process was the Human Resources Manager. In this case, the vacancy required management tasks. Five competences were evaluated in five candidates by four experts (the Human Resources Manager and other managers). Different results were provided according to different levels of exigency. The result was satisfactory and useful for making the final decision (Canós-Darós and Liern 2010).

3.2 Competence Encouragement in Smart Organizations

Achievement needs, defined as the desire to excel and achieve success in a competitive environment, consist of taking responsibility for problem solving in order to increase employee satisfaction. To put individual skills into practice it is necessary to leave them free to act by avoiding the pressures of higher levels and being free to create ideas as well as practice creativity. These factors are essential for generating intangible assets such as continuous learning (the pillar of a smart organization) among the members of the organization by working together. Employees need jobs that serve as a challenge, they need to feel that they are part of the company and a manager can meet these needs through empowerment, helping people in developing their competences by giving them more responsibilities and opportunities for new learning. The same happens if we are talking about the performance of a team. Summarizing: people need to be motivated.

We define motivation as the will to make an effort with a given goal, through the behavior (Dieleman et al. 2006; Tabassi and Bakar 2009). This definition seems simple, but if we analyze the motivation in depth we discover that it is a complex and multidimensional phenomenon, because there are many factors motivating and not all of them have the same influence to all the people or in all the geographical areas. An exhaustive list of motivation factors can be seen in Canós-Darós (2013). Because of the subjectivity underlying motivation processes, managers and employees may have different beliefs, expectations, and viewpoints about the same reality. So, managers face the challenge of measuring the influence of different motivating factors on their employees. On the one hand, this can be a problem because measures can be deficient, including personal opinions, deliberate distortions and other errors. On the other hand, the measures are useful for decision-making in the field of human resources. Whether it is intrinsic or extrinsic motivation, different factors can be measured (Canós-Darós 2012).

Thus, we can use motivation factors as the raw material of an algorithm to identify the most motivated employees in an organization, that is, the employees more identified with the policies and practices of motivation designed by the managers (Canós-Darós 2013). We have applied this algorithm in the private and public sector in: a Spanish service company, a sports association and a department of a town hall (Bautista Banacloy et al. 2016; Canós-Darós et al. 2015). In all the cases the motivation factors chosen by managers were similar to that chosen by employees, although the differences were significant because they confirm that both groups can have different points of view about the same fact (in this case, motivation policy). These differences were the basis to improve motivation policies through an improvement of communication policies.

3.3 Competence Development in Smart Organizations

As important as the assessment of existing competences for the performance of a job, it's the capacity to develop other competences based on experience, training and interaction within a work team. The proper assignment of employees to the corresponding jobs according to their competences facilitates their development to carry out the work in a more satisfactory way and to share knowledge. Although some times the low performance of a worker leads him or her to be fired, he or she can be valuable to the company. In these cases, a rellocation is the appropriate procedure to obtain the maximum contribution of staff knowledge. This reassignment can be done in many different ways, for example, by using the Hungarian algorithm as it is proposed in Fúller et al. (2012).

On another hand, there is no consensus on the high performance practices of human resources in companies (Marin-Garcia 2013); we consider as example training or reward policies. A general methodology used to choose the set of human resources practices in the area of operations that lead to the combination of desired operational performance according to the previous state of the organization is presented in Marin-Garcia et al. (2013). Previously, 27 practices related to participation, training, remuneration of workers and knowledge management were presented to analyze the practices that must be carried out to introduce production systems. Because a unified questionnaire about these practices did not exist, one is proposed and applied in 98 companies.

References

Badoo N, Hall T, Jagielska D (2006) Software developer motivation in a high maturity company: a case study. Softw Process Improv Pract 11:219–228

Bautista Banacloy A, Canós-Darós L, Guijarro Tarradellas E, Babiloni Griñón E (2016) Análisis de factores motivadores en una organización pública de ámbito local. XXX AEDEM Annual Meeting, Las Palmas de Gran Canaria

Canós-Darós L (2012) An algorithm to identify the most motivated employees in knowledge-based companies. Global Innovation and Knowledge Academy (GIKA 2012), Valencia

Canós-Darós L (2013) An algorithm to identify the most motivated employees. Manag Decis 51(4):813–823

Canós-Darós L, Casasús Estellés T, Liern Carrión V, Pérez Cantó JC (2014) Soft computing methods for personnel selection based on the valuation of competences. Int J Intell Syst 29:1079–1099

Canós-Darós L, Santandreu Mascarell C, Guijarro E, Babiloni E (2015) La medida de la motivación: Aplicación en una organización deportiva. XXIX Annual Conference of the Academy of Management and Business Economics, San Sebastian

Canós-Darós L, Liern V (2008) Soft computing-based aggregation methods for human resource management. Eur J Oper Res 189(3):669–681

Canós-Darós L, Liern V (2010) An application of goal programming in personnel selection decisions. In: 24th European conference on operational research—EURO XXIV, Lisboa (Portugal)

Canós-Darós L, Casasús T, Crespo E, Lara T, Pérez JC (2011) Personnel selection based on fuzzy methods. Revista de Matemática. Teoría y Aplicaciones 18(1):177–192

Canós-Darós L, Valdés J, Zaragoza PC (2002) Competency management as a fundamental part of knowledge management. IX Congreso Internacional de la Asociación Europea de Dirección y Economía de la Empresa, París (Francia)

Casasús Estellés T, Canós-Darós L, Pla-Ferrando ML (2012) Application of fuzzy techniques to Human Resources Management. Lambert Academic Publisher

Dieleman M, Toonen H, Touré H, Martineau T (2006) The match between motivation and performance management of health sector workers in Mali. Human Resour Health 4:2–9

Fúller R, Canós-Darós L, Canós-Darós Darós, MJ (2012) Transparent fuzzy logic based methods for some human resources problems. Rect@ 13(1):27–41

Marin-Garcia JA (2013) What do we know about the relationship between High Involvement Work Practices and performance? Working Papers on Operations Management 4(2):1–15

Marin-Garcia JA, Perelló Marín MR, Canós-Darós L, Valero Herrero M (2013) How to choose the best order when implanting HIWP in Operations Area. XVII Congreso de Ingeniería de Organización. In: 7th international conference on industrial engineering and industrial management (CIO 2013). XIX international conference on industrial engineering and operations management (ICIEOM 2013), Valladolid

Nonaka I, Takeuchi H (1995) The knowledge-creating company. How japanese companies create the dynamics of innovation. Oxford University Press, New York

Ordóñez M (1995) La nueva gestión de los recursos humanos. Ediciones Gestión 2000
Osterloh M, Frey BS (2000) Motivation, knowledge transfer and organizational forms. Organ Sci 11(5):538–550
Senge P (1992) La quinta disciplina. Cómo impulsar el aprendizaje en la organización inteligente, Granica, Barcelona
SubbaNarasimha PN (2001) Salience of knowledge in a strategic theory of the firm. J Intellect Capital 2(3):215–224
Tabassi AA, Bakar AH (2009) Training, motivation, and performance: the case of human resource management in construction projects in Mashhad, Iran. Int J Project Manage 27:471–480
Ulrich D (1998) Intellectual capital = competence x commitment. Sloan Manag Rev 39(2):15–26
Valverde M, Ryan G, Soler C (2006) Distributing HRM responsibilities: a classification of organisations. Personnel Review 35(6):618–636
Valverde M, Scullion H, Ryan G (2013) Talent management in Spanish médium-sized organisations. The Int J Human Resour Manag 24(9):1832–1852

When to Stop? A New Stop Criterion for Combinatorial Optimization Search Techniques

Yuval Cohen, Roy Gelbard and Marlene Amorim

Abstract This research draws on theories in discrete event simulation, order statistics and record-breaking statistics to develop a methodology for deciding when to stop a combinatorial optimization search. During the first optimization period, the objective function improves rapidly with the iterations, and the improvement slows gradually until it almost stalls. We adopt a popular method to detect the period of rapid improvements of the "warm-up period", and then we propose a special control chart technique to identify with a given certainty reaching a steady state. Then, we suggest using the theory of record-breaking to decide on a stopping criterion. In addition, the paper develops estimates for the optimum bounds and estimates for value and timing of the next expected improvement. The advantages of this approach are discussed.

Keywords Stop criterion · Convergence · Evolutionary · Swarm · Solution's quality · Bound

1 Introduction

The convergence towards the optimum of genetic algorithms and other related search techniques is well established (Holland 1992; Pandey et al. 2014). The objective value

Y. Cohen (✉)
Department of Industrial Engineering, Tel-Aviv Afeka College of Engineering, 38 Mivtsa Kadesh, 69988 Tel Aviv, Israel
e-mail: yuvalc@afeka.ac.il

R. Gelbard
Department of Business Administration, Bar-Ilan University, 5290002 Ramat-Gan, Israel
e-mail: r_gelbard@yahoo.com

M. Amorim
Department of Economics, Management and Industrial Engineering, University of Aveiro, 3810-193 Aveiro, Portugal
e-mail: mamorim@ua.pt

Á. Ortiz et al. (eds.), *Engineering Digital Transformation*, Lecture Notes in Management and Industrial Engineering,
https://doi.org/10.1007/978-3-319-96005-0_26

of all search techniques shows a typical initial rapid improvement period followed by continual decrease in improvement rate, as the search goes-on (Rawlins and Sushil 2014).

There are several popular stopping criteria such as: (1) number of generations (Liu et al. 2014), (2) number of generations without minimal improvement (Ha et al. 2014), (3) maximum optimization time (Glorieux 2015, p. 9), (4) optimization threshold value (Glorieux 2015, p. 9), (5) a threshold value for the difference between the objective best value and average value, over the generations (Toledo et al. 2014). However, setting the stop criterion in each case is more an art than a science. Martí et al. (2016) tried to integrate science into this decision, but ended up with complex models based on progress indicators with a Kalman filter which is used for data gathering. This paper suggests a more structured and reason-based strategy. It first identifies an initial "warmup period" (Rossetti et al. 2005), of significant improvement, then it tests when the objective function reaches a plateau (Robinson 2007), finally, it adopts the statistical inference of a record breaking process from a stationary random process (Glick 1978; Krug and Jain 2005).

2 Objectives

The objective of the research is to propose an effective and efficient stopping criterion (Hermadi et al. 2014; Kim 2013) that could be applicable to most search techniques (for example, Genetic Algorithms (GA), Simulated Annealing (SA), Ant Colony (AC), Particle Swarm Optimization (PSO), etc.). The suggested stopping technique should also be based on current scientific tools and on reason. Additionally, the proposed technique should be easy to apply and easily understood. Most importantly, the technique should offer ways to control its reliability.

To fulfil the objectives we set three different goals to be achieved:

1. The identification of the initial period of rapid improvement in the objective function.
2. The identification of the second period of the objective function convergence to a plateau.
3. The identification of a sufficiently good solution by assuming a stationary homoscedastic process.

3 Identifying the End of the Intensive Warm-up Period

The first step is to identify the end of the initial steep descent or "warmup period" (Rossetti et al. 2005). For that purpose, we adopt the following Definitions:

F_{max} Largest value of the fitness function
F_{min} Smallest value of the fitness function

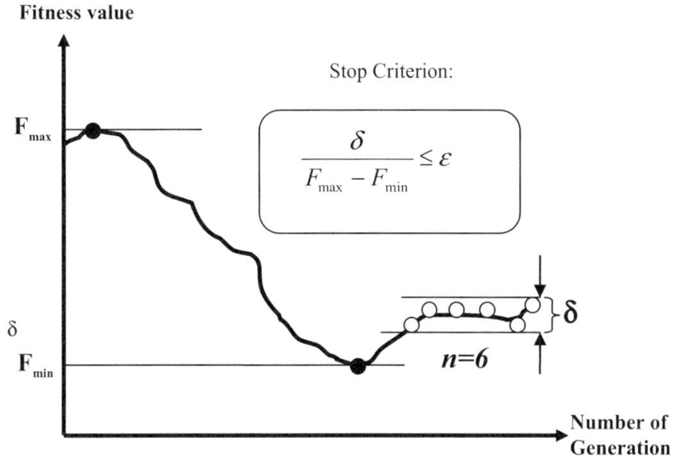

Fig. 1 Example of identifying the warm-up period in evolutionary optimization

Total improvement $= F_{max}$-F_{min}

n The number of last observations of the fitness function for consideration.
δ The range of fitness function value of the last n observations.
e Threshold value for the ratio between δ and the total improvement

First we need to decide on the value of n (the number of last observations we consider). Once n is know (say $n = 6$), δ could be computed continuously as follows:

$$\delta = \mathrm{Max}(X_t, X_{t-1}, \ldots X_{t-n}) - \mathrm{Min}(X_t, X_{t-1}, \ldots X_{t-n}) \tag{1}$$

Note that δ is analogous to the total improvement, but is only related to the last n observations. Thus, the ratio between δ and the total-improvement is the ratio between the local improvement (of last n points) and the global improvement. This ratio is expected to decline through time and using an epsilon as a threshold is utilized as a stopping rule as described in Fig. 1.

Reaching the threshold described in Fig. 1, is telling us that the process finished its steep progress towards the optimum. However, it does not mean that the process reached a stationary condition or hit a plateau. It still remains up to a second procedure to ascertain the steady state of the process. This step is described in Sect. 4.

4 Identifying the Steady State

Once the initial period of intensive improvement is curtailed, a quest for steadiness begins. A second procedure is applied to make sure that the process is stationary with unchanging variance. A popular quality control based technique was adopted for this

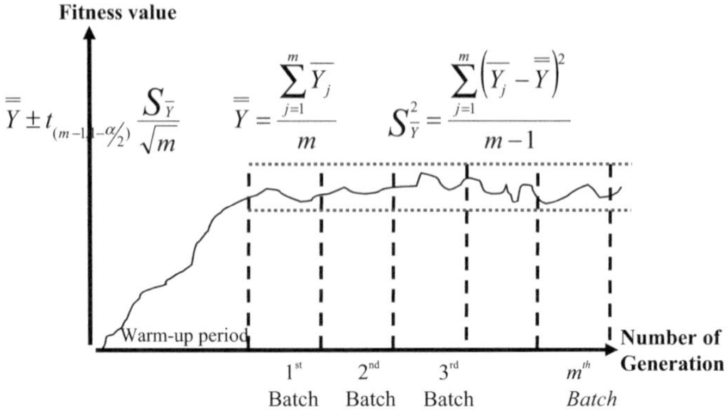

Fig. 2 An example for the method of batch means

purpose. The technique is widely used by the discrete event simulation community. The method is named the method of "Batch Means" (Robinson 2007). This method is sometimes used outside the world of discrete event simulations: for example, Rossetti et al. (2005).

In this scheme we define:

n the number of observations in each batch.
m the number of required batches.
$\bar{Y}_j$ the mean of batch j.
$\bar{\bar{Y}}$ the total mean of all observations.
$S^2_{\bar{Y}}$ the variance of the batch means.

Setting values to n, and m (typically greater than 6) generates m groups of n observations each. The central limit theorem (CLT) ensures that the batch means are distributed approximately as Normal IIDs, and thus by the procedure tests for outliers from the following range:

$$\bar{\bar{Y}} \pm t_{(m-1,1-\alpha/2)} \frac{S_{\bar{Y}}}{\sqrt{m}} \tag{2}$$

If there are no outliers, we adopt the end-point of last batch (*mth* batch) to the beginning of the third step of the overall procedure. Otherwise, a repeated procedure of m-means will be applied where the first batch will start right after the last outlier (Fig. 2).

Table 1 Record number *r* and its simulated statistical parameters (Glick 1978)

r	2	3	4	5	6	7	8
$E(n)$	4	14	40	109	291	781	2097
$Median = M_r$	4	10	26	69	183	490	1316
M_r/M_{r-1}		2.50	2.60	2.65	2.65	2.68	2.69

5 Identifying Sufficiently Satisfying Record

Once the second step of batch means is done, a steady state process could be assumed, for the observations that follow the last batch. In that region, a third step is taken using record-breaking theory for deciding when to stop the search. For presenting the record-breaking theory and our proposed stopping rule, the following definition are given.

Definitions

r	Record number (r = 2: second record, r = 3: third record, etc.)
n	Number of generations/observations
$E(n)$	mean number of trials/observations until a certain record number.
R_n	number of records in n generations/observations.
M_r	Median number of trials/observations related to the *rth* record
M_r/M_{r-1}	The ratio of current median to median of last record

Table 1 shows results of an extensive simulation trial for a process of drawing n iid variables, until the 8th record is achieved. Table 1, shows for each *rth* record: the mean average observations ($E(n)$). However $E(n)$ approaches infinity as n grows, while median is stable. Therefore, we prefer dealing with medians. Median is denoted (M_r), and the ratio of current median to last one (M_r/M_{r-1}) is based on (Glick 1978). This ratio approaches $e = 2.718$ as r grows.

Thus:

$$\frac{Median\{n_{r+1}\}}{Median\{n_r\}} = e = 2.718 \qquad (3)$$

This means that the median number of observations between the current record to next record is 2.718 times the total number of observations so far (until current record). Setting a threshold for number of generations without improvement is easy. For example, limiting the search for achieving next record by maximum of additional million generations has a median of last record: at, or after, a total of: 1,000,000/2.718 = 367,918 generations.

Additional way to set a threshold is to use the number of records directly. In that case, the threshold could be determined using the Chebyshev's bound for ensuring that the solution is in a certain percentile of the total population of solutions. Table 2

Table 2 Expectation, variance, and standard deviation for n i.i.d. observations (Glick 1978)

Number observations n	$E(R_n)$	$V(R_n)$	$\sigma(R_n)$	$E(R_n)+3\sigma(R_n)$
100	5.19	3.55	1.88	10.84
200	5.88	4.24	2.06	12.06
300	6.28	4.64	2.15	12.74
400	6.57	4.93	2.22	13.23
500	6.79	5.15	2.27	13.60
600	6.97	5.33	2.31	13.90
700	7.13	5.49	2.34	14.16
800	7.26	5.62	2.37	14.37
900	7.38	5.74	2.40	14.57
1000	7.49	5.8564	2.42	14.75
1,000,000	14.39	12.75	3.57	25.10

shows results for several thousand repetitions of generating n iid observations and keeping R_n at each repetition.

For each n, the mean $E(R_n)$ and variance $V(R_n)$ number of records was computed. This enables to find a number of records that ensures with high probability that the best of n observations was found. The rightmost column of Table 2, uses 3 standard deviations above the mean. Using Chebyshev's one-sided bound ensures that the best solution among n is found with a probability of 94.4%:

$$\Pr\{\text{no better records exist for } n|k = 3\} = 1 - (1/(2(\mathrm{k}^2)))$$
$$= 1 - ((1/2)(1/3^2)) = 17/18 = 0.944 \quad (4)$$

Thus, a process of waiting for the next 15 records is equivalent to waiting at the minimum 1200 generations, and waiting for 25 records is equivalent to observing 1,000,000 generations.

6 Results

We compared the results of the proposed GA stopping technique with optimal values of the benchmark RCPSP solutions on PSPLIB (Kolisch and Sprecher 1997; Kolisch et al. 1999). We used 30 solutions for 4 different network sizes: 120, 90, 60 and 30 activities. The stopping parameters we used were: (1) For the intensive improvement stage: $n=8$, $\varepsilon=0.005$; (2) for the batch means $n=10$, $m=5$ and $\alpha=0.05$; (3) setting the start point at the end of step-2 and waiting for 15 more records. The results are summarized in Table 3.

Thus, preliminary results (summarized in Table 3) show that the suggested scheme ensures high quality solutions, albeit in some cases adding some computational time.

Table 3 Results of the GA stopping rules vs. benchmark optimum (Kolisch et al. 1999)

No. of activities	Benchmark cases	# reach optimum	% reach optimum (%)	Within 1% of the optimal value (%)
30	30	28	93	100
60	30	24	80	100
90	30	18	60	98
120	30	14	47	93

7 Conclusion

The paper suggests a new approach for stopping criterion for search techniques. First, the length of the warmup period with intensive improvement is identified. Then, batch means method is applied for reaching a flat part of the fitness function—to be approximated as a static distribution process. This approximation enables using record breaking statistics for stopping inference rules. Preliminary results show high quality solutions could be obtained using this scheme. It is left for future research to examine the impact of most model parameters, and the trade-off between solutions quality and its time-performance. For that purpose far more experimentation is needed, as well as sensitivity analysis for each parameter and their combinations.

References

Glick N (1978) Breaking records and breaking boards. Am Math Monthly 2–26

Glorieux E (2015) Constructive cooperative coevolution for optimising interacting production stations. Doctoral dissertation, University West, Trollhättan: Sweden

Ha MP, Kumar L, Ananthapadmanabha T (2014) A novel approach for optimal allocation of a distributed generator in a radial distribution feeder for loss minimization and tail end node voltage improvement during peak load. Int Trans Electr Comput Eng Syst 2(2):67–72

Hermadi I, Lokan C, Sarker R (2014) Dynamic stopping criteria for search-based test data generation for path testing. Inf Softw Technol 56(4):395–407

Holland J (1992) Adaptation in natural and artificial systems. The MIT Press; Reprint edition (originally published in 1975)

Kim JL (2013) Genetic algorithm stopping criteria for optimization of construction resource scheduling problems. Constr Manag Econ 31(1):3–19

Kolisch R, Sprecher A (1997) PSPLIB-a project scheduling problem library: OR software-ORSEP operations research software exchange program. Eur J Oper Res 96(1):205–216

Kolisch R, Schwindt C, Sprecher A (1999) Benchmark instances for project scheduling problems. Int Ser Oper Res Manag Sci 197–212

Krug J, Jain K (2005) Breaking records in the evolutionary race. Physica A 358(1):1–9

Liu HL, Gu F, Zhang Q (2014) Decomposition of a multi-objective optimization problem into a number of simple multi-objective sub-problems. IEEE Trans Evol Comput 18(3):450–455

Martí L, García J, Berlanga A, Molina JM (2016) A stopping criterion for multi-objective optimization evolutionary algorithms. Inf Sci 367:700–718

Pandey HM, Chaudhary A, Mehrotra D (2014) A comparative review of approaches to prevent premature convergence in GA. Appl Soft Comput 24:1047–1077
Robinson S (2007) A statistical process control approach to selecting a warm-up period for a discrete-event simulation. Eur J Oper Res 176(1):332–346
Rossetti MD, Li Z, Qu P (2005) Exploring exponentially weighted moving average control charts to determine the warmup period, In: Proceedings of the 2005 winter simulation conference, 771–780
Rawlins GJE, Sushil JL (2014) Syntactic analysis of convergence in genetic algorithms. Found Genet Algorithms (FOGA 2), 2:141
Toledo CFM, Oliveira L, França PM (2014) Global optimization using a genetic algorithm with hierarchically structured population. J Comput Appl Math 261:341–351

Another Way to Teach Operation Management

Sofía Estelles-Miguel, José Fernando Garrigos-Simon,
Marta Elena Palmer Gato and José Miguel Albarracin Guillem

Abstract In the context of the Business Management degrees Operation Management subject provides a wide variety of tools related with skill like decision making, solving problems and information management. Usually this subject is taught by master classes for the theoretical part of the subject and problem solving for the practical one. The present paper presents a new methodology used for teaching theoretical part in this subject. For the authors the students should learn Operation Management doing Operation Management. In this way the teachers have develop a Project-Based Learning for this subject. This paper present also the motivation for this teaching innovation, the problems encountered in its implementation and the results obtained.

Keywords Project-Based learning · Competences · Skills · Teaching-learning process

1 Introduction

The nowadays educational environment requires the application and transformation of new teaching and learning methods and tools (Garrigos-Simon et al. 2015). This

S. Estelles-Miguel (✉) · M. E. Palmer Gato · J. M. Albarracin Guillem
Dpto. de Organización de Empresas. Facultad de Administración y Dirección de Empresas,
Universitat Politècnica de València, Camino de Vera S/N, 46022 Valencia, Spain
e-mail: soesmi@omp.upv.es

M. E. Palmer Gato
e-mail: marpalga@doe.upv.es

J. M. Albarracin Guillem
e-mail: jmalbarr@omp.upv.es

J. F. Garrigos-Simon
Dpto. de Organización de Empresas. Escuela Técnica Superior de Informática,
Universitat Politècnica de València, Camino de Vera S/N, 46022 Valencia, Spain
e-mail: fgarrigos@doe.upv.es

Á. Ortiz et al. (eds.), *Engineering Digital Transformation*, Lecture Notes
in Management and Industrial Engineering,
https://doi.org/10.1007/978-3-319-96005-0_27

becomes even more evident when we visit those companies that subsequently hire our students. The biggest complaint is that students lack skills such as teamwork, leadership, social skill, dealing with staff, negotiation, corporate culture, etc. Perhaps we are not focusing the issue properly. The question should be "What do the staffing companies that they hire really need?"

Possibly we are not acting in the right way when this is happening. This is why we ask ourselves what to change and how to change it. Although the classic form has been the keynote presentations due to overcrowded classrooms (Barbera Ribera et al. 2015). Thinking about it, we detected that the theoretical part that we teach in master classes, later the students memorized it, they "vomited it" in the exams and they forgot it quickly. Then, memorizing the theoretical part does not bring anything meaningful. The question became: how to make them learn and fix these concepts? Is it possible to develop those skills that they will need in their working life at the same time?

With these questions in mind, a project-based learning approach was proposed in the Operation Management subject. In this paper we summarize the author's experience during the course 2015/2016 in this subject at the Faculty of Business Administration in the Universitat Politècnica de València (UPV).

2 Project Based Learning

Adaptation to the European Higher Education Area (EHEA) implies that the subjects are designed in competences and in learning objectives, this affects not only the teaching-learning methodology but also the evaluation process (Estelles-Miguel et al. 2016). EHEA also introduces not just focus on learn the concepts of the subject, but also in develop skills and competences to prepare students for future integration into the working world. But it is not easy the management of skills and generates numerous difficulties (Levy-Leboyer 2003; Estelles-Miguel et al. 2013).

The teachers need to change the classes based on rote learning of knowledge to another where students are an active part becoming responsible and protagonists for their own learning rather than passive subjects (Zabala Videla and Arnau Belmonte 2007).

Project Based Learning (PBL) is a teaching method in which students gain skills and knowledge by working for a period to investigate and respond to and autentic, engaging and complex question, problem or challenge.

Projects are complex tasks, based on challenging questions or problems, that involve students in design, problem-solving, decision making, or research activities; giving students the opportunity to work relatively autonomously over extended periods of time; and culminate in realistic products or present actions (Jones et al. 1997; Thomas et al. 1999). There are a great diversity of PBL and the idea of assigning projects to students is not a new one. But in PBL the project is the central teaching strategy, students learn the central concepts of the subject via the project. The project help the students to understand the concepts and principles of the subject. The central activities of the project must involve the construction of knowledge (Bereiter and Scardamalia 1999).

Projects must be realistic in their characteristics and give a feeling of authenticity to students.

This work try to achive the following goals:

1. To get that students learn by doing and catch up on the subject.
2. To promote responsibility, teamwork and autonomy in the students.
3. To improve the students qualifications.

2.1 Methodology

Exist ideas for enhancing students' ability to benefit from PBL, primarily through the introduction of so-called "Scaffolding" (learning aids, models, training strategies) intended to help students become proficient at conducting inquiry activities. Learning is maximized if the context for learning resembles the real-life context in which the to-be-learned material will be used. While learning is minimized if the context in which learning occurs is dissimilar to the context in which the learning will be used. And it is important for the students to be able to apply what they learn to solve problems and make decisions, instructions be carried out in a problem-solving context. Learning that occurs in the context of problem solving is more likely to be retained and applied. Such learning is also seen as being more flexible than the inert knowledge that is acquired as a result of more traditional didactic teaching methods.

According to Aranda (2009), this type of learning consists of proposing a real problem to students team, for which solution they will have to work collaboratively in a project that they will have to design following initial guidelines marked by the teachers, and where each student has an individualized role with some goals to achieve. Taking advantage of this the authors of the present paper developed a PBL that use all these proposals.

2.1.1 Innovation Development

To carry out the project satisfactorily it will be necessary to establish certain elements:

- Objectives of the project:
  - The overall objective.
  - And each participant objectives.
- The initial instructions of each participant.
- The final evaluation of the project.
  The students have to create a company that makes pasties (PBL).
- First teachers deliver the same problem for all groups. This problem consists of a fictitious company created by students who make pasties.

- They are indicated the ingredients and quantities of them, also the tools, machines, restrictions and processing times.
- They are asked to calculate some data related to this productive process.
- If calculations are performed in the right way all groups should get the same result in this first step.

- In the next session the team must decide what will be the role of each member of the group. Each group will have 5 components. The roles are:
  - Director, Product Manager, Process Manager, Layout Manager and Location Manager.
- In the following sessions they must make decisions about their company:
  - Each company can make a different product (all are pasties, but the flavors, masses, shapes and others have to decide each group).
  - Each company will have a different manufacturing process depending on the type of product that they have decided to manufacture.
  - Each company will have a different layout depending on the manufacturing process they have decided.
  - And finally each company will have a different location.
  - The decisions taken at each previous step have a significant influence on the decisions to be taken in the subsequent step and must be consistent and logical.
  - In this part of the process each team will obtain a different solution to the PBL. But it must be logical and respectful with all the previous decisions taken.

The PBL has a value of 15% on the final grade of the subject. The rating of the PBL is given by the teacher. On the other hand, the director should prepare a report by rating each of his/her colleagues, reporting how each performed the assigned work. The value can be A, B or C:

- A: If his/her performance has been outstanding: it will add 1 point to the score.
- B: If his/her performance has been remarkable.
- C: If his/her performance has been lower than the rest of the group; this will subtract 1 point to the score.
- The director's note will depend on the quality of his/her final report about the work of his/her peers.

3 Results

In this section we present the results obtained during the academic year 2015/16 comparing them with the results of the previous year. We compare the final results and not only the grades of this part of the subject (Tables 1 and 2).

In the results it can be observed that the average mark has increased by almost one point (0.86). It can be seen that the maximum grade has been reduced by 0.70 points and the minimum grade of the course has slightly increased.

Table 1 Comparison of results of the final marks of the two consecutive courses

Course	2015/16		2014/15	
	Students	%	Students	%
Outstanding	7	4.40	2	1.19
Remarkable	57	35.85	23	13.69
Approved	72	45.28	122	72.62
Suspense	12	7.55	19	11.31
Notsubmitted	11	6.92	2	1.19
Total	159		168	

Table 2 Comparison of results of the average marks of the two consecutive courses

Course	2015/16	2014/15
Average grade	6.57	5.71
Standard deviation	1.44	1.15
Maximum grade	9.30	10.00
Minimum grade	1.10	1.00

Although the number of students approved has dropped from 87.50% in 2014/2015 to 85.53% in 2015/2016, the number of students, has increased the number of approved students with respect to submitted students going from 88.55 to 91.89% (Fig. 1).

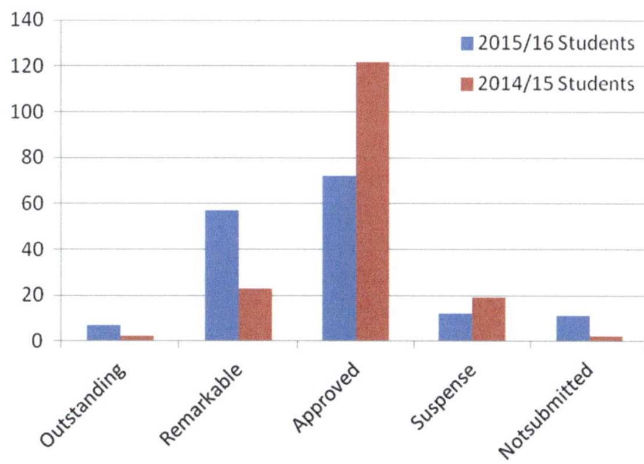

Fig. 1 Marks of the subject courses 2014/2015 and 2015/2016

4 Conclusions

This paper presents the experience realized in the implantation of a PBL in the Business Administration Faculty of the UPV. At this time are still little few data of the experience.

Boaler (2002) and Cohen et al. (2004) talk about the effects of the PBL on the quality of the knowledge acquired by students. Although the authors of this paper agree with this, a validation of this claim should be made.

Marx et al. (1997), Thomas (2000) and Kuhn (2005) talk about the challenges faced by both teachers and students in the implementation of the PBL. We agree with this statement on the one hand teachers face a new challenge, a new way of teaching and evaluation. On the other hand, this is also completely new for the students, while it is a challenge for them. Students can learn much more, but they must also strive much more.

PBL is relatively challenging to enact and plan (Thomas 2000) in fact, most teachers will find some difficult aspects in the implementation of PBL.

There is evidences that some students have difficulty self-directing, especially in complex projects (Thomas 2000), especially with time and project management. This can be resolved as long as there is adequate support from teachers.

Although there has been an improvement in the marks with the change of procedure we will have to continue to monitor whether this process is maintained. We will have to improve those problems that have been detected.

Even with all the failures observed, we consider that it has been a good experience in general for both students and teachers.

References

Aranda SR (2009) Aprendizaje Basado en Proyectos. Revista Innovación Experiencias Educativas 24:1–6

Barbera Ribera MT, Estelles-Miguel S, Barbera Ribera MT, Dema Perez CM (2015) Student opinion on the application of active methodologies. In: Peris-Ortiz M, Merigó Lindahl JM (eds) Sustainable learning in higher education: developing competencies for the global marketplace. Springer, Switzerland, pp 157–167

Bereiter C, Scardamalia M (1999) Process and product in PBL research. Ontario Institutes for Studies in Education. University of Toronto, Toronto

Boaler J (2002) Experiencing school mathematics: traditional and reform approaches to teaching and their impact on student learning. Routledge

Cohen L, Manion L, Morrison K (2004). A guide to teaching practice. Psychology Press

Estelles-Miguel S, Barbera Ribera MT, Dema Perez CM (2013) Aplicación del portafolio grupal en la asignatura de dirección de producción y logística: Resultados de una Prueba Piloto, Educade. Revista de educación en contabilidad, finanzas y administración de empresas, no 4, pp. 124–138. Available from http://www.educade.es/numero04.html. 9 Jan 2016

Estelles-Miguel S, Albarracín Guillem JM, Palmer Gato, ME et al (2016) Aprendizaje y evaluación basada en proyectos en una asignatura de Dirección de Operaciones, Libro de Actas del Congreso

Nacional de Innovación Educativa y de Docencia en Red, Valencia, pp. 1–8. Available from https://riunet.upv.es/handle/10251/70954. 15 Jan 2016
Garrigos-Simon FJ, Oltra JV, Montesa-Andres JO et al (2015) The use of facebook and social networks to improve education. Dirección y Organización, 55:4–10
Jones BF, Rasmussen CM, Moffit MC (1997) Real-life problem solving: A collaborative approach to interdisciplinary learning. American Psycological Association, Washington, DC
Kuhn D (2005) Education for thinking. Harvard University Press
Levy-Leboyer C (2003) Gestión de las Competencias. Ediciones Gestión 2000. Barcelona
Marx RW, Blumenfeld PC, Krajcik JS, Soloway E (1997) Enacting project-based science: challenges for practice and policy. Elementary Sch J 97:341–358
Thomas JW, Mergendoller JR, Michaelson A (1999) Project-based learning: a handbook for middle and high school teachers. The Buck Institute for Education, Novato, CA
Thomas JW (2000) A review of research on project-based learning
Zabala Videla A, Arnau Belmonte L (2007) 11 ideas clave. Cómo aprender y enseñar competencias, Colección Ideas Clave. Serie Didáctica/Diseño y Desarrollo Curricular. Editorial GRAÓ, de IRIF, S.L. Barcelona

Qualitative Patents Evaluation Through the Analysis of Their Citations. Case of the Technological Sectors in the Basque Country

J. Gavilanes-Trapote, Ernesto Cilleruelo-Carrasco, I. Etxeberria-Agiriano, Gaizka Garechana and Alejandro Rodríguez Andara

Abstract Patents are an output of the level of innovation of a company or region. Patent quantitative studies are performed by simply counting the number of these documents. For the qualitative evaluation, there is a certain consensus among the authors to consider the citations as the most adequate indicator. However, this indicator presents several problems regarding its correct interpretation. In the present study, in order to avoid the typical citation interpretation biases, a precise methodology is presented. As an illustrative example, we present a comparative study of the quality of patents in technological sectors of the Basque Country region over the period 1991–2011.

Keywords Patent indicators · Citations · Patents · Regional analysis Intellectual property analysis

J. Gavilanes-Trapote (✉) · A. Rodríguez Andara
Industrial Organization and Management Engineering Department, Faculty of Engineering Vitoria-Gasteiz, University of the Basque Country (UPV/EHU), Nieves Cano 12, 01006 Vitoria-Gasteiz, Spain
e-mail: javier.gavilanes@ehu.eus

E. Cilleruelo-Carrasco
Industrial Organization and Management Engineering Department, Faculty of Engineering Bilbao, University of the Basque Country (UPV/EHU), Alameda de Urquijo, 48013 Bilbao, Spain

I. Etxeberria-Agiriano
Computer Languages and Systems Department, Faculty of Engineering Vitoria-Gasteiz, University of the Basque Country (UPV/EHU), Nieves Cano 12, 01006 Vitoria-Gasteiz, Spain

G. Garechana
Industrial Organization and Management Engineering Department, Faculty of Economics and Business Bilbao, University of the Basque Country (UPV/EHU), C/ Elcano, 48008 Bilbao, Spain

Á. Ortiz et al. (eds.), *Engineering Digital Transformation*, Lecture Notes in Management and Industrial Engineering,
https://doi.org/10.1007/978-3-319-96005-0_28

1 Introduction

In the last decade, the use of citations from both patents and non-patent literature (NPL) as indicators of innovation has increased significantly (OCDE 2009; Higham et al. 2017). Since they indicate the scientific and technological background of inventions, they allow monitoring the evolution of knowledge (Acs et al. 2002; Gavilanes et al. 2015; You et al. 2017). It has been proved that the number of citations received by a patent reflects, on average, its technological and commercial importance and, thus, it helps addressing the problem of heterogeneity in the value of different patents (Hall et al. 2005; Squicciarini et al. 2013; Bakker 2017).

There are basically two types of citations. On the one hand, patent references are citations of relevant technology previously protected by other patents applied for anywhere in the world, at any time and in any language (*backward* and *forward* citations). On the other hand, references classified as non-patent literature (NPL) are scientific publications, conference proceedings, books, database guides, technical manuals, standard descriptions, and so on.

In principle, it is possible to discriminate the quality of patents through the changes of citation intensity associated with various effects (Callaert et al. 2006). However, there are several phenomena that are inherent to patent citation data that hinder this task (Hall et al. 2001; Messinis 2011; Roach and Cohen 2013). Firstly, the number of citations received by a particular patent is *truncated* because only those received to date are known. Secondly, differences in patent examination practices over time may produce differences in the intensity of patents and are unrelated to the true impact (Alcacer and Gittelman 2006). Thirdly, the problem created by the increase in the number of patent citations is exacerbated by the fact that the number of patents granted has also been increasing in various patent offices. Finally, the number of citations that are made (and received) by patent varies considerably depending on the technological field or the maturity of the technology.

2 Objectives

The main objective of this paper is to translate a concrete methodology to help a better interpretation of patent citations as indicators of quality of industrial property documents.

This methodology is applied to the study of the quality of patents in the Basque Country among the different technological sectors throughout the study period.

Due to the laborious nature of data collection, an ad hoc designed program is also presented to automate the collection of citations and the calculation of technological intensity ratios.

3 Methods

The patent citations in the Basque Country have been collected for the 1991-2011 period through the PATSTAT database of the European Patent Office (EPO). Vantage Point software has been used for data cleaning and processing. A new patent classification criterion developed by Kay et al. (2014) has been used to address the *population* problem of certain sectors defined through the International Patent Classification (IPC). An ad hoc program has been developed for counting and calculating the ratios of technological intensity of each type of citation and for each of the 35 technological sectors under study.

Two types of technological intensity ratios are presented to determine the quality of patents: one that takes into account all patents (TI) and a second ratio that only counts patents that have citations (TI*).

Results are presented in different tables, one for each type of citation and for each type of technological intensity ratio. In each table data are disaggregated by sector and for periods of 5 years.

In order to solve the citation *truncation* problem, the study has been limited to patents of 2011. This makes it possible to ensure that practically all the patents that have been applied for so far have been published and therefore have been taken into account. For the differences in patent examination practices over time, the study is presented by comparing citation groups over the same period of time. To solve the patent *inflation* problem, the average citation value for each period is included to be compared with the partial value of each sector and see the variation of the deviation over time. Finally, the citation differences among patents from different sectors are resolved by comparing technology sectors.

4 Results

In the provided results tables, the value in parentheses represents the number of patents per period and sector. For each period we include the average value of the ratio to allow comparing with the partial value of each sector and reduce the *inflation* bias of the citations. The technological intensity ratio of forward citations is also included with a temporary analysis window limited to 5 years after the study period.

In Table 1, through the TI ratio, it is observed that the technological sector that has increased the number of *backward* citations in the study period is "Cosm & Med Chem". If we study this sector through the TI* ratio (Table 2) we observe that the increase in the number of citations in the first half of the period is due to an increase in the number of patents in the sector and not to the number of citations of each patent. However, in the second half, there is an increase in the number of citations per patent and, consequently, the quality of the citations.

In the case of *forward* citations, there are no significant differences between Tables 3 and 4. This indicates that the evolution of the sectors of the Basque Country

Table 1 Technological knowledge capture by industrial sector, TI (backward citations)

Sectors	I.T. (1992-96) Average: 3.4	I.T. (1997-01) Average : 3.6	I.T. (2002-06) Average : 3.2	I.T. (2007-11) Average : 3.8
Vehicles	3.9 (84)	3.7 (135)	3.8 (196)	4.1 (153)
Vehicle parts	3.4 (34)	3.5 (52)	3.7 (89)	4.4 (97)
TV, Imaging & Comm	2.6 (7)	4.3 (4)	2.9 (10)	3.4 (5)
Turbines & Engines	5.1 (24)	3.5 (38)	4.7 (58)	4.6 (66)
Textiles	3.0 (15)	4.3 (30)	2.8 (44)	4.4 (21)
Telephone Comm	2.4 (5)	5.0 (3)	2.5 (2)	2.2 (5)
Semiconductors	3.9 (26)	3.4 (33)	3.5 (42)	3.3 (31)
Recording	0.0 (0)	4.5 (2)	1.3 (3)	2.2 (5)
Radio, Comm	(2) 6.5	3.5 (11)	1.5 (2)	3.3 (4)
Plastics & Wheels	4.2 (33)	2.8 (24)	3.2 (44)	3.8 (52)
Photolithography	5.0 (1)	2.0 (1)	4.3 (3)	5.0 (3)
Optics	2.5 (11)	2.9 (19)	3.6 (13)	3.8 (10)
Metals	3.4 (45)	2.3 (55)	3.6 (66)	3.8 (59)
Medical devices	2.9 (10)	3.2 (17)	3.1 (12)	2.7 (6)
Med Instruments	5.4 (20)	4.6 (27)	3.1 (33)	4.5 (29)
Measurement	4.1 (34)	3.8 (20)	3.2 (52)	3.8 (33)
Machine Tools	4.0 (35)	2.5 (58)	3.3 (84)	4.4 (39)
Lighting	4.2 (6)	4.6 (13)	4.5 (10)	4.9 (13)
Lab equipment	3.5 (19)	4.1 (19)	3.4 (25)	3.6 (33)
Info Transmission	3.2 (21)	3.3 (9)	3.2 (27)	2.9 (15)
Heating & Cooling	4.3 (21)	3.4 (30)	3.8 (66)	4.1 (45)
Furnace	3.8 (27)	3.9 (31)	3.7 (39)	4.3 (22)
Food	2.1 (17)	3.9 (28)	4.2 (15)	4.6 (17)
Electric Power	3.3 (14)	3.0 (12)	3.7 (21)	3.4 (29)
Drugs, Med Chem	1.4 (28)	1.8 (21)	1.5 (30)	2.5 (35)
Domestic applicances	3.5 (66)	3.3 (84)	3.1 (107)	4.1 (76)
Data Commerce	3.3 (15)	4.1 (25)	3.0 (33)	3.9 (29)
→ Cosm & Med Chem	0.3 (9)	1.3 (21)	1.4 (12)	3.6 (20)
Copying & Printing	4.9 (7)	(6) 6.2	4.3 (6)	3.0 (1)
Construction	3.5 (76)	3.2 (112)	4.3 (113)	5.1 (115)
Computing	4.1 (10)	3.1 (8)	2.7 (20)	3.6 (13)
Combustion Engines	3.8 (4)	(8) 5.8	2.5 (8)	5.0 (1)
Chem & Polymers	2.6 (23)	2.2 (29)	2.7 (22)	2.9 (21)
Catalysis & Separation	3.6 (26)	4.1 (20)	2.9 (16)	4.3 (36)
Biologics	2.6 (14)	3.6 (25)	2.4 (38)	2.9 (47)

Table 2 Technological knowledge capture by industrial sector, TI* (backward citations)

Sectors	I.T.* (1992-96) Average : 4.1	I.T.* (1997-01) Average : 4.5	I.T.* (2002-06) Average : 4.3	I.T.* (2007-11) Average : 4.3
Vehicles	4.3 (77)	5.0 (100)	4.9 (151)	4.7 (132)
Vehicle parts	4.3 (27)	4.8 (38)	4.6 (71)	5.0 (86)
TV, Imaging & Comm	3.6 (5)	4.3 (4)	3.2 (9)	(3) 5.7
Turbines & Engines	5.1 (24)	5.1 (26)	4.8 (56)	4.7 (64)
Textiles	4.1 (11)	5.1 (25)	4.0 (31)	5.1 (18)
Telephone Comm	4.0 (3)	5.0 (3)	5.0 (1)	3.7 (3)
Semiconductors	4.3 (24)	4.5 (25)	4.7 (31)	3.7 (28)
Recording	0.0 (0)	4.5 (2)	2.0 (2)	3.7 (3)
Radio, Comm	6.5 (2)	4.2 (9)	3.0 (1)	3.3 (4)
Plastics & Wheels	4.6 (30)	4.7 (14)	4.8 (29)	4.6 (42)
Photolithography	5.0 (1)	2.0 (1)	(2) 6.5	5.0 (3)
Optics	3.4 (8)	4.2 (13)	4.3 (11)	3.8 (10)
Metals	4.2 (37)	4.1 (31)	4.2 (56)	4.7 (47)
Medical devices	3.6 (8)	(9) 6.0	4.6 (8)	4.0 (4)
Med Instruments	(15) 7.1	5.6 (22)	4.7 (22)	5.0 (26)
Measurement	4.5 (31)	4.2 (18)	3.9 (42)	4.1 (31)
Machine Tools	4.1 (34)	3.3 (44)	4.0 (69)	4.7 (36)
Lighting	4.2 (6)	4.6 (13)	5.0 (9)	(11) 5.8
Lab equipment	3.9 (17)	4.5 (17)	3.7 (23)	4.4 (27)
Info Transmission	3.8 (18)	3.8 (8)	4.1 (21)	3.3 (13)
Heating & Cooling	4.8 (19)	5.1 (20)	5.4 (47)	4.9 (38)
Furnace	4.5 (23)	4.4 (28)	5.1 (28)	4.5 (21)
Food	3.3 (11)	5.0 (22)	4.2 (15)	4.9 (16)
Electric Power	3.8 (12)	3.6 (10)	4.1 (19)	3.8 (26)
Drugs, Med Chem	2.9 (14)	4.1 (9)	2.8 (16)	2.9 (31)
Domestic applicances	4.1 (56)	4.7 (59)	5.0 (67)	5.0 (62)
Data Commerce	3.8 (13)	4.7 (22)	4.1 (24)	4.4 (26)
→ Cosm & Med Chem	1.5 (2)	4.7 (6)	4.3 (4)	3.6 (20)
Copying & Printing	4.9 (7)	(6) 6.2	4.3 (6)	3.0 (1)
Construction	3.9 (69)	4.2 (84)	5.2 (94)	5.4 (108)
Computing	4.1 (10)	3.6 (7)	3.6 (15)	3.9 (12)
Combustion Engines	5.0 (3)	(8) 5.8	4.0 (5)	5.0 (1)
Chem & Polymers	3.8 (16)	3.8 (17)	3.8 (16)	3.4 (18)
Catalysis & Separation	4.4 (21)	4.6 (18)	3.9 (12)	4.4 (35)
Biologics	3.7 (10)	4.7 (19)	3.4 (27)	3.6 (37)

Table 3 Technological knowledge emission by industrial sector, TI (forward citations)

Sectors	I.T. (1992-96) Average : 0.2		I.T. (1997-01) Average : 0.3		I.T. (2002-06) Average : 0.5	
Vehicles	0.2	(84)	0.3	(135)	0.4	(196)
Vehicle parts	0.0	(34)	0.3	(52)	0.4	(89)
TV, Imaging & Comm	0.0	(7)	0.0	(4)	0.5	(10)
Turbines & Engines	0.1	(24)	0.2	(38)	0.7	(58)
Textiles	0.1	(15)	0.6	(30)	0.7	(44)
Telephone Comm	0.4	(5)	0.0	(3)	0.0	(2)
Semiconductors	0.4	(26)	0.4	(33)	1.5	(42)
Recording	0.0	(0)	0.0	(2)	0.0	(3)
Radio, Comm	0.0	(0)	0.1	(11)	0.0	(2)
Plastics & Wheels	0.1	(33)	0.5	(24)	0.5	(44)
Photolithography	1.0	(1)	0.0	(1)	0.0	(3)
Optics	0.0	(11)	0.6	(19)	0.7	(13)
Metals	0.0	(45)	0.3	(55)	0.4	(66)
Medical devices	0.0	(10)	0.2	(17)	0.5	(12)
Med Instruments	0.4	(20)	0.4	(27)	0.5	(33)
Measurement	0.1	(34)	0.3	(20)	0.6	(52)
Machine Tools	0.1	(35)	0.1	(58)	0.5	(84)
Lighting	0.0	(6)	0.6	(13)	0.3	(10)
Lab equipment	0.1	(19)	0.2	(19)	0.4	(25)
Info Transmission	0.3	(21)	0.0	(9)	0.6	(27)
Heating & Cooling	0.2	(21)	0.6	(30)	0.6	(66)
Furnace	0.2	(27)	0.4	(31)	0.7	(39)
Food	0.8	(17)	0.6	(28)	0.3	(15)
Electric Power	0.4	(14)	0.8	(12)	2.3	(21)
Drugs, Med Chem	0.1	(28)	0.6	(21)	0.3	(30)
Domestic applicances	0.1	(66)	0.3	(84)	0.4	(107)
Data Commerce	0.1	(15)	0.4	(25)	0.7	(33)
Cosm & Med Chem	0.0	(9)	0.4	(21)	1.0	(12)
Copying & Printing	0.1	(7)	0.2	(8)	0.0	(6)
Construction	0.2	(76)	0.4	(112)	0.3	(113)
Computing	0.1	(10)	0.0	(8)	0.5	(20)
Combustion Engines	0.0	(4)	0.1	(8)	0.9	(8)
Chem & Polymers	0.5	(23)	0.6	(29)	0.0	(22)
Catalysis & Separation	0.1	(26)	0.4	(20)	0.5	(16)
Biologics	0.1	(14)	0.1	(25)	0.3	(38)

as agents transmitting technological knowledge to the rest of the world is due to both quantitative and qualitative aspects, that is, not only the number of cited patents but also the number of patent citations.

Finally, studying NPL citations we highlight the "Chemicals & Polymers" sector, which presents a fourth period with a considerable increase compared to the previous period and the average of all patents of that period (Table 5). Analyzing Table 6 we see that the number of patents with NPL citations increases by 50% (from 8 to 12) and the average number of patent citations increases by 100%, from 1.1 to 2.2. So we can conclude that the increase is due in greater measure to qualitative aspects.

5 Conclusion

This work shows the difficulty of correctly analyzing patent citations. There are several phenomena that must be taken into account in order to avoid bias in the interpretation of data: *truncation* effect, lack of standardization among different patent offices, *inflation* effect or differences in citation number according to technological sector.

Table 4 Technological knowledge emission by industrial sector, TI* (forward citations)

Sectors	I.T.* (1992-96) Average : 1.1		I.T.* (1997-01) Average : 1.5		I.T.* (2002-06) Average : 2.3	
Vehicles	1.4	(9)	1.5	(22)	2.2	(40)
Vehicle parts	1.0	(1)	1.8	(8)	2.1	(15)
TV, Imaging & Comm	0.0	(0)	0.0	(0)	1.7	(3)
Turbines & Engines	1.0	(2)	1.5	(6)	2.6	(16)
Textiles	1.0	(1)	2.1	(8)	2.9	(11)
Telephone Comm	1.0	(2)	0.0	(0)	0.0	(0)
Semiconductors	1.7	(6)	2.0	(7)	4.5	(14)
Recording	0.0	(0)	0.0	(0)	0.0	(0)
Radio, Comm	0.0	(0)	1.0	(1)	0.0	(0)
Plastics & Wheels	1.0	(3)	2.6	(5)	2.1	(10)
Photolithography	1.0	(1)	0.0	(0)	0.0	(0)
Optics	0.0	(0)	2.0	(6)	2.3	(4)
Metals	0.0	(0)	1.8	(10)	2.6	(11)
Medical devices	0.0	(0)	1.3	(3)	2.0	(3)
Med Instruments	3.5	(2)	1.8	(6)	2.5	(6)
Measurement	1.0	(2)	1.3	(4)	1.9	(16)
Machine Tools	1.5	(2)	1.0	(7)	3.4	(13)
Lighting	0.0	(0)	2.0	(4)	3.0	(1)
Lab equipment	1.0	(2)	1.5	(2)	1.4	(7)
Info Transmission	3.0	(2)	0.0	(0)	2.1	(7)
Heating & Cooling	1.0	(4)	2.1	(8)	3.2	(13)
Furnace	1.3	(4)	1.9	(7)	2.9	(10)
Food	1.9	(7)	2.3	(8)	1.0	(4)
Electric Power	3.0	(2)	4.5	(2)	8.2	(6)
Drugs, Med Chem	1.0	(3)	3.0	(4)	2.0	(5)
Domestic applicances	1.2	(6)	2.4	(12)	2.1	(23)
Data Commerce	1.0	(2)	1.7	(6)	1.9	(12)
Cosm & Med Chem	0.0	(0)	2.3	(4)	3.0	(4)
Copying & Printing	1.0	(1)	1.0	(1)	0.0	(0)
Construction	1.5	(11)	1.8	(25)	1.5	(26)
Computing	1.0	(1)	0.0	(0)	3.3	(3)
Combustion Engines	0.0	(0)	1.0	(1)	7.0	(1)
Chem & Polymers	4.0	(3)	2.1	(8)	1.0	(1)
Catalysis & Separation	1.0	(2)	1.6	(5)	2.0	(4)
Biologics	1.0	(1)	1.0	(2)	2.2	(6)

Table 5 Science-technology knowledge transfer by industrial sector, TI (NPL citations)

Sectors	I.T. (1992-96) Average: 0.2		I.T. (1997-01) Average : 0.5		I.T. (2002-06) Average : 0.5		I.T. (2007-11) Average : 0.6	
Vehicles	0.1	(84)	0.2	(135)	0.2	(196)	0.2	(153)
Vehicle parts	0.0	(34)	0.3	(52)	0.3	(89)	0.4	(97)
TV, Imaging & Comm	0.3	(7)	1.0	(4)	0.2	(10)	0.0	(5)
Turbines & Engines	0.0	(24)	0.2	(38)	0.3	(58)	0.3	(66)
Textiles	0.0	(15)	0.1	(30)	0.2	(44)	0.1	(21)
Telephone Comm	0.0	(5)	1.3	(3)	0.0	(2)	0.2	(5)
Semiconductors	0.3	(26)	0.5	(33)	1.0	(42)	0.5	(31)
Recording	0.0	(0)	1.5	(2)	0.3	(3)	1.0	(5)
Radio, Comm	1.0	(0)	0.2	(11)	1.5	(2)	0.0	(4)
Plastics & Wheels	0.1	(33)	0.4	(24)	0.3	(44)	0.3	(52)
Photolithography	1.0	(1)	1.0	(1)	2.0	(3)	0.0	(3)
Optics	0.1	(11)	0.2	(19)	0.2	(13)	1.2	(10)
Metals	0.2	(45)	0.3	(55)	0.3	(66)	0.4	(59)
Medical devices	0.0	(10)	0.1	(17)	0.3	(12)	1.0	(6)
Med Instruments	0.3	(20)	0.1	(27)	0.6	(33)	0.6	(29)
Measurement	0.3	(34)	0.4	(20)	0.3	(52)	0.1	(33)
Machine Tools	0.1	(35)	0.2	(58)	0.2	(84)	0.1	(39)
Lighting	0.0	(6)	0.7	(13)	0.3	(10)	0.5	(13)
Lab equipment	0.8	(19)	0.5	(19)	0.5	(25)	1.2	(33)
Info Transmission	0.2	(21)	0.3	(9)	0.4	(27)	0.6	(15)
Heating & Cooling	0.1	(21)	0.3	(30)	0.1	(66)	0.2	(45)
Furnace	0.1	(27)	0.3	(31)	0.3	(39)	0.2	(22)
Food	0.9	(17)	1.1	(28)	1.3	(15)	0.9	(17)
Electric Power	0.1	(14)	1.0	(12)	0.9	(21)	0.8	(29)
Drugs, Med Chem	0.2	(28)	0.5	(21)	1.0	(30)	2.3	(35)
Domestic applicances	0.1	(66)	0.1	(84)	0.2	(107)	0.3	(76)
Data Commerce	0.2	(15)	0.2	(25)	0.1	(33)	0.3	(29)
Cosm & Med Chem	0.2	(9)	0.2	(21)	0.3	(12)	1.3	(20)
Copying & Printing	0.3	(7)	1.8	(8)	0.5	(6)	1.0	(1)
Construction	0.2	(76)	0.2	(112)	0.2	(113)	0.3	(115)
Computing	0.0	(10)	0.5	(8)	0.5	(20)	0.2	(13)
Combustion Engines	0.5	(4)	0.0	(8)	0.0	(8)	0.0	(1)
→ Chem & Polymers	0.1	(23)	0.7	(29)	0.4	(22)	1.2	(21)
Catalysis & Separation	0.0	(26)	0.5	(20)	1.1	(16)	0.5	(36)
Biologics	0.2	(14)	0.9	(25)	2.4	(38)	2.3	(47)

Table 6 Science-technology knowledge transfer by industrial sector, TI* (NPL citations)

Sectors	I.T.* (1992-96) Average : 1.1	I.T.* (1997-01) Average : 1.6	I.T.* (2002-06) Average : 1.8	I.T.* (2007-11) Average : 2.0
Vehicles	1.1 (8)	1.3 (21)	1.4 (33)	1.3 (20)
Vehicle parts	0.0 (0)	1.3 (10)	1.4 (20)	2.3 (15)
TV, Imaging & Comm	1.0 (2)	2.0 (2)	2.0 (1)	0.0 (0)
Turbines & Engines	1.0 (1)	1.2 (5)	1.9 (9)	1.8 (10)
Textiles	0.0 (0)	1.0 (2)	1.4 (5)	1.0 (2)
Telephone Comm	0.0 (0)	2.0 (2)	0.0 (0)	1.0 (1)
Semiconductors	1.5 (6)	1.6 (11)	3.1 (14)	2.3 (7)
Recording	0.0 (0)	3.0 (1)	1.0 (1)	5.0 (1)
Radio, Comm	2.0 (1)	1.0 (2)	3.0 (1)	0.0 (0)
Plastics & Wheels	2.0 (2)	1.3 (7)	1.2 (9)	1.3 (10)
Photolithography	1.0 (1)	1.0 (1)	(1) 6.0	0.0 (0)
Optics	1.0 (1)	1.3 (3)	1.0 (2)	2.0 (6)
Metals	2.0 (4)	1.2 (15)	1.3 (15)	2.6 (9)
Medical devices	0.0 (0)	1.0 (2)	1.3 (3)	3.0 (2)
Med Instruments	1.7 (3)	1.0 (3)	2.2 (9)	2.3 (7)
Measurement	1.4 (7)	1.4 (5)	1.3 (11)	1.3 (3)
Machine Tools	1.0 (2)	1.5 (8)	1.3 (12)	1.3 (3)
Lighting	0.0 (0)	1.5 (6)	1.5 (2)	2.0 (3)
Lab equipment	2.5 (6)	1.5 (6)	1.7 (7)	3.4 (12)
Info Transmission	1.3 (4)	3.0 (1)	1.3 (8)	2.3 (4)
Heating & Cooling	1.0 (3)	1.3 (7)	1.3 (3)	2.0 (4)
Furnace	1.5 (2)	1.3 (6)	4.3 (3)	5.0 (1)
Food	1.9 (8)	3.1 (10)	2.0 (10)	2.1 (7)
Electric Power	1.0 (2)	2.4 (5)	2.1 (9)	2.0 (12)
Drugs, Med Chem	1.5 (4)	1.8 (6)	1.6 (19)	3.1 (26)
Domestic applicances	1.6 (5)	1.0 (10)	1.3 (19)	1.5 (13)
Data Commerce	1.0 (3)	2.0 (3)	1.0 (3)	3.3 (3)
Cosm & Med Chem	2.0 (1)	1.0 (5)	1.0 (3)	2.9 (9)
Copying & Printing	1.0 (2)	2.8 (4)	1.5 (2)	1.0 (1)
Construction	1.8 (10)	1.1 (16)	1.4 (14)	2.2 (15)
Computing	0.0 (0)	1.0 (4)	1.7 (6)	1.0 (2)
Combustion Engines	2.0 (1)	0.0 (0)	0.0 (0)	0.0 (0)
→ Chem & Polymers	1.0 (2)	1.7 (11)	1.1 (8)	2.2 (12)
Catalysis & Separation	1.0 (1)	1.7 (6)	1.8 (10)	1.7 (10)
Biologics	1.5 (2)	2.8 (8)	4.0 (23)	3.3 (33)

The solutions that are proposed require a count of a considerable number of citations in a disaggregated manner by sectors, periods of time and in the case of forward citations is complicated even more having to operate with windows of time of equal duration in each one of the periods of study. To simplify this task and avoid errors, a program has been developed that automates counting. This program is available through the authors, together with instructions for its use.

With respect to the analyzed ratios we can conclude that the TI ratio is useful to determine the evolution of patent citations in a sector from a quantitative point of view. This analysis requires comparing the ratio of the sector to the average of all sectors in the same period of time. On the other hand, the TI* ratio is used to determine the evolution of patent quality in a technology sector. This analysis will likewise require the comparison with the average of the ratio of all sectors of the study period.

The field of study of patent citations requires further investigation to determine, for example, a composite indicator of patent quality that discriminates the importance of each type of citation.

Acknowledgements The authors want to thank Jon Zabala for making and sharing the program allowing the analysis of data.

Calculation program

The Excel calculation program together with the associated documentation is available through the corresponding author.

References

Acs ZJ, Anselin L, Varga A (2002) Patents and innovation counts as measures of regional production of new knowledge. Res Policy 31:1069–1085

Alcacer J, Gittelman M (2006) Patent citations as a measure of knowledge flows: the influence of examiner citations. Rev Econ Stat 88:774–779. https://doi.org/10.1162/rest.88.4.774

Bakker J (2017) The log-linear relation between patent citations and patent value. Scientometrics 110:879–892

Callaert J, Van Looy B, Verbeek A et al (2006) Traces of prior art: an analysis of non-patent references found in patent documents. Scientometrics 69:3. https://doi.org/10.1007/s11192-006-0135-8

Gavilanes-Trapote J, Río-Belver RM, Cilleruelo E, Garechana G, Larruscain-Sarasola J (2015) Patent overlay maps: Spain and the Basque Country. Int J Technol Manage 69:261–274. https://doi.org/10.1504/ijtm.2015.072976

Hall BH, Jaffe AB, Trajtenberg, M (2001) The NBER patent citation data file: lessons, insights and methodological tools. National Bureau of Economic Research

Hall BH, Jaffe A and Trajtenberg M (2005) Market value and patent citations. Rand J Econ 36 (Spring)

Higham KW, Governale M, Jaffe AB et al (2017) Fame and obsolescence: Disentangling growth and aging dynamics of patent citations. Phys Rev E 95:042309—Published 13 April 2017

Kay L, Newman N, Youtie J, Porter AL, Rafols I (2014) Patent overlay mapping: visualizing technological distance. J Assn Inf Sci Tec 65:2432–2443. https://doi.org/10.1002/asi.23146

Messinis G (2011) Triadic citations, country biases and patent value: The case of pharmaceuticals. Scientometrics 89:813–833

Organisation for Economic Cooperation and Development (OECD) (2009) OECD Patent Statistics Manual

Roach M, Cohen W (2013) Lens or prism? Patent citations as a measure of knowledge flows from public research. Manage Sci 59:504–525

Squicciarini M, Dernis H, Criscuolo C (2013) Measuring patent quality: indicators of technological and economic value. OECD science, technology and industry working papers, 2013/03. OECD Publishing. https://doi.org/10.1787/5k4522wkw1r8-en

You H, Li M, Hipel KW et al (2017) Development trend forecasting for coherent light generator technology based on patent citation network analysis. Scientometrics 111:297. https://doi.org/10.1007/s11192-017-2252-y

Optimization for Fulfilling Education Requirements of Refugee Children

Macit Betul and Gokcen Hadi

Abstract Complex humanitarian emergencies caused by war disaster necessitates fulfillment of shelter, health and education requirements of refugees. Developing countries hosting refugees are tackling with challenges and overstretching of available resources in meeting sheltering and education requirements of refugees. This study proposes a mathematical model that minimizes costs while deciding on the location of refugee camp sites, and allocation of refugee education demand to the existing and additional shifts in schools. Proposed model aims to help policy makers for decision making in camp site selection and school capacity building. A computational study using data of Nizip town in Gaziantep/Turkey is presented for model application.

Keywords Complex humanitarian emergencies · Refugee camp site selection
Refugees · Emergency education

1 Introduction

Events jeopardizing or damaging human life, property and environment are defined as disasters (Nikbakhsh et al. 2011). Disasters are classified as natural disasters, technological disasters and man-made disasters (Cappola 2011). Shelter location and education requirement fulfillment problems elaborated in this study are related with war disaster which is classified as a man-made disaster by scholars.

Man-made disasters like war, civil war and terrorism are defined as "complex humanitarian emergencies" in disaster management literature. These emergencies generate large number of refugees, internally displaced populations and are one of

M. Betul (✉) · G. Hadi
Department of Industrial Engineering, Gazi University, Eti Mh., Yukselis Sk. No: 5, Maltepe, Ankara, Turkey
e-mail: betul.macit@gazi.edu.tr

G. Hadi
e-mail: hgokcen@gazi.edu.tr

Á. Ortiz et al. (eds.), *Engineering Digital Transformation*, Lecture Notes in Management and Industrial Engineering,
https://doi.org/10.1007/978-3-319-96005-0_29

the most frequently occurring disasters in last 20 years (Sharp and Beadling 2012). Nature of these emergencies leads to complex and critical decision making processes related with preparation of emergency response plans and planning of humanitarian relief activities. Therefore Operations Research (OR) methods are useful for testing alternative emergency response plans for these emergencies (Van Wassenhove 2012). In this study, a complex humanitarian emergency case is presented and a mathematical model is proposed for fulfilling sheltering and education requirements of refugees.

2 Literature Review

Many studies in literature investigated facility location for emergency logistics (Caunhye et al. 2012). However shelter site location problems are commonly studied for post disaster sheltering problems in major natural disasters. Alcada-Almeida et al. (2009), Li et al. (2012) and Kilci et al. (2015) are among scholars studied sheltering problems for disaster victims. Alcada-Almeida et al. (2009) proposed a multi objective facility location-routing model that decides shelter locations for disaster victims and identifies evacuation roads for fire disasters. Li et al. (2012) presented a scenario based bi-level programming model for shelter location selection in a range of hurricane events. Kilci et al. (2015) studied locating tent cities by Red Crescent in Turkey for earthquake victims and improved current method by calculating efficient usage of tent cities.

To the best of our knowledge, there is a lack of research on refugee camp site selection and emergency education in OR literature. Cetinkaya et al. (2016) proposed a GIS based multi-criteria decision making model for refugee camp selection. This model considered 19 sub-criteria that are used by Turkish Disaster and Emergency Management Authority for camp site selection. 16 sub-criteria used in model are related with physical and infrastructural conditions and 3 of them are related with social aspects. The social criterion which is related with "proximity to local community" has been one of the most considerable factors.

This study addresses "spatial relation with local education network" as a criterion for refugee camp site selection in addition to the criteria defined in literature. Uncertainty about the postwar process and the length of the stay of refugees necessitates discussing sustainability of education in camp sites. Long term separation of refugee children from local community in camp areas does not support integration of refugee children with hosting societies. Besides it is a huge cost to the hosting country, since quality issues regarding refugee education in camps will cause low earning potential in future. In this study proposed model selects candidate refugee camp sites according to their spatial relation with local education network and allocates the education demand in and out of the camps according to spatial relation between camp sites and schools, and between urban district centroids and schools. By handling two problems together, it is expected to generate a synchronized solution for both.

3 Problem Definition and Mathematical Model

Conflict in Syria entering its seventh year, resulted by 6.3 million internally displaced people and 5.05 million registered refugees according to the UN Refugee Agency figures (UNHCR 2017). Neighbouring countries namely Egypt, Iraq, Jordan, Lebanon and Turkey opened their borders for humanitarian aid. Among them, Jordan, Lebanon and Turkey are hosting the largest populations of Syrian refugees (UNHCR 2017). These countries are struggling to fulfil sheltering, health and education requirements of refugees. Due to the development challenges, lack of resources and poor planning, an education crisis rise in the hosting countries (Culbertson et al. 2015).

UN figures indicate that 10% of registered Syrian refugee population are living in camps and remaining 90% of refugees are living in urban, peri-urban and rural areas in hosting countries (UNHCR 2017). According to Regional Refugee & Resilience Plan (3RP) 2017–2018, there are more than 1.6 million school age children in 3RP countries—namely Egypt, Iraq, Jordan, Lebanon and Turkey—and 48% of Syrian children are out of school (UN 2017). Actions towards sheltering and education needs differ from country to country. In Jordan and Lebanon, second shifts are added to the government's schools for meeting education requirements of refugees (UN 2016a). In Turkey education facilities established in camping areas cover education demand in camps (AFAD 2013). However 90% Syrian refugee population remain outside of the camps in Turkey (EC 2017) and by December 2016 Ministry of National Education reported that 60% of Syrian children enrolled in formal education (MONE 2016). For covering refugee education demand outside the camps, double shift policy is applied along with establishment of temporary education centers (UN 2016b).

Meeting education requirement of refugee children in developing countries necessitates utilizing existing education facilities of hosting country and investing on new education facilities and alternative education approaches. Available spaces in existing shifts needs to be identified and schools that can open double shifts needs to be selected. Identification of available capacity is not sufficient, as Culbertson and Constant (2015) stated, in Lebanon's Akkar region it was possible to find available spaces but there was no matching between children needing education and the available spaces in schools. The issue is the lack of spatial association of refugee children with schools having available spaces. In this study proposed mathematical model handles sheltering and education needs by spatial relation of refugee children with schools. This model is expected to enable policy makers to examine alternative scenarios for planning humanitarian relief activities regarding sheltering and education. The sets, parameters, scalars, decision variables and mathematical formulation are presented below.

Sets

i set of candidate locations for establishing refugee camps
g set of schools that have additional spaces in existing shifts

l set of schools that can engage in double shifting
b set of districts hosting refugees living outside of the camps
s set of grades in education system

Parameters

c_i	cost of establishing refugee camp at candidate location i
cap_i	capacity of refugee camp i
cpc_{gs}	capacity of existing shifts at school g for grade s
cp_{ls}	capacity of additional shift at school l for grade s
dis_{gi}	distance between school g and refugee camp i
db_{gb}	distance between school g and district b
de_{li}	distance between school l and refugee camp i
da_{lb}	distance between school l and district b
$demand_{is}$	education demand at refugee camp i for grade s
$dema_{bs}$	education demand at district b for grade s

Scalars

rcost	annual rent and living cost per person living outside of camps
cusc	artificial socioeconomic cost which is greater than education costs
cext	annual cost of education in existing shift per student
cadd	annual cost of education in additional shift per student
M	a big number
KA	number of refugee camps that will be established
TNR	total number of refugees
dmaxi	critical distance threshold for camps and schools
dmaxb	critical distance threshold for districts and schools

Decision Variables

x_i	if candidate location i is selected for establishing refugee camp 1, else 0
pr_{gi}	if school g serves for refugee camp i 1, else 0
pr_{gb}	if school g serves for district b 1, else 0
ser_{li}	if school l serves for refugee camp i 1, else 0
sr_{lb}	if school l serves for district b 1, else 0
sc_{gis}	number of refugee students at school g, living in camp i, enrolled in grade s
sb_{gbs}	number of refugee students at school g, living at district b, enrolled in grade s
stc_{lis}	number of refugee students at school l, living in camp i, enrolled in grade s

stb_{lbs}	number of refugee students at school l, living in district b, enrolled in grade s
rout	number of refugees living outside of the camps
unschooled	number of refugee children who cannot access education

Mathematical Formulation

$$\text{minimize}\,\Sigma_{i=1}^{I} c_i x_i + (\text{rcost} \times \text{rout}) + \left(\text{cext}\left(\Sigma_{g=1}^{G}\Sigma_{i=1}^{I}\Sigma_{s=1}^{S} sc_{gis} + \Sigma_{g=1}^{G}\Sigma_{b=1}^{B}\Sigma_{s=1}^{S} sb_{gbs}\right)\right)$$
$$+ \left(\text{cadd} \times \left(\Sigma_{l=1}^{L}\Sigma_{i=1}^{I}\Sigma_{s=1}^{S} stc_{lis} + \Sigma_{l=1}^{L}\Sigma_{b=1}^{B}\Sigma_{s=1}^{S} stb_{lbs}\right)\right) + (\text{cusc} \times \text{unschooled}) \quad (1)$$

$$\Sigma_{i=1}^{I} x_i \le KN \quad (2)$$

$$\Sigma_{i=1}^{I} cap_i x_i + \text{rout} \ge TNR \quad (3)$$

$$\Sigma_{g=1}^{G}\Sigma_{i=1}^{I}\Sigma_{s=1}^{S} sc_{gis} + \Sigma_{l=1}^{L}\Sigma_{i=1}^{I}\Sigma_{s=1}^{S} stc_{lis} + \Sigma_{g=1}^{G}\Sigma_{b=1}^{B}\Sigma_{s=1}^{S} sb_{gbs} + \Sigma_{l=1}^{L}\Sigma_{b=1}^{B}\Sigma_{s=1}^{S} stb_{lbs}$$
$$+ \text{rout} \ge \Sigma_{i=1}^{I}\Sigma_{s=1}^{S}(\text{demand}_{is} \times x_i) + \Sigma_{b=1}^{B}\Sigma_{s=1}^{S} \text{dema}_{bs} \quad (4a)$$

$$\Sigma_{g=1}^{G}\Sigma_{i=1}^{I}\Sigma_{s=1}^{S} sc_{gis} + \Sigma_{l=1}^{L}\Sigma_{i=1}^{I}\Sigma_{s=1}^{S} stc_{lis} \le \Sigma_{i=1}^{I}\Sigma_{s=1}^{S} \text{demand}_{is} x_i$$
$$\Sigma_{g=1}^{G}\Sigma_{b=1}^{B}\Sigma_{s=1}^{S} stb_{gbs} + \Sigma_{l=1}^{L}\Sigma_{b=1}^{B}\Sigma_{s=1}^{S} stb_{lbs} \le \Sigma_{b=1}^{B}\Sigma_{s=1}^{S} \text{dema}_{bs} \quad (4b)$$

$$\Sigma_i sc_{gis} + \Sigma_b sb_{gbs} \le cpc_{g,s} \quad \text{for } \forall(gs) \quad (5a)$$

$$\Sigma_i stc_{lis} + \Sigma_b stc_{lbs} \le cp_{l,s} \quad \text{for } \forall(ls) \quad (5b)$$

$$pr_{gi} \times dis_{gi} \le dmaxi \quad \text{for } \forall(g, i) \quad (6a)$$

$$pb_{gb} \times db_{gb} \le dmaxb \quad \text{for } \forall(g, b) \quad (6b)$$

$$ser_{li} \times de_{li} \le dmaxi \quad \text{for } \forall(l, i) \quad (6c)$$

$$sr_{lb} \times da_{lb} \le dmaxb \quad \text{for } \forall(l, b) \quad (6d)$$

$$pr_{gi} \le x_i \quad \text{for } \forall(g, i) \quad (7a)$$

$$ser_{li} \le x_i \quad \text{for } \forall(l, i) \quad (7b)$$

$$sc_{gis} \le M \times pr_{gi} \quad \text{for } \forall(g, i, s) \quad (8a)$$

$$sb_{gbs} \le M \times pb_{gb} \quad \text{for } \forall(g, b, s) \quad (8b)$$

$$stc_{lis} \le M \times ser_{li} \quad \text{for } \forall(l, i, s) \quad (8c)$$

$$stb_{lbs} \le M \times sr_{lb} \quad \text{for } \forall(l, b, s) \quad (8d)$$

$$x_i, pr_{gi}, pb_{gb}, ser_{li}, sr_{lb} \in \{0, 1\} \quad (9)$$

$$\text{rout, unschooled}, sc_{gis}, sb_{gbs}, stc_{lis}, stb_{lbs} \ge 0 \quad (10)$$

Notation (1) is the objective function that minimizes the cost of refugee camp establishment, the cost of rent and living outside of the camps and education costs. Constraint (2) controls maximum number of candidate locations to be selected for refugee camp site establishment. Constraint (3) helps to calculate number of refugees living outside of camps and balances the demand. Total number of refugees needing shelters is greater than the available capacity in camps as it is in reality. Constraint (4a) and (4b) calculates refugee children who cannot access education and balances demand. Constraint (5a) and (5b) balances student assignments and school capacities. Total education demand is greater than total capacity of schools. Constraint (6) ensures that maximum distance rule is applied. Constraint (7) controls that schools serve for a refugee camp site if it is opened. Constraint (8) controls that student assignments are done if the maximum distance constraint is ensured. Constraint (9) states that site selection, "site to school" and "district to school" serving relations are binary decisions. Constraint (10) states that number of unschooled children, number of refugees outside of the camps and the number of students assigned to schools are positive variables.

4 Computational Studies

Data of Nizip town in Gaziantep/Turkey is used for model implementation. Gaziantep is one of the top five cities hosting large number of Syrian refugees in Turkey (GIGM 2017). Nizip town is currently hosting refugee population in urban area and two camps established in 2012 and 2013. Children living in refugee camps access education facilities in camp areas. During 2012 and 2013 refugee children living outside of the camps mostly accessed education in temporary education centers. After adoption of temporary protection legislation for Syrian refugees in 2014 (Official Gazette 2014), the number of refugee children accessing education in governmental schools grown (Emin 2016). In this computational study, camp site selection scenario with spatial relation to education network of Nizip is assessed, allocation of education demands to the schools and the capacity gaps are presented. Demand and cost values are assumed to be deterministic, since historical data is available regarding refugee population and related costs.

Centroids of districts outside of Nizip town center are calculated by ARCGIS 10.3 and ten centroids are selected as candidate camp site locations. The centroids of districts in Nizip town center are used for calculation of distance between refugee children living in urban area and schools. Travel distance data is obtained by Google Maps API. As Pearce (2000) stated it is not possible to acquire postal code of each student for allocation and therefore methods of allocating centroids to schools is preferred. Maximum distance between schools and camp site centroids is determined as 15 km, as maximum distance a child can travel for accessing education. Maximum distance between schools and urban district centroids is determined as 2 km. International Organization of Migration (2017) provides transportation funding for children living 2 km away from schools. Therefore 2 km is accepted as limit for cost-efficient allocation in urban area.

For schools; number of teachers, number of classes, number of students enrolled, shift policy and geographic coordinates are collected. Selection of schools is done based on student-teacher criteria. In Nizip average student-teacher ratio for primary and secondary schools are 22 and 19.8. 13 out of 20 primary, 17 out of 25 secondary and 4 out of 21 upper-secondary schools have school-teacher ratios that exceed the average town student-teacher ratio. Therefore these schools are excluded from new assignments to the existing shifts. Additional spaces in existing shifts are determined by not exceeding regional average ratio. In Southeastern region average student-teacher ratio in primary and secondary school are 23 and 18 respectively (Ministry of National Education 2016).

Double shift schools are selected within single shift schools. 6 primary schools and 5 secondary schools are suitable for opening second shifts. Upper-secondary schools have full-time schooling; it is not possible to apply double shifting at this level of education. Double shift capacities are calculated by ensuring not to exceed the average regional student-teacher ratios for primary and secondary schools. And 2014 population figures of refugees in Nizip are taken into account for identifying the sheltering and education demands in case study (GKK 2014).

In this case study, education demand in camp and local area are 42% and 58% of the total refugee education demand respectively and two camps are opened. According to the model results, 25% of the total education demand can be met by capacity building in existing shifts and 36% of the demand can be met by adding second shifts in available schools, hence in total 61% of the total refugee education demand can be met. 39% of the total education demand cannot be met due to the insufficient capacity; therefore investment on new education facilities is needed. In this case study, spatial relation of candidate camp locations with government schools in local area is considered and camp sites are selected accordingly. By this consideration only 25% of the education demand in camps can be met by government schools in local area. If decision makers plan to meet education demand in camp sites via government schools in local area, location selection for new education facilities should be performed by considering the spatial relation between camp sites and new education facilities.

5 Conclusion and Future Work

This study proposed a mathematical model that considers spatial relation while deciding on camp site locations and refugee education demand allocation. Proposed model aims to help national and international stakeholders for decision making on sheltering and education demands of refugees by considering alternative cases. For future studies spatial relation with health institutions and humanitarian aid warehouses can also be considered for refugee camp site selection. Regarding the emergency education part of the problem, location selection for new schools and transportation planning for refugee children can be examined.

References

AFAD (2013) Republic of Turkey Prime Minister Disaster and Emergency Management Authority, Quarterly Magazine of AFAD, November 2013, Issue 2, p 19

Alcada-Almeida L, Tralha L, Santos L, Coutinho-Rodrigues J (2009) A multiobjective approach to locate emergency shelters and identify evacuation routes in urban area. Geogr. Anal 41:9–29

Cappola DP (2011) Chapter 1: The management of disasters, introduction to international disaster management (2nd ed, p 2–35). Elsevier Inc
Caunhye AM, Nie X, Pokharel S (2012) Optimization models in emergency logistics: a literature review. Socio-Econ Plann Sci 46:4–13
Cetinkaya C, Ozceylan E, Erbas M, Kabak M (2016) GIS-based fuzzy MCDA approach for siting refugee camp: a case study for south-eastern Turkey. Int J Disaster Risk Reduction 18:218–231
Culbertson S, Constant L (2015) Education of Syrian refugee children: managing the crisis in Turkey, Lebanon and Jordan, RAND Research Reports, No. 859, RAND Corporation, Santa Monica, Calif
EC (2017) Turkey: refugee crisis, ECHO factsheet, European Commission, Civil Protection and Humanitarian Aid Operations, April 2017. Accessed 12 May 2017
Emin MN (2016) Turkiye'deki Suriyeli Cocuklarin Egitimi, Temel Egitim Politikaları, SETA
GIGM (2017) Republic of Turkey, Ministry of Interior, Directorate General of Migration Management, Temporary Protection Statistics. Accessed 14 May 2017
GKK (2014), Gaziantep Kent Konseyi, Suriyeli Misafirlere Iliskin Rapor
International Office of Migration (2017), Regional Response to Syria Crisis Report
Kilci F, Kara BY, Bozkaya B (2015) Locating temporary shelter areas after an earthquake: a case for Turkey. Eur J Oper Res 243:323–332
Li ACY, Nozick L, Xu N, Davidson R (2012) Shelter location and transportation planning under hurricane conditions. Transp Res Part E 48:715–729
Ministry of National Education (2016), National Education Statistics Report 2015–16
Nikbakhsh E, Farahani RZ (2011) Chapter 15: Humanitarian logistics planning and disaster relief operations. In: Logistics operations and management. Elsevier Inc., pp 291–326
Official Gazette (2014) Gecici Koruma Yonetmeligi, No: 2014/6883, November 22, 2014
Pearce J (2000) Techniques for defining school catchment areas for comparison with census data. Comput Environ Urban Syst 24:282–303
Sharp TW, Beadling CW (2012) Chapter 18: Disasters, complex emergencies and population displacement. In: Magill AJ, Hill, David R, Solomon T, Ryan ET (eds) Hunter's tropical medicine and emerging infectious disease (9th ed.). Elsevier Inc., pp 148–155
UN (2016a) Regional refugee & resilience Plan 2016–2017, 2016 Annual Report. Accessed 12 May 2017
UN (2016b) 3RP Regional strategic overview 2016–2017
UN (2017) 3RP Regional strategic overview 2017–2018
UNHCR (2017) Information sharing portal. Accessed 12 May 2017
Van Wassenhove LN (2012) Using OR to adapt supply chain best practices to humanitarian logistics. Int Trans Oper Res 19:307–322

Using Twitter as a Tool to Foster Social Resilience in Emergency Situations: A Case of Study

María Martínez Rojas, María del Carmen Pardo-Ferreira, Antonio López-Arquillos and Juan Carlos Rubio-Romero

Abstract In this paper, the behaviour of Twitter activity under an emergency situation is explored. Concretely, the chemical accident occurred on February 8, 2017 in Paterna (Valencia) is analyzed. Then, we analyse how the information regarding this event is propagated through the twitter network in the hours and days following this disaster.

Keywords Twitter · Emergency · Social resilience · Social media analytics

1 Introduction

Twitter is a micro-blogging platform that allows to publish and exchange short messages, known as *tweets*, making it an ideal environment for the real-time propagation of breaking-news (Mendoza et al. 2010; Twitter 2017). In recent years, there is a growing interest in the use of twitter in emergency response and recovery to foster social resilience, which is defined as the "capacity of social groups and communities to recover from, or respond positively to, crises" (Reuter and Spielhofer 2016).

In this paper, we analyse the event occurred at 9:24 am in the Industrial area named "Fuente del Jarro" in Paterna (Valencia). At this time, there were a strong explosion which caused an immense smoke in the area. Then, the industrial area was evacuated and more than 150 companies and 1000 workers were affected.

2 Objectives

The objective of this paper is to analyze the information spread by twitter social networks in the hours and days immediately following the mentioned event.

M. M. Rojas (✉) · M. del C. Pardo-Ferreira · A. López-Arquillos · J. C. Rubio-Romero
Dept. de Economía y Administración de Empresas, Universidad de Málaga, Málaga, Spain
e-mail: mmrojas@uma.es

Á. Ortiz et al. (eds.), *Engineering Digital Transformation*, Lecture Notes in Management and Industrial Engineering,
https://doi.org/10.1007/978-3-319-96005-0_30

3 Methods

To achieve the objective, firstly, user activity data between February 8 and February 13 is collected. To do this, we use an open source data analytics platform (Knime 2017), which makes possible the communication to the twitter API. The analysis is focused on tweets which include the hashtags #Explotion and #Paterna. Secondly, we perform a data pre-processing consisting in gather tweets, eliminate duplicate ones and clean some fields which are very heterogeneous due to the freedom offered by this social network. After these steps, we proceed to analyse the obtained data.

4 Results

As a result of the first step, a total of 4615 tweets have been obtained. From these, 790 tweets are unique which are the basis of other users that have retweeted them. The impact was considerably high both in the first day and in the hours close to the explosion. In addition to users who witnessed the explosion, twitter was used by the local government and the office of emergencies to warn of the fire and report that firefighters and police were on their way. Additionally, based on the extracted information, the location of the users who echoed the news can be analysed. In this regard, users from different countries retweeted about the explosion, as for example, Moscow, Venezuela, London, Netherlands, etc. However, some original tweets were written in French and English, as for example, those spread by *FranceNews* the same day of the explosion.

5 Conclusion

It has been found a large amount of noise in data due to the freedom that users have when they post a tweet. It makes the task of extracting information very laborious. Nowadays, it is a reality that the information spread in twitter provide timely insight into events as they are developed. It makes an opportunity to foster social resilience from emergency situations. To achieve this end, the development of methods that allow the automatically identification of information through the tweets that serve for the management of emergencies are necessary.

Acknowledgements This work has been supported by the Spanish Ministry of Economic, Industry and Competitiveness (Juan de la Cierva – Formación FJCI-2015-24093).

References

Knime (2017) [online] Knime.org. https://www.knime.org/. Accessed 18 Feb 2017
Mendoza M, Poblete B, Castillo C (2010) Twitter under crisis: can we trust what we RT? In: Proceedings of SOMA '10. ACM, New York, NY, USA, pp 71–79
Reuter C, Spielhofer T (2016) Towards social resilience: a quantitative and qualitative survey on citizens' perception of social media in emergencies in Europe. Technological Forecasting and Social Change
Twitter (2017) [online] https://www.twitter.com. Accessed 18 Feb 2017

Managing Volunteer Assignment in a Sport Event

Manuel Mateo Doll, Joaquín Bautista Valhondo and Rocio de la Torre Martínez

Abstract Any sport event can count with volunteers to support the rest of participants (organizers, officials…). This group usually concentrates an important number of collaborators who work selflessly, but with heterogeneity of profiles. The organizational success of these events depends to a large extent on their efficiency in the assigned tasks. Here, it is designed a methodology that generates an assignment of volunteers to tasks, taken into account the suitability of each person to the operations to be performed. One of the stages in the methodology is based in the resolution of a Mixed Integer Linear Program. This methodology is evaluated and applied to data from a world event.

Keywords Assignment · Mixed integer linear programming · Sports event

1 Introduction

Some sport events, in addition to hired personnel, require volunteering for their correct functioning. Volunteers form an essential collective in the organization and development of any sport event. They collaborate in the great majority of tasks necessary for the accomplishment of the event. Therefore, that the event can be carried out and its quality will depend to a great extent on the work that they perform.

M. Mateo Doll (✉) · J. Bautista Valhondo
Dpto. de Organización de Empresas. Institut Organització i Control, Escuela Técnica Superior Ingeniería Industrial, Universitat Politècnica de Catalunya, Av. Diagonal, 647, 7°, 08028 Barcelona, Spain
e-mail: manel.mateo@upc.edu

J. Bautista Valhondo
e-mail: joaquin.bautista@upc.edu

R. de la Torre Martínez
Dpto. de Organización de Empresas, Escuela Técnica Superior Ingeniería Industrial, Universitat Politècnica de Catalunya, Av. Diagonal, 647, 7°, 08028 Barcelona, Spain
e-mail: maria.rocio.de.torre@upc.edu

Á. Ortiz et al. (eds.), *Engineering Digital Transformation*, Lecture Notes in Management and Industrial Engineering,
https://doi.org/10.1007/978-3-319-96005-0_31

In addition, beyond its influence in the accomplishment of tasks, the group of volunteers have direct contact with sportspeople and public, and therefore, they are the direct reference of these in any circumstance. It is necessary to decide which tasks can be done by a volunteer (v.gr., health support) and find, among the candidates, the best prepared people to perform them. For decisions in sport events, the main problem faced in the literature is the tournament scheduling (Pinedo 2009), specially related to optimize logistic or profit issues (Kendall et al. 2010). The mixed integer linear programming has been used, but basically in this sense to schedule matches in a competition (Della Croce and Oliveri 2006; Drexl and Knust 2007; Recalde et al. 2013). This is not the case for assignment problem referred to people, which is introduced in Llovet (2016) for a world basket championship.

This paper presents a procedure for solving the Volunteer Assignment Problem and a mathematical model for this. The organization of the rest of the paper is as follows: Sect. 2 describes briefly the problem. Section 3 presents the methodology. Section 4 includes the model of the problem. Section 5 describes the case study in which the model has been applied. Finally, conclusion is detailed in Sect. 6.

2 The Volunteer Assignment in Sport Events

An Assignment Problem (Kunh 1955) is a particular case of linear programming (Dantzig and Thapa 1997). This problem has a lot of extensions. Among them, the Generalized Assignment Problem (GAP) has been treated by several authors (Chu and Beasley 1997; Öncan 2007; Krumke and Thielen 2013). As in any Assignment Problem, the volunteer assignment in sport events is based on two sets of elements: each one of the candidates to be volunteers and each one of the positions to be occupied by these. Of course, if a campaign presents properly an event and the capitation of interested people, the number of applications should be greater than the number of positions.

The information required to the candidate should be approximately the following one: source of capitation (web, sponsor…); personal data (age, studies, nationality, driving license and job situation, among others); knowledge of the candidate about what is done in the event (for instance, relation with this sport); skills in certain tasks (for instance, relation with health); participation in other similar events or even non-profit organizations; time availability (it there will be different shifts) and geographical availability (in case of different venues); knowledge of languages (local language and English), and finally, the preferences of the candidate for each one of the different work areas or tasks in which he/she can participate.

On the other hand, in a sport event the work areas or tasks to be done can be very varied: accreditation, competition, access control, logistics, press, marketing, match entertainment, VIP attention, information points, health services, technology, transport… Once all the necessary operations to be done in each area are correctly defined, it will determine the number of positions, i.e. the number of people involved per each area or task, venue and, if it is required, shift.

The objective is to develop a procedure capable of treating the huge amount of data about potential volunteers, in order to obtain an optimal allocation of the people registered to the tasks without any remuneration that can be covered by them.

3 Methodology for the Volunteer Assignment

As a result of the elements to be considered in this problem, a methodology with at least five stages is proposed, to which a sixth optional stage is added:

1. Determine the different kinds of volunteer needs, once studied the competition, the calendar, the venues… Analyse the list of operations for each task position (Rogalksy et al. 2016).
2. According to the dimension of the event at each venue, determine the number of positions for each task at each of the venues (and/or shifts, matches…).
3. Obtain the list of volunteer candidates and their associated information.
4. Evaluate each of the candidates for each of the possible positions.
5. Solve the model for the assignment of candidates to volunteer positions. As a result, the assignment of candidates to volunteer positions is obtained.
6. Optionally, sensitivity analysis may be added to consider qualitative aspects omitted in the previous stages.

4 Mixed Integer Linear Programming (MILP) Model for the Volunteer Assignment

As it has seen in the previous Chapter, in the fifth stage, a model is developed to assign a candidate to each one of the positions, considering the candidate's preferences for tasks and venues previously indicated. Therefore, the objective will be the maximization of the overall score (fitness) which takes into account capabilities and preferences of the people.

The total number of volunteer candidates is n; the number of different tasks is m and the number of venues is l.

4.1 Basic Nomenclature

Parameters:

I	Number of volunteers (i.e. volunteer candidates): $i = 1, \ldots, n$ $(n \equiv \|I\|)$
J	Number of tasks: $j = 1, \ldots, m$ $(m \equiv \|J\|)$
K	Number of venues (or matches, periods, shifts…): $k = 1, \ldots, l$ $(l \equiv \|K\|)$
$n_{j,k}$	Maximum number of volunteers that can develop the task $j \in J$ in the venue $k \in K$. It is equivalent to the number of job positions available to develop the task $j \in J$ in the venue $k \in K$
$p_{i,j}$	Score in points obtained by the volunteer $i \in I$ to develop the task $j \in J$ based on his/her skills and aptitudes. It is equivalent to the fitness of volunteer $i \in I$ to develop the task $j \in J$
$a_{i,j}$	Degree of satisfaction obtained by the volunteer $i \in I$ when he/she develops the task $j \in J$. It also expresses the volunteer's wish to realize a certain task. Generally it is stated that $0 \leq a_{i,j} \leq 1$. In this work we will suppose that values $a_{i,j}$ are binary: $a_{i,j} \in \{0, 1\}$
$b_{i,k}$	Degree of satisfaction obtained by the volunteer $i \in I$ when he/she is assigned to the venue $k \in K$. It also expresses the volunteer's wish to be present in a venue. Generally it is stated that $0 \leq b_{i,k} \leq 1$. In this work we will suppose that values $b_{i,k}$ are binary: $b_{i,k} \in \{0, 1\}$
$f_{i,j,k}$	Fitness for the volunteer $i \in I$ to develop the task $j \in J$ in the venue or match $k \in K$. The value is obtained as: $f_{i,j,k} = a_{i,j} \cdot b_{i,k} \cdot p_{i,j}$

Variables:

F	Overall fitness of all the volunteers to the tasks and the venues
$x_{i,j,k}$	Binary variable that is equal to 1 if the volunteer $i \in I$ is assigned to the task $j \in J$ in the venue $k \in K$, and equal to 0 otherwise

4.2 MILP Model

MILP-1: $a_{i,j} \in \{0, 1\}$; $b_{i,k} \in \{0, 1\}$

$$\max F = \sum_{i=1}^{n}\sum_{j=1}^{m}\sum_{k=1}^{l} f_{i,j,k}x_{i,j,k} \tag{1}$$

Subject to:

$$\sum_{k=1}^{l} x_{i,j,k} \leq a_{i,j} \quad \forall i = 1, \ldots, n \;\; \forall j = 1, \ldots, m \tag{2}$$

$$\sum_{j=1}^{m} x_{i,j,k} \leq b_{i,k} \quad \forall i = 1, \ldots, n \;\; \forall k = 1, \ldots, l \tag{3}$$

$$\sum_{i=1}^{n} x_{i,j,k} \leq n_{j,k} \quad \forall j = 1, \ldots, m \;\; \forall k = 1, \ldots, l \tag{4}$$

$$x_{i,j,k} \in \{0, 1\} \quad \forall i = 1, \ldots, n \;\; \forall j = 1, \ldots, m \;\; \forall k = 1, \ldots, l \tag{5}$$

In the model (MILP-1: $a_{i,j} \in \{0, 1\}$; $b_{i,k} \in \{0, 1\}$), the objective function (1) represents the maximization of the fitness of the set of volunteers to the tasks and venues; the constraints (2) guarantee that none of the volunteers is assigned to a task that he/she does not want, while (3) guarantee that none of the volunteers must go to a venue or match in which he/she does not want to participate; the constraints (4) limit the assignment of volunteers to the available job positions, according to tasks and venues or matches; finally, the constraints (5) impose that the decision variables ($x_{i,j,k}$) are binary.

4.3 Data Source

The values given to the parameters are obtained by displaying the stages 1 to 4 of the proposed methodology. In effect:

- Stage-1: On one hand, when the set of tasks to develop during the event is determined in this stage, the set J is defined. On the other hand, the set of venues, matches, etc. (it can be any division in the whole competition) for the event is determined and the set K is defined.
- Stage-2: As the number of job positions associated to each task $j \in J$ in each venue $k \in K$ is determined, in this stage the values $n_{j,k}$ are established.
- Stage-3: When the list of volunteers is obtained, in this stage the set I is defined. Their preference on tasks ($a_{i,j}$) and venues ($b_{i,k}$) are collected. Moreover, the rest

of information for them (including personal data, skills…) is obtained, but it will be used in the next stage.

- Stage-4: To evaluate to the elements of the set I according to the suitability on the elements of the set J, a score, $p_{i,j}$, is given to each volunteer $i \in I$ for each task $j \in J$, in function of the information provided by the volunteer in his/her application form to opt for the volunteer positions.

5 Case Study

5.1 Data

The model was tested by the data provided in the volunteer recruitment for the World Basketball Championship 2014. The number of applications to volunteer tasks is $n = 14{,}774$. The number of tasks (positions to be occupied) is $m = 27$. The number of venues is $l = 6$ (in this case, each different pavilion). As each candidate is asked for a two task preferences, the addition of $a_{i,j}$ for all the volunteers and tasks is 29,548, greater to the requirements at each task (see Table 1). Each candidate can choose a single venue ($k = 1,\ldots,6$) or none (see Table 2). The total number of task positions, in all the venues, is 2148 ($j = 1,\ldots,27$).

Table 1 Number of positions versus applications for each task j

j	1	2,3	4,5	6,7	8,9	10,11	12,13
$\sum_{k=1}^{l} n_{j,k}$	48	92	262	420	66	310	46
$\sum_{i=1}^{n} a_{i,j}$	6907	2122	6798	1599	999	1588	457

(continued)

Table 1 (continued)

j	14,15	16,17	18,19	20,21	22,23	24,25	26,27
$\sum_{k=1}^{l} n_{j,k}$	156	76	76	90	122	264	120
$\sum_{i=1}^{n} a_{i,j}$	1808	2184	2103	761	962	762	498

Table 2 Number of positions versus applications for each venue k

k	1	2	3	4	5	6	None
$\sum_{j=1}^{m} n_{j,k}$	397	357	365	353	326	350	0
$\sum_{i=1}^{n} b_{i,k}$	868	1086	2973	3043	3907	1084	1813

5.2 *Experiment*

In this case, some tasks are, among others, logistics, marketing, press support, accreditations, information points, transport, technology or health services. Each candidate can reach a maximum score of 100 points, but it is very difficult to obtain them. We provide, as an example, the values of three volunteers (Table 3).

The model has 487,704 constraints and 2,393,389 variables. It is solved with IBM ILOG CPLEX Optimization Studio. A personal computer, Intel® CoreTM i7 CPU at 3.4 GHz and 8 GB RAM, is used to perform the experiment.

The upper bound for the objective function would be 214,800. After the resolution of the model, the value of the objective function for the assignment is 176,866 points. The improvement evaluated with respect to the mean values of the volunteer candidates is 38.32%. The CPU time is 2.25 s. It can be seen that the average score of all the volunteers in a task and a venue usually is above 70 points in nearly all the cases (with an exception of a task in a venue). Moreover, the results are greater than 80 points in many tasks. Another aspect to note from the results is that the three locations with the highest score are the ones with the highest number of applications.

6 Conclusion

A methodology is designed to assign a group of volunteers to a set of positions to support the organization of a sport event. It is necessary to collect data from all the people that applied to volunteer and the tasks they are going to perform. Each person is scored on each task. Thanks to a mixed integer linear programming, people are assigned to tasks trying to reach a maximum overall score. This procedure is tested on the volunteer assignment of a world sport event, whose data were obtained. It could be adapted to other similar events.

Respect to future research, we want to develop the following studies: (1) analyse the impact, on the volunteer assignment and the global fitness, generated by the partial and total exclusion the degrees of satisfaction $a_{i,j}$ and $b_{i,k}$, for the assigned task and venue; (2) consider $a_{i,j}$ and $b_{i,k}$ as continuous parameters; (3) include time constraints (work shifts) in the assignment models and also space constraints (capacity in the venue sites); and (4) formulate and exploit new assignment models that incorporate the typology of volunteering, tasks and venues.

Acknowledgements This work was funded by the Ministerio de Economía y Competitividad (Spanish Government) through the TIN2014-57497-P (FHI-SELM2) project.

Table 3 Score for 3 of the 14,774 volunteers (denoted by i, i', i'') for each task j

j	1	2,3	4,5	6,7	8,9	10,11	12,13	14,15	16,17	18,19	20,21	22,23	24,25	26,27
$p_{i,j}$	44	50	69	77	82	0	79	56	85	61	50	80	77	0
$p_{i',j}$	78	93	75	92	91	93	93	75	92	78	93	73	77	92
$p_{i'',j}$	72	64	69	62	55	67	64	50	62	56	64	0	46	0

References

Chu PC, Beasley JE (1997) A genetic algorithm for the generalised assignment problem. Comput Oper Res 24:17–23

Dantzig GB, Thapa MN (1997) Linear programming 1: introduction. Springer, New York

Della Croce F, Oliveri D (2006) Scheduling the Italian football league: an ILP-based approach. Comput Oper Res 33:1963–1974

Drexl A, Knust S (2007) Sports league scheduling: graph- and resource-based models. Omega 35:465–471

Kendall G, Knust S, Ribeiro CC, Urrutia S (2010) Scheduling in sports: an annotated bibliography. Comput Oper Res 37:1–19

Krumke SO, Thielen C (2013) The generalized assignment problem with minimum quantities. Eur J Oper Res 228:46–55

Kunh HW (1955) The Hungarian method for the assignment problem. Naval Res Logistics 2:83–97

Llovet I (2016) Diseño de una herramienta de soporte en la gestión de voluntarios en una competición deportiva global. Proyecto Final de Carrera, ETS Enginyeria Industrial Barcelona, Barcelona

Öncan T (2007) Survey of the generalized assignment problem and its applications. INFOR Inf Syst Oper Res 45:123–141

Pinedo ML (2009) Planning and scheduling in manufacturing and services. Springer, New York

Recalde D, Torres R, Vaca P (2013) Scheduling the professional Equadorian football league by integer programming. Comput Oper Res 40:2478–2484

Rogalksy K, Doherty A, Paradis KF (2016) Understanding the sport event volunteer experience: an investigation of role ambiguity and its correlates. J Sport Manag 30:453–469

Optimal Tax Planning with Mathematical Programming Models

E. Parra

Abstract Tax planning is a matter of increasing interest to large corporations. Complex tax systems must be optimized using mathematical programming methods. As an example, this paper considers the special tax system in the Canary Islands (REF), which includes incentives for investments in the region and can be used by companies to improve the return of their investments plans. A model for tax planning is implemented and solved with standard software.

Keywords Tax planning · Optimization models

1 Introduction

Optimization techniques are currently used in numerous business areas: production and transport planning, portfolio selection and others (a good example of models for business optimization can be found from Kallrath 1997 to Baker 2011). These techniques have also been used for financial planning since the very beginnings of mathematical programming in business.

Although tax planning is widely used in large corporations, with special emphasis on transfer price optimization (see Klassen et al. 2017), there are very few works on practical tax planning using mathematical programming. Even in one of the foremost collections of articles on tax in Europe (GQTR 2017), it is difficult to find papers on corporate optimization models for tax planning. In a recent work (Dinh 2014), a specific tax planning problem is solved theoretically. Even general manuals on tax planning, (i.e. Schanz and Schanz 2011) do not include optimization models.

This paper describes the implementation of a tax planning optimization model for the Canary Islands special tax system (REF, "Regimen Económico y Fiscal") using standard software to illustrate the advantages of using optimization models in tax

E. Parra (✉)
Department of Economics, University of Alcala, Madrid, Spain
e-mail: enrique.parra@uah.es

Á. Ortiz et al. (eds.), *Engineering Digital Transformation*, Lecture Notes in Management and Industrial Engineering,
https://doi.org/10.1007/978-3-319-96005-0_32

planning. This example is used to demonstrate a general approach to tax planning optimization problems.

2 The Tax-Financial Model

2.1 *Canary Islands Special Tax System*

As an example of tax planning optimization, and to establish the general concepts, a multi-period model for REF ("Regimen Económico y Fiscal") is implemented for the Canary Island autonomous region.

The aim of this model is to maximize the cash flow generated (with or without discount, as preferred) in the entire period, making use of the various incentives in the Canary Islands special tax system, mainly "Reserva de Inversiones" (reserves for investments, RIC) and "Deducción por inversiones" (investment deductions, DI). These incentives are regulated by Spanish and European Union laws (Ley 30/1972, 20/1991, 19/1994; RDL 15/2014).

The sample model is built to determine the optimal amounts of RIC and DI each year in the planned period.

The main reasons for using optimization are:

1. RIC is direct-deductible from the taxable profit, thus reducing the tax to pay (general tax rate for firms is 25% in Spain).
2. RIC can reach a maximum equivalent amount of 90% of the undistributed accounting profit which, after taxes, depends on the Corporate Tax rate and the RIC and the investment deduction.
3. Undistributed After-Tax Earnings depend on voluntary reserves and the dividends paid to shareholders.
4. RIC is a reserve that must be materialized in the following three years in various types of investments (Ley 19/1994).
5. DI cannot exceed the ceiling of the corporate taxable profit.

The conjunction of these conditions is perfect for applying optimization models, rather than using trial and error or the more sophisticated simulation procedures commonly used.

2.2 *The Mathematical Model*

This type of model can be used in budgeting or strategic planning. The model can begin with some pre-tax profit (for example a EBITDA base series for each tax year calculated before including the investment plan), and a pre-defined depreciation for previous and planned investments, could be included in a horizon of several years in a complete financial-tax planning model. Other data are the investment plan for each year that produces a profit after one year (r), corporate tax rate (t), tax deduction for investments (tdr(i)).

For simplicity, we develop the model without interest costs, and assume all the investments are working throughout the whole fiscal year, and that their effects are included in the pre-tax earnings for the year following the investment (of course the model can be extended by adding more periods (i.e. quarters or months) to improve precision).

To explain the model, let us assume a company established in the Canary Islands (Spain) is deciding its strategic plan for the coming years. It has determined the investments to be made each year, and projected the before-tax profits and the amortizations it will have in the following year. The calculations of these concepts can be included in our optimization model; this is not done here in order to simplify and address the core problem. The model can easily be extended to include debt and its cost, as will be shown below.

Making use of the tax advantages offered by the Canary Islands, the model calculates the optimal amount to be entered as the RIC for each year of the strategic plan. These will be the decision variables for the optimization model. The model includes the investment deductions, an incentive compatible with the previous one. The conditions for the RIC are that it cannot exceed 90% of the undistributed profit. Profits before the tax base series do not include the profitability of the cash for each year, which we will assume to be remunerated at an interest rate i.

The model can be modeled as a linear programming problem as explained below, and solved in any of the available systems. This paper presents one possibility combining Excel with What's Best. This solution allows a combined use of simulation/optimization, and tests different options as shown below.

Model TaxCan

Data:

For i = 1,…, N number of tax years planned

t	Corporate tax rate
r	Interest rate for cash
tdr (i)	Deduction from Investment Tax rate
INV (i)	Investment plan
D (i)	Depreciation
EBT (i)	Earnings before taxes (base)
DIV (i)	Dividends paid in tax year i

(optional parameter that can be imposed by shareholders, considered paid at the end of tax year i)

Variables

RIC (i) Reserve for Investment in Canary Islands

(year i in which RIC is accounted and allows tax deduction)

Dinv (i) Direct investment (not from RIC)

(investments this year that have not been "reserved" in previous years)

TMatRIC (j,i) Materialization of RIC

(the amount to be used as real investment from RIC (i) in tax year j must be only 3 years before)

Objective function:

$$\text{MAX} \sum_{i=1}^{n} \text{CF(i)}$$

Subject to:
(Calculations/constraints for each i, tax year)

$$\text{EFC (i)} = \text{r} * \text{Cash (i} - 1) \tag{1}$$

(Earnings from cash in fiscal year i)

$$\text{EBTC (i)} = \text{EBT (i)} + \text{EFC (i)} \tag{2}$$

(Earnings before taxes computed)

$$\text{INV\_Ded (i)} = \text{DINV (i)} + \text{TMatRIC (i)} \tag{3}$$

(Tax deductible investment)

$$\text{Td (i)} = \text{tdr (i)} * \text{INV\_Ded (i)} \tag{4}$$

(Tax deduction from investment)

$$\text{CCash (i)} = \text{CCash (i} - 1) + \text{EBTC (i)} + \text{D(i)} - \text{Ctax (i)} - \text{Inv (i)} \tag{5}$$

(Cumulative cash at the end of year i)

$$\text{TBD (i)} = \text{t} * [\text{EBTC (i)} - \text{RIC (i)}] \tag{6}$$

(Tax before deduction from investment. Note RIC (i) reduces the tax base)

$$\text{Ctax (i)} = \text{TBD (i)} - \text{TD (i)} \tag{7}$$

(Final Corporate Tax)

$$\text{EAT (i)} = \text{EBTC (i)} - \text{Ctax (i)} \tag{8}$$

(Earnings after taxes, must be positive)

$$\text{CF (i)} = \text{EAT (i)} + \text{D (i)} \tag{9}$$

(Cash flow computation, usually D depends on the investment plan, not considered in this version)

$$\text{INV C (i)} = \text{Dinv (i)} + \text{TMatRIC (i)} \tag{10}$$

(Investment computation: direct investment plus the materialization of previous years' RIC in year i)

$$\text{INV C (i)} = \text{INV (i)} \tag{11}$$

(Investment plan control equation: computed investment must be equal to investment plan)

$$\text{RIC (i)} = \text{SUM for j} = \text{i} + 1 \text{ to i} + 3 \text{ of MatRIC (i, j)} \tag{12}$$

(Entire amount of RIC(i) must be used in the following 3 years)

$$\text{TMatRIC (i)} = \text{SUM for j} = \text{i} - 3 \text{ to i} - 1 \text{ of MatRIC (j, i)} \tag{13}$$

(Actual investment in year i is computed from the previous years' RICs)

$$\text{NDE (i)} = \text{EAT (i)} - \text{Dividends (i)} \tag{14}$$

(Non-distributed earnings, not including legal reserves)

$$\text{RICLimit (i)} = 0.9 * \text{NDE (i)} \tag{15}$$

(RIC limit −90% of non-distributed earnings is computed)

$$\text{RIC(i)} <= \text{RICLimit(i)} \tag{16}$$

(Limit constraint formulation)

There are other objective functions that could be considered, such as CCash at last year and others.

2.3 *Implementation of the Optimization Model*

This type of model is solved in this case with the What's Best spreadsheet optimization (LINDO 2017) due to its ease for integrating in Excel, but the model can be

	A	B	D	E	F	G	H	I
1	**Model TaxCan**							
2			**2.018**	**2.019**	**2.020**	**2.021**	**2.022**	**2.023**
3								
4	**I. Data (k€)**							
5		Investment plan	600	2.305	1.204	2.712	666	561
6		Depreciation	1.346	1.639	1.870	1.990	2.261	2.328
7		Earnings before taxes (Base)	5.076	2.781	2.790	2.880	2.974	3.076
8		Corporate tax	25%	25%	25%	25%	25%	25%
9		Interest rate for cash	**2%**					
10		Deduction from Investment Tax rate	25%	25%	25%	25%	25%	25%

Fig. 1 Model data

	A	B	D	E	F	G	H	I
1	**Model TaxCan**							
2			**2.018**	**2.019**	**2.020**	**2.021**	**2.022**	**2.023**
3								
12	**II. Equations**							
13		Earnings from Cash		94	41	62	43	81
14		Earnings before taxes Computed	5.076	2.875	2.831	2.942	3.018	3.157
15		Tax deductible Investment	600	2.305	1.204	2.712	666	561
16								
17		Tax Deduction from Investment	150	576	301	678	166	140
18		Cumulative Cash	4.703	2.067	3.090	2.163	4.025	4.275
19								
20		*Corporate tax computation*						
21		Reserve for Investment in Canary Is.	0	0	0	0	0	0
22		Tax before deduction from Investment	1.269	719	708	735	754	789
23								
24		Final Corporate tax	1.119	143	407	58	588	649
25		Earnings after taxes	3.957	2.733	2.424	2.884	2.430	2.508
26		(Must be positive)	>=	>=	>=	>=	>=	>=

Fig. 2 Model equations

formulated in other commercial optimization software such as Frontline (2017) or any other. See Fourer (2015) for a recent survey.

The model can be formulated in an Excel spreadsheet as can be seen below (the underlying equations are explained above), in Figs. 1 (data) and 2 (equations).

Used as a simulator, the Excel model allows all the accounting of the investments to be allocated in the year in which they are going to be implemented: 600 k€ (2018), 2.305 k€ (2019) and so on. No dividends are paid. To aid visualization, only row 33 is shaded in Fig. 3.

In this base case, total cash flow is 28.370 k€ (cell J30).

For the model to give a better solution, we enter new variables related to RIC (shaded cells in rows 34, 36, 37…) in Fig. 4.

Cell J30 is the objective, and the optimal value is now 30.260 k€ when the model uses different levels of RIC every tax year. For example, in 2018 the company will post ("will reserve") 4.476 k€, 593 k€ in 2019 and so on.

(It should be noted that the shaded cells are the model variables.)

	A	B	D	E	F	G	H	I	J	K	L
1	**Model TaxCan**										
2			2.018	2.019	2.020	2.021	2.022	2.023	Total		
3											
27											
28	**III. Decisions**										
29											
30		Cash Flow	5.303	4.372	4.294	4.874	4.691	4.835	28.370		
31		Investment plan (control equation)	600	2.305	1.204	2.712	666	561	8.047	=	8.047
32		Must be equal to plan	=	=	=	=	=	=			
33		Direct investment (non from RIC)	600	2.305	1.204	2.712	666	561	8.047		
34		RIC accounting							0		
35		Materialization of RIC	0	0	0	0	0	0		Constraint	
36		RIC from year 2018 in year 2019, 2020 or 2021		0	0	0			0	=	0
37		RIC from year 2019 in year 2020, 2021 or 2022			0	0	0		0	=	0
38		RIC from year 2020 in year 2021, 2022 or 2023				0	0	0	0	=	0
39		RIC from year 2021 in year 2022 or 2023					0	0	0	=	0
40		RIC from year 2022 in year 2023						0	0	=	0
41											
42		Dividends	0	0	0	0	0	0	0	=>=	0

Fig. 3 Variables and objective function

	A	B	D	E	F	G	H	I	J	K	L
1	**Model TaxCan**										
2			2.018	2.019	2.020	2.021	2.022	2.023	Total		
3											
28	**III. Decisions**										
29											
30		Cash Flow	6.422	4.537	4.704	4.940	4.819	4.837	30.260		
31		Investment plan (control equation)	600	2.305	1.204	2.712	666	561	8.047	=	8.047
32		Must be equal to plan	=	=	=	=	=	=			
33		Direct investment (non from RIC)	600	0	0	0	0	0	600		
34		RIC accounting	4.476	593	1.630	238	509		7.447		
35		Materialization of RIC	0	2.305	1.204	2.712	666	561		Constraint	
36		RIC from year 2018 in year 2019, 2020 or 2021		2.305	611	1.560			4.476	=	4.476
37		RIC from year 2019 in year 2020, 2021 or 2022			593	0	0		593	=	593
38		RIC from year 2020 in year 2021, 2022 or 2023				1.152	427	52	1.630	=	1.630
39		RIC from year 2021 in year 2022 or 2023					238	0	238	=	238
40		RIC from year 2022 in year 2023						509	509	=	509

Fig. 4 Optimal solution without dividends

The 2018 RIC will be used for the real investments in year 2019 (2.305 k€), 2020 (611 k€), and 2021 (1.560 k€), but the 2019 RIC will be completely used in 2020 (593 k€).

The optimization model has improved the cumulative cash flow by nearly 2 M€.

Of course, this model can be used by imposing more conditions. For example, if the dividends request is changed, the model will recalculate the new optimal solution. Let us assume the shareholders claim 50% of the profit after taxes. The new solution is shown in Fig. 5.

The cash flow is worse: 29.402 k€, practically 1 M€ less than the optimum without dividends.

The question is: would be this be the best solution for the shareholders?

The model can be managed in various ways to obtain different solutions.

	A	B	D	E	F	G	H	I	J	K	L	M
1	**Model TaxCan**											
2			**2.018**	**2.019**	**2.020**	**2.021**	**2.022**	**2.023**	Total			
3												
28	**III. Decisions**											
29												
30		**Cash Flow**	**5.805**	**4.480**	**4.579**	**4.911**	**4.810**	**4.818**	**29.402**			
31		Investment plan (control equation)	600	2.305	1.204	2.712	666	561	8.047	=	8.047	
32		Must be equal to plan	=	=	=	=	=	=				
33		**Direct investment (non from RIC)**	**600**	**299**	**668**	**1.492**	**457**	**0**	3.516			
34		**RIC accounting**	2.007	536	1.219	209	561		4.531			
35		**Materialization of RIC**	0	2.007	536	1.219	209	561		Constraint		
36		**RIC from year 2018 in year 2019, 2020 or 2021**		2.007	0	0			2.007	=	2.007	
37		**RIC from year 2019 in year 2020, 2021 or 2022**			536	0	0		536	=	536	
38		**RIC from year 2020 in year 2021, 2022 or 2023**				1.219	0	0	1.219	=	1.219	
39		**RIC from year 2021 in year 2022 or 2023**					209	0	209	=	209	
40		**RIC from year 2022 in year 2023**						561	561	=	561	
41												
42		**Dividends**	2.229	1.420	1.355	1.460	1.274	1.245	8.984	>=	0	

Fig. 5 Optimal solution with dividends

3 Conclusion

Tax planning optimization models can improve the cash flow generated by corporations. This paper presents an example for the special tax system in Canary Islands (REF).

Acknowledgements I would like to thank to Ms. Prudence Brooke-Turner for her revision of the English manuscript.

References

Baker K (2011) Optimization modeling with spreadsheets. Wiley-Blackwell; 2nd Revised edn

Dinh H (2014) Optimization in finance: approaches for modeling and solving the multi-period loss offset problem in German income tax system. http://scidok.sulb.uni-saarland.de/volltexte/2015/6154/pdf/Dissertation_Dinh.pdf

Fourer R (2015) Linear programming: software survey. OR MS Today. INFORMS. 42:3

Frontline (2017). Solver for Excel. http://www.solver.com/excel-solver-help

GQTR (2017) Group on quantitative tax research. http://www.arqus.info/

Kallrath J, Wilson J (1997) Business Optimisation: Using Mathematical Programming. Palgrave

Klassen KJ, Lisowsky P, Mescall D (2017) Transfer pricing: strategies, practices, and tax minimization. Contemp Account Res 34:455–493. https://doi.org/10.1111/1911-3846.12239

Ley 30/1972. https://www.boe.es/boe/dias/1972/07/24/pdfs/A13287-13292.pdf

Ley 20/1991. https://www.boe.es/boe/dias/1991/06/08/pdfs/A18795-18820.pdf

Ley 19/1994. https://www.boe.es/boe/dias/1994/07/07/pdfs/A21719-21736.pdf

LINDO (2017) What's best. http://www.lindo.com

RDL 15/2014. https://www.boe.es/boe/dias/2014/12/20/pdfs/BOE-A-2014-13248.pdf

Schanz D, Schanz S (2011) Business taxation and financial decisions. Springer, Berlin

Lessons for Technological Innovation Analysis. A Case of Study Based on McLuhan Tetrad Applied to Laser Cleaning Machines

José Luis Roca-González, Germán Rodríguez-Bermúdez and Antonio Juan Briones-Peñalver

Abstract One of the main purposes for a Project Management course under the Industrial Engineering grade is to lead the students by early applying their multidisciplinary competences to a professional framework, in order to develop their active profile to encourage them to always look for the best available techniques to brings organizations up to date. This paper summarizes how the McLuhan Tetrad help the students to handle the laser cleaning technique as a case of study.

Keywords Laser cleaning machine · McLuhan's Tetrad · Innovation analysis

1 Introduction

One of the main features linked to accomplish innovation processes goals is to materialize technological innovations as a wide extended advance in order to update the industrial activity and therefore the industry efficiency. This way the Industrial Organizations are able to start life cycles of activity one after one, facing changes of competitive environment by applying the Best Available Techniques (BAT) at each time.

However there are several cases of study where already successfully tested innovations, under a commercialization phase, do not reach the expected implementation to revolutionise an specific industrial sector or an industrial process and so the inno-

J. L. Roca-González (✉) · G. Rodríguez-Bermúdez
Dpto. de Ingeniería y Técnicas Aplicadas, Centro Universitario de la Defensa en la Academia General del Aire, C/Coronel Lopez Peña sn, 30720 San Javier, Murcia, Spain
e-mail: jluis.roca@cud.upct.es

G. Rodríguez-Bermúdez
e-mail: german.rodriguez@cud.upct.es

A. J. Briones-Peñalver
Dpto. de Economía de la Empresa. Facultad de Ciencias de la Empresa, Universidad Politécnica de Cartagena, C/Real nº3, 30201 Cartagena, Colombia
e-mail: aj.briones@upct.es

Á. Ortiz et al. (eds.), *Engineering Digital Transformation*, Lecture Notes in Management and Industrial Engineering,
https://doi.org/10.1007/978-3-319-96005-0_33

vation main purpose fail to success. Perhaps the availability studies of Research and Development increases the financial requirements each time and more often to a higher level to assure its future success but most of the time there still being in addition some exogenous factors that are hard to take into consideration when an specific innovation may be able to be used for a certain industrial activity.

Therefore the aim of this paper is to share how some of those exogenous factors can be identify when a "Design Thinking Tool" such as McLuhan Tetrad is use while working in a R&D report. This chapter analyzes as an example for future lesson extrapolation a real Case of Study that summarizes the application of this tool. The interest of the case of study introduced in this paper was selected in order to make focus in R&D lessons that may help students to acquire a best understand of how the exogenous factors may affect future implementations when developing innovation analysis. Such lessons were extracted from experiences achieved by the authors by teaching Industrial Engineering and Project Management courses for the last year students of the grade in Industrial Engineering at the University Centre of Defence for the Spanish Air Force Academy.

This chapter follows two different phases/steps that were use while teaching lesson in relationship with the understanding of the theoretical framework that defines the R&D applied to real cases of study. The first step was set up with the common innovation analysis under a professional point of view, what requires a technological and economic full-understand of an innovation, defining specific technical capabilities, industrial uses, real cost of implementation, pricing ratios of performance among other aspects. The second step forced the students to applied what is call the McLuhan Tetrad as an important tool to develop the creativity of future engineers thanks to the "design thinking" processes linked to this tool.

2 The Case of Study: Cleaning Laser Machines (CLM)

Although laser technology applications have been widely known and studied in order to remove layers of rust, chemical particles, dust, paints an any other type of particles (Veiko et al. 2013) nowadays new possible applications still being developed like the archaeological heritage maintenance or graffiti removing from public buildings, mould surface cleaning, surface tolerance improvement, smooth end for coating surfaces, medical skin treatment like tattoo remove among others (Wang et al. 2017) what opens new ways to improve several industrial uses. This section shows the first step in innovation analysis through the technological and economic points of view applied to CLM as a case of study.

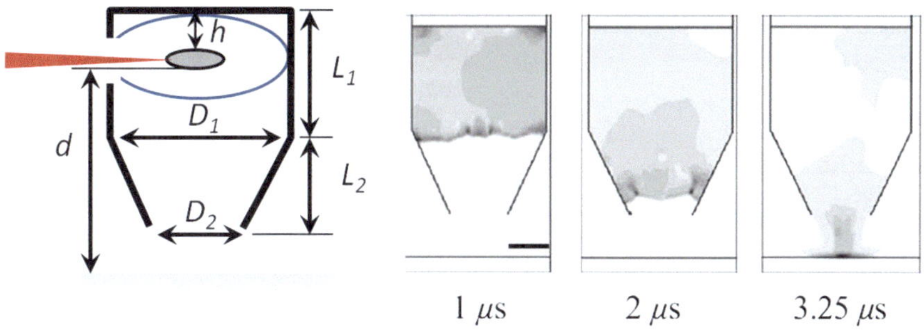

Fig. 1 Plasma confinement depending on convergent tubular typologies and its pressure distribution. *Source*: Jang et al. (2009)

2.1 Technological Analysis

The Laser technique is based on the physic phenomenon of directional discharge of plasma in a continuous flow state, created by several pulses of high energy laser what generates a spherical shock wave while the plasma is expanding (See Fig. 1) and makes that the laser action over the target surfaces to be directly proportional to the intensity and power of the mentioned shock wave, what finally helps to classify the industrial uses of this technique in relationship within the rated power of the equipment commercialized.

The effectiveness of the laser technique takes shape through two different ways that may happen at the same time, first because of the high energy impact over the target what volatilize any remains out of the external layer and after that because of the thermal expansion of the layer below, what helps to release any undesired layer out of the surface (AlShaer et al. 2014). So the versatility about the possible accuracy that can be reached by this technique has been a matter of interest for several case of study publications in where some technical information can be extracted to have a full view about the advantages of this technique. Among the most recent publications of high performance cases of study, there is the example of how this technique was used to clean metallic mirrors that were designed to be use at the International Thermonuclear Experimental Reactor (ITER), in where excellent results were obtained for laser pulses within a range of 1064 nm at 350 mj/cm^2 and 266 nm for 150 mj/cm^2 of power (Maffini et al. 2015) that secure that no damage could be done over the mirrors what was confirmed by an analysis with scanning electron microscope.

In relationship with metallurgical studies it is estimated that around the 30–50% of energy that is used in these industrial processes is lost by friction (Sasi et al. 2017) what reveals how useful the laser technique may be just because it removes the friction between the tool and the target and therefore the loss of energy involved (Uthaijunyawong et al. 2017). Other applications as an example, makes able low cost processes to recovery the polycarbonate pieces from old CDs/DVDs in order to

Table 1 Comparison of present cleaning surface techniques

Technique	Laser	Chemical	Mechanical	Dry ice	Ultrasound
Application	No contact	Surface contact	Abrasive contact	No contact	Surface contact
Superficial damage	None	Yes	Yes	None	None
Efficiency per pass	High just with 1–2 passes	Low. Needs several passes	Low. Bad uniformity	Middle in 2–3 passes	Middle
Consumable	Electric energy	Chemical detergent	Sand, balls, abrasive paper	Dry ice	Detergent and liquid
Results	Homogeneous High accuracy Easy to control	Irregular Low accuracy Hard to control	Irregular Middle accuracy No control	Homogeneous Low accuracy	Homogeneous
Emissions	Not contaminant	Chemical residues	Residual dust can be recovered	Not contaminant	Not contaminant
Investment	High. Low operation cost	Low. High operation costs	Middle. High operation costs	Middle. High cost of consumables	Low

Source Herolaser Ltd.

develop new processes that may take advantage of this technique in a way that was not possible to apply before (Leone et al. 2013).

The accuracy of this technique beside the need of preserving the integrity of surfaces below external layers, allows it to be applied for historic heritage conservation (Ersoy et al. 2014) better than any other methodology known up to nowadays in a similar that happens for plenty of industrial uses (See Table 1).

2.2 *Economic Analysis Summary*

Regarding the economic impact of the LCM uses, the price of purchasing equipment with this technology fluctuates from 50.000€ to 450.000 € for a power range between 20 W up to 1000 W and for other options like renting it can hired from 4.000€ to 7.000€/week for middle to high power units (200–1000 W) and with a capability of processing of about 20 m^2/h what can be traduced as a operating cost of 4€/h or 30c€/m^2 for low-medium powers and 5€/h for higher what may counteract the high price of acquisition when it is compared with the operating cost of other techniques such as mechanical applications (Careddu and Akkoyun 2016) that may cost about 70€/m2 when cleaning surfaces with sand. The profitability therefore depends on each case of study beside other aspects that may reject the economic point of view in order to accomplish with technical/process criteria. However these cost estimations may be use to provide a fast economic framework for a general purpose study.

3 McLuhan's Tetrad

The four questions that conform what is called "The McLuhan's Tetrad" in relationship with communications media where originally wrote to answer how new ways of communication could affect the Society and therefore to an incoming Information and Technology Society just by wondering when applying a new communication media these questions: *What could be left obsolete, What could be retrieved, What new ways could be extended or enhance and finally What negative consequences could exist* (McLuhan and Powers 1989). This Tetrad has often being updated to new frameworks (Hempell 1996) to provide a full view and understanding about any matter that required a self critic process before customer acceptance, most recently it has been applied to innovations development processes in order to analyse all circumstances or dimensions of influence attached to the innovation cycle (Sameshima 2006), so nowadays these questions have been reformulated by some authors as follow (Verstraete 2011) for this purpose:

- What does the innovation enhance? The purpose of this question is to answer to what capabilities can be developed by its implementation under an operative and practical point of view, trying to find out new solutions to the industrial needs.
- What does the innovation retrieves? Any innovation may carry out the recovery of old knowledge that could have been forgotten and thanks to a new development it could be of interest to redefine thanks to any new domain of use for the new innovation.
- What will make the innovation to become obsolete? As innovations shall provide competitive advantages for industrial processes it will make competitors to abandon some industrial processes or techniques to update their industries with the new developments.
- What negative aspect will show up? The action-reaction principle will probably materialize negatively in a direct or indirect way, creating new problems that may force to work out some possible solutions, to prevent these consequences.

The most interesting features of these questions show up when are used to develop design thinking processes by project management students (Steinbeck 2011) as it is defined in the design thinking theoretical background, these processes help students to Understand, Observe, Think up, Judge, Draft, Solve by designing and to Validate the work under a wide view of the problem analyzed (Hasso-Plattner-Institut 2009).

The students of Project Management at the University Centre of Defence at the Spanish Air Force Academy (grade in Industrial Engineering, class of 2017) applied these question to the case of study of Laser Cleaning Machines by summarizing the answers as follow:

1. LCM extend or enhance the capability of removing any contaminant out from any type of surface without damaging the external layer beyond the contaminants and saving the work time needed for that purpose.
2. The chance of using this technology with portable equipments retrieves the chance of repairing on site some surfaces that were rejected to treat and expected

to become worse until being replaced. Examples for these surfaces are the graffiti and rust remove treatment from cultural heritage and industrial facilities like high voltage towers, industrial piping among others and the maintenance processes at confined spaces like other cases in where the cost of dismantling and transport were not able to be assumed without applying this technique.
As a matter of fact the marketing of laser protection equipment is also retrieved but under a transportable set up in order to protect the working environment while applying the laser. Out of industrial purpose, the LCM can be used also for tattoo removal treatments what may become as a business opportunity for existing aesthetic clinic o new ones.
3. What may be left obsolete are some mechanical treatment over surfaces such as grit/shot blasting or chemical processes like picking or soda blasting that may be more expensive and what creates more toxic residues.
4. In return, these techniques may have a negative influence as it could be the lost of competitiveness of some industrial business when they do not adapt this technology to their processes what may lead to their end of activity in a last term. Also it may retrieve negatively in labour accidents due these laser equipments are easily to handle and use what may lead to unauthorized workers to manage inappropriately the machine what may caused burns most often on eyes, even permanent blindness, due to direct or indirect exposure to laser beam or its reflection.

After answering these questions, the student ended the "design thinking" process by redacting a short survey (See Table 2) addressed to maintenance head of department for three types of industries: Group one for heavy factories located at "Región de Murcia", Group two for middle and small business and Group 3 for some business belonging to service sector, cultural heritage maintenance among other groups of interest for this innovation. Finally the survey itself summarizes the full understanding about the innovation requirements that may help any head of department to decide its acquisition what accomplished the course goal in relationship this matter.

4 Conclusions

In Spain the grade of implementation of LCM has not reached an expected ratio in relationship with the technical profit that could be obtained for several industrial applications (See Fig. 2) what shows how the main manufacturers of this equipment shall increase their commercializing efforts to reach the full industrial market. The case of study analyzed for Spain reported that there are, among others possible equipment providers, at least three main business organizations that could cover all the territory of Spain, these are: "*Materias Primas Abrasivas SL (MPA)*" located at Barcelona being the only authorized partner of the German Company "*Cleanlasersysteme (Gmbh)*" to commercialise their patent at Spain, "*Tech Iberica SL*" at Madrid and "*Josdan Aplicaciones Laser*" at Burgos that offer the specialist support

Table 2 Survey proposed by students after the "Design Thinking Process"

In relationship with your maintenance labours, answer the next questions

1-Are most of surfaces maintenance operations applied on site?	□ Yes	□ No
2-Are the surfaces maintenance done with own resources?	□ Yes	□ No
3-Have the surfaces maintenance a high cost for your budget (more than 8.000€/week)?	□ Yes	□ No
4-Is the surface rust a constant problem at your facilities?	□ Yes	□ No
5-Do you apply any rust countermeasure or prevention technique?	□ Yes	□ No
6-Classify your industry in relationship with the maintenance ratio, as follow:		
□ Small: *It is limited to small pieces or parts which are easily removed when needed and whose surface area does not exceed 0.063 m²/unit (DIN A4)*		
□ Middle: It is applied to surfaces of area near to 0.6 m² *(1000 mm × 500 mm) or shafts with a length lower than 2 m*		
□ High: It is applied to surfaces of area higher than 0.6 m² *(>1000 mm × 500 mm) or shafts with a length higher than 2 m*		
7-Classify your industry in relationship with the maintenance complexity:		
□ Very easy: The surfaces ar very smooth and are very easy to move		
□ Normal: There are not smooth surfaces or they have plenty of curves. The surfaces can be accessible even they require any mechanical handling or any specific position for the operator		
□ Very difficult: Irregular surfaces, corners that are hardly to access, installation over 2 m of high, heavy parts that cannot be unassembled or that requires uncomfortable position for the operators		
8-Score under your experience the grade of utility of the next surfaces cleaning techniques (U: Unknown, 1 if It is of no use, 2 poor useful, 3 useful and 4 very useful)		

Laser	U	1	2	3	4
Chemical	U	1	2	3	4
Grit/shot blast	U	1	2	3	4
Dry ice	U	1	2	3	4
Ultrasonic	U	1	2	3	4

to successfully apply this technology. Although it may exist some other providers for this technology, they makes focus on the laser implementation for an specific application rather than a multi-sector purpose.

The high acquisition cost and its initial economic investment beside the rate of use for some industries suggest that better than buying this technology it could more effective to rent the equipment what requires to manage and plan the labours of surfaces treatment in order to group the highest number of surfaces to be treated within the hiring hours of work. Even some business may develop cooperative strategies of

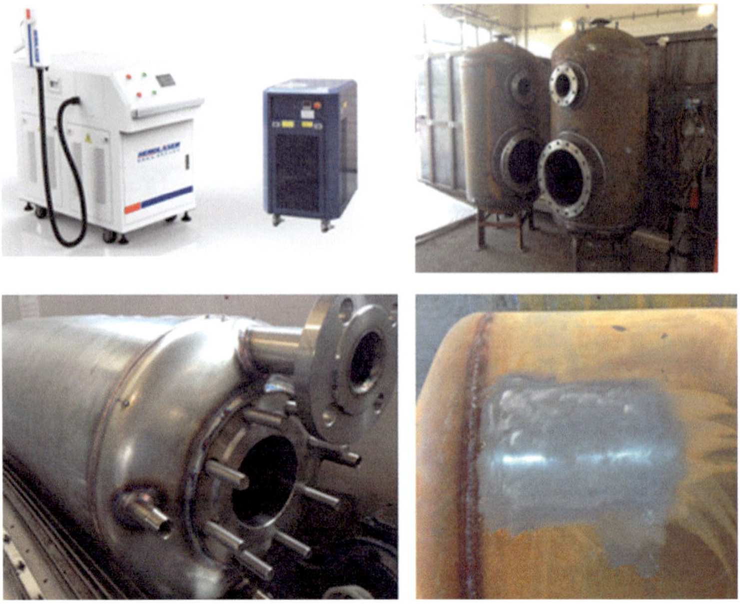

Fig. 2 Applications examples over mechanical tanks. *Source* Herolaser Ltd.

synergy by creating new "spin-off" business to decentralise the laser services what may offer as a reward higher flexibility in order to contract the services just when are needed.

Under the academic framework, the experience achieved shared how students increased their motivation to participate in real innovation analysis processes what accomplished the course main unit description and the academic grade expectations for future engineers what opened a new research line to apply future cases of study in relationship with R&D where innovation processes can take advantage of design thinking theory trough the McLuhan Tetrad.

References

AlShaer AW, Li L, Mistry A (2014) The effects of short pulse laser surface cleaning on porosity formation and reduction in laser welding of aluminium alloy for automotive component manufacture. Opt Laser Technol. 64:162–171. ISSN: 0030-3992. https://doi.org/10.1016/j.optlastec.2014.05.010

Careddu N, Akkoyun O (2016) An investigation on the efficiency of water-jet technology for graffiti cleaning. J Cult Herit 19:426–434. ISSN: 1296-2074. https://doi.org/10.1016/j.culher.2015.11.009

Ersoy T, Tunay T, U˘guryol M, Mavilii G, Akturk S (2014) Femtosecond laser cleaning of historical paper with sizing. J Cult Herit 15:258–265. ISNN: 1296-2074. https://doi.org/10.1016/j.culher.2013.07.002

Hasso-Plattner-Institut (2009) School of design thinking. Available: http://www.hpi.uni-potsdam.de/d_school/home.htm [2011]
Hempell A (1996) The resonating interval: exploring the process of the Tetrad (S. F. University, Ed.). No longer available at http://www.anthonyhempell.com must be requested at https://www.linkedin.com/in/ahempell/
Jang D, Oh JH, Lee JM, Kim D (2009) Enhanced efficiency of laser shock cleaning process by geometrical confinement of laser-induced plasma. J Appl Phys 106(1):014913
Leone C, Genna S, Caggiano A (2013) Compact disc laser cleaning for polycarbonate recovering. In: 2nd CIRP global web conference, pp 73–78
Maffini A, Ucello A, Dellasega D, Russo V, Perisinotto S, Passoni M (2015) Laser cleaning of diagnostic mirrors from tokamak-like carbon contaminants. J Nucl Mater 463:944–947. ISSN: 0022-3115. https://doi.org/10.1016/j.jnucmat.2014.10.016
McLuhan M, Powers BR (1989) The global village: transformations in world life and media in the 21st century. Oxford University Press, New York
Sameshima P (2006) Household at the shore: A Marshall McLuhan metaphor. J Can Assoc Curriculum Stud 4(1):51–58
Sasi R, Subbus K, Palani IA (2017) Performance of laser surface textured high speed steel cutting tool in machining of AL17075-T6 aerospace alloy. Surf Coat Technol 313:337–346. ISSN: 0257-8972. https://doi.org/10.1016/j.surfcoat.2017.01.118
Steinbeck R (2011) Building creative competence in globally distributed courses through design thinking. Comunicar XIX(37):27–34. ISSN: 1134-3478
Uthaijunyawong T, Siriwattanayotin S, Viriyarattanasak C, Tangwarodomnukun V (2017) Laser cleaning performance and PAHs formation in the removal of roasting marinade stain. Food Bioprod Process 102:81–89. ISSN: 0960-3085. https://doi.org/10.1016/j.fbp.2016.12.002
Veiko V, Samohvalov A, Ageev E (2013) Laser cleaning of engraved roll scoupled with spectroscopic control. Opt Laser Technol 54:170–175. ISSN: 0030-3992. https://doi.org/10.1016/j.optlastec.2013.05.015
Verstraete G (2011) The Tetrad: a dialectical model of media ecology theory. University of Alberta published at HASTAC, Humanities, Arts, Science and Technology Alliance and Collaborative. https://www.hastac.org/sites/default/files/documents/the_tetrad_dialectical_model_-_publication_copy.pdf
Wang X, Han P, Giovannini M, Ehmann K (2017) Modeling of machined depth in laser surface texturing of medical needles. Precis Eng 47:10–18. ISSN: 0141-6359. https://doi.org/10.1016/j.precisioneng.2016.06.012

Using a Serious Game for the Practical Learning of Productive and Environmental Efficiencies

Javier Santos García, Andrés Muñoz-Villamizar, Elisabeth Viles and Marta Ormazabal

Abstract The increase in environmental and social awareness has triggered numerous green initiatives at companies. Traditionally, companies' training related to environmental management focuses on technical aspects and theoretical concepts. In addition to these theoretical contents, it is necessary to carry out practical activities to impart the desired behaviors and skills. Serious games have been found to improve employees understanding of practical environmental sustainability challenges, by offering opportunities to obtain first-hand experiences that may otherwise be too costly and difficult to reproduce in reality. This work presents Pizz@green, a serious game which simulates the productive and environmental efficiencies generated by the improvements of processes in a pizza shop. The purpose of the game is the practical learning of the environmental impact caused by the production processes of a company.

Keywords Serious games · Sustainable development · Lean · Efficiency · OEE

J. Santos García · A. Muñoz-Villamizar (✉) · E. Viles · M. Ormazabal
Dpto. de Organización Industrial. Tecnun, Universidad de Navarra, Pº Manuel Lardizabal, 13, 20018 San Sebastián, Spain
e-mail: andres.munoz1@unisabana.edu.co; afmvillamizar@tecnun.es

J. Santos García
e-mail: jsantos@tecnun.es

E. Viles
e-mail: eviles@tecnun.es

M. Ormazabal
e-mail: mormazabal@tecnun.es

A. Muñoz-Villamizar
Escuela Internacional de Ciencias Económicas y Administrativas, Universidad de La Sabana, km. 7, Autopista Norte de Bogotá. Chía, Cundinamarca, Colombia

Á. Ortiz et al. (eds.), *Engineering Digital Transformation*, Lecture Notes in Management and Industrial Engineering,
https://doi.org/10.1007/978-3-319-96005-0_34

1 Introduction

Increasingly, companies are taking initiatives to mitigate the negative social and environmental impacts to address the emerging needs of society. The assumption that natural resources are infinite and that the regenerative capacity of the environment is able to compensate for all human action is no longer acceptable. Thus, organizations are becoming progressively more aware of their operations' impacts on people, the planet and profits (triple bottom line perspective), and are under increasing pressure to account for their resource consumption and environmental footprint (Alayón et al. 2017).

Jabbour (2015) concludes that managers must invest in environmental training if they want to improve their company's environmental management, and industry associations and universities may help companies plan and offer environmental training or improve their actual environmental training activities. Consequently, serious environmental management games have been found to improve understanding of practical environmental sustainability challenges by offering opportunities to obtain first-hand experiences that may otherwise be too costly, difficult or dangerous to reproduce in reality (Madani et al. 2017).

The objective of this paper is twofold. First, it aims to describe a conceptual framework for sustainability and environmental training, leading to a better understanding of the benefits of using serious games in this field. Second, it will present a specific serious game, called Pizz@green, which was developed as part of a larger research project.

This paper is organized as follows. In Sect. 2, we review the relevant literature from three angles: first, from the general window of sustainable production and consumption; second, from the window of employee training in companies; and third, from the window of using serious games as an effective learning/training tool. Section 3 sets out the proposed serious game, Pizz@green. Finally, Sect. 4 presents some concluding remarks and research perspectives.

2 Literature Review

2.1 Sustainable Production and Consumption

The term sustainability integrates social, environmental and economic responsibilities. Kleindorfer et al. (2005) use the term to include environmental management, supply chains and a broad perspective on triple bottom line thinking that integrates profit, people and the planet into corporate culture, strategy and operations. Sustainable development can be defined as "development that meets the needs of the present without compromising the ability of future generations to meet their own needs" (Brundtland 1987).

Therefore, the concept of sustainable consumption and production has emerged as an overarching objective and as an essential requirement for a better understanding of sustainable development (Ülkü and Hsuan 2017). Sustainable consumption and production is about promoting resource and energy efficiency, sustainable infrastructure, and providing access to basic services, green and decent jobs and a better quality of life for all. Its implementation helps to achieve overall development plans, reduce future economic, environmental and social costs, strengthen economic competitiveness and reduce poverty. Sustainable consumption and production aims at 'doing more and better with less' (UNSDG12 2016).

Thus, it is important that companies considerer both classical performance objectives and the sustainability requirements of the market. The concept of lean production, has been implemented worldwide to manage competitive businesses, could be considered the best way to create opportunities for developing resource efficient manufacturing systems (Andersson and Bellgran 2015). A well-known concept for improving production performance is Total Productive Maintenance (TPM). Developed by Nakajima (1988), it is one of the most used lean tools. TPM is a production-driven improvement methodology that is designed to optimize equipment reliability and ensure efficient management of plant assets through the use of employee involvement, linking manufacturing, maintenance and engineering (Ahuja and Khamba 2008). As defined by Nakajima (1988), Overall Equipment Effectiveness (OEE) is an important measure within the concept of TPM. The OEE measure is traditionally used by practitioners as an operational measure to monitor production performance, but it can also be used as an indicator for process improvement activities in a production context (Andersson and Bellgran 2015). The challenge is to correctly implement OEE and sustainable development at the same time. These tools, applied together, should considerably improve both production and environmental efficiencies.

2.2 Company-LED Training

Aligning sustainability goals with employee and market incentives can be difficult (Kleindorfer et al. 2005). As environmental considerations must be integrated into corporate culture and business planning at all levels of design, manufacturing, distribution and disposal (O'Brien 1999), employees need to take pride in their work and need to believe that their companies operate in a prudent and responsible manner and care about employee health and safety. Thus, there is a need for training built on the cultural and scientific bases that allow the acquisition of these new knowledge and skills, which are increasingly complex (Hernandez 2014). In-company training is a task of utmost importance for the future of the company itself.

Gimenez et al. (2012) concluded that environmental action programs (internal and external) have a positive impact on each dimension of the triple bottom line (environmental, social and economic performance). This means that the implementation of environmental action initiatives (e.g., using technological tools to evaluate the environmental impact generated by companies' processes) leads to improvements in environmental, social and economic performance.

In this context, the technological revolution has generated new challenges, but it has also provided new solutions. E-learning has emerged as a new training alternative that enables workers' ongoing training and qualification needs to be met properly. E-learning constitutes a flexible methodology that makes it easier to combine work and training (Batalla-Busquets and Martínez-Argüelles 2014). Companies could use e-learning methods (i.e.,) to obtain useful experiences, knowledge, and expertise to help sustainable development and to increase productive efficiency, customer value and satisfaction.

2.3 Serious Games

Batalla-Busquets and Martínez-Argüelles (2014) state that most of the teaching community accepts e-learning as a valid and efficient training strategy that provides a set of benefits for on-the-job training. Benefits include reducing the amount of time allocated for training, improving student support, allowing for personalized interaction and better follow-up on each student's participation and progression, increasing flexibility in learning by making training available whenever needed and allowing students to study at their own pace.

Game-based learning is a method that has been found to increase soft skills such as critical thinking, creative problem solving and teamwork; it also improves cognitive development, learning retention and social learning (Madani et al. 2017). A principal argument for using games in education and training is the engaging nature of gaming and the motivational power that games display: the ability of hooking and absorbing players in such a way that they can hardly stop playing (Westera 2017).

In this context, serious games are video games used in a professional environment to make training more exciting and immersive (Zyda 2005). There are numerous definitions of serious games. However, a common theme in most definitions is that serious games are not designed solely for entertainment purposes, but as a tool to educate, train and inform users. That is not to say that serious games are not entertaining or fun for players, only that entertainment is not the primary focus of such games (Madani et al. 2017). Serious games allow for safe experimentation in realistic environments, stimulate problem ownership by role adoption, and allow for learning-by-doing approaches, which support the acquisition of tacit and contextualized knowledge, such as sustainable development in companies.

3 Pizz@green

3.1 Background

Pizz@green is a serious game jointly developed by the University of Navarra and Ikasplay, a company dedicated to the development of educational games. In 2013, these two organizations collaborated on the development of another serious game: Jolaslean (see Fig. 1). Jolaslean is the result of a research project that studied the technical feasibility of a wireless sensor network to diagnosis the productive efficiency of industrial equipment. The goal of Jolaslean was to maximize total benefit and efficiency using lean thinking tools, such as OEE. It is being used in master's degree courses and in specialized courses for companies. Real user play showed the effectiveness of the game as a learning tool to enhance lean thinking.

In 2016, Jolaslean was adapted to handle environmental considerations. However, the results were not as expected. Users were not aware of improvements' environmental impacts: there were not specific scenarios to evaluate these environmental impacts and the improvements were a bit far-fetched to handle environmental considerations. In addition, some areas for improvement were identified (e.g., the development of an environment that was more familiar to users than car racing tracks). Hence the idea of design a new game focused on both environmental and productive efficiencies.

3.2 Game Description

Our game attempts to evaluate the productive and environmental effects of the actions carried out in a common production process, such as preparing pizzas. The game

Fig. 1 Jolaslean and Pizz@green main screens

simultaneously measures productivity and environmental efficiency, as generated by improvements made to the pizza production process. These improvements are chosen by the players as the game progresses.

Pizz@green is a single-player game, where the player must prepare pizzas according to a list of 3 different orders over the course of 3 rounds, each lasting 10 min (see Fig. 1). The order list will be updated every time a correct pizza is made. The player has 16 different ingredients (e.g., ham, bacon, beef, chicken, pepperoni, cheese, pineapple, mushroom, tomato, red pepper, green pepper, etc.), 8 of which are available in each round; there are also 2 types of dough (i.e., thin and tick) and 3 types of sauces (i.e., tomato, BBQ and creamy garlic). Random combination of all these components will generate the list of 3 different orders.

Pizzas are baked in an oven. The oven has problems in the variability and loss of heat, and pizzas are burned if they are not removed in time. As mentioned before, when any order is correctly made, a new one will immediately appear. If the pizza is not correct, it needs to be made again. Incorrect pizzas are thrown away, and the order remains active until it is correctly made. Finally, there is an economic penalty for trying to deliver an incorrect pizza.

The player is able to invest the money generated by making correct pizzas in some upgrades at the end of each of the 3 rounds. These improvements include better tools, such as a modern oven, a trash separator, biodegradable packaging, setup time reduction, etc. The game has a total of 27 upgrades. Some upgrades only affect production efficiency, some only affect environmental efficiency, and others affect both.

Additionally, Pizz@green has 3 different mini-games that are activated during various stages of the game. First, after every 20 correctly made pizzas, ingredients are changed via a mini-game in the product store (see Fig. 2). Second, pizzas will occasionally be dirty when they come out of the oven, and the oven must be cleaned via a cleaning mini-game. This game can be avoided by doing preventive cleaning. Finally, a third mini-game comes up every time that 3 pizzas are burnt. In this case, the oven thermostat is damaged and it must be changed using a third mini-game.

At the end of each round, total profit and production and environmental efficiencies are calculated. Incomes are computed according to the number of correctly made pizzas, while outcomes are calculated according to ingredients used (including those wasted and/or burned) and selected upgrades. As mentioned before, production efficiency is computed according to OEE, which measures equipment efficiency with 3 indicators: availability (the equipment is operational when needed), performance (an established production speed is maintained) and quality (only correct products are made) (Nakajima 1988). Environmental efficiency is computed using materials consumed (i.e., ingredients, water, energy) and residues generated (i.e., wasted ingredients, CO_2 emissions, wastewater, packing waste).

The user's performance data is stored as a telemetric of the pizza shop. The user sees this report at the end of the game (see Fig. 2). This allows users to understand that continuously monitoring equipment facilitates decision-making processes. Besides evaluating the consequences, in general performance, of the decisions they have made by analyzing the evolution of all results.

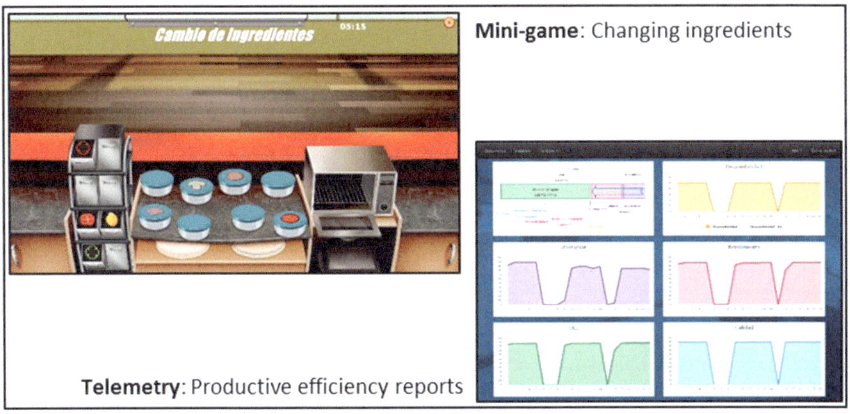

Fig. 2 Pizz@green screenshots

Pizz@green can be combined with a theoretical presentation of concepts and numerical problems to show the effect of environmental and production improvement tools. In the initial playing session, participants use only their intuition and/or their previous knowledge. Then several games sessions with ad hoc scenarios (which can be easily prepared) can be added to separately improve OEE and sustainability indicators. If necessary, a final session can be run in which the performance of all players is evaluated and the best performance is rewarded. It is expected that results of the final session will be much better than the first session's results, showing the training to have been effective. What players learn in the game can be easily transferred to their real work. Players can start monitoring their production (real equipment) and improve their performance using the same indicators and improvement tools presented in the game.

4 Conclusions and Perspectives

The aim of this study was to describe a conceptual framework for sustainability and environmental training for a better understanding of the benefits of using serious games in this field. A new serious game, Pizz@green, was proposed and described as an example of applying practical learning tools to help workers acquire behaviors that are appropriate for managing and improving environmental and production aspects in their jobs.

The main argument for the proposal is that training related to environmental management traditionally focuses on theoretical concepts rather than to practical experiences. Thus, serious games are relevant not only for evaluating the improvements of processes but also environmental performance and its evolution within companies.

Further research should quantify the impact of using serious games in training. Statistical methods can be applied to evaluate the performance of two groups of workers (i.e., one that uses serious games and one that does not), and thus determine the effectiveness of the proposed method.

Acknowledgements This project has been funded by the Basque Government, UE2015-15.

References

Ahuja IPS, Khamba JS (2008) Assessment of contributions of successful TPM initiatives towards competitive manufacturing. J Qual Maintenance Eng 14(4):356–374

Alayón C, Säfsten K, Johansson G (2017) Conceptual sustainable production principles in practice: do they reflect what companies do? J Clean Prod 141:693–701

Andersson C, Bellgran M (2015) On the complexity of using performance measures: enhancing sustained production improvement capability by combining OEE and productivity. J Manufact Syst 35:144–154

Batalla-Busquets JM, Martínez-Argüelles MJ (2014) Determining factors in online training in companies. Int J Manufact Educ 12(2):68–79

Brundtland GH (1987) World commission on environment and development: our common future. Report of the world commission on environment and development. Oxford University

Gimenez C, Sierra V, Rodon J (2012) Sustainable operations: their impact on the triple bottom line. Int J Prod Econ 140(1):149–159

Hernandez R (2014) Training in companies, permanent training and adult learning: an epistemological approach. Procedia - Soc and Behav Sci 139: 434–440

Jabbour CJC (2015) Environmental training and environmental management maturity of Brazilian companies with ISO14001: empirical evidence. J Clean Prod 96:331–338

Kleindorfer PR, Singhal K, VanWassenhove L (2005) Sustainable operations management. Prod Oper Manag 14(4):482–492

Madani K, Pierce TW, Mirchi A (2017) Serious games on environmental management. Sustain Cities Soc 29:1–11

Nakajima S (1988) Introduction to total productive maintenance TPM. Productivity Press, Cambridge, MA

O'Brien C (1999) Sustainable production - a new paradigm for a new millennium. Int J Prod Econ 60:1–7

Ülkü MA, Hsuan J (2017) Towards sustainable consumption and production: competitive pricing of modular products for green consumers. J Clean Prod 142:4230–4242

UNSDG12 (2016) (United Nations Sustainable Development Goals #12) Goal 12: Ensure sustainable consumption and production patterns. Available at: http://www.un.org/sustainabledevelopment/sustainable-consumption-production. Accessed: 24 Jan 2017

Westera W (2017) How people learn while playing serious games: a computational modelling approach. J Comput Sci 18:32–45

Zyda M (2005) From visual simulation to virtual reality to games. Computer 38(9):25–32

Training Methodology for the Design of Robust Processes Based on Design of Experiments. Case Study, Launcher

Gorka Unzueta Aranguren and Jose Alberto Eguren Egiguren

Abstract Rapid and precise modelling to design robust industrial processes is a key element in the competitiveness of companies. This paper describes Methodology of design and modelling of robust processes based on Design of Experiments, and applies it to a case. The concrete case in which the methodology has been applied has been used to train people from industrial environments in the use of tools of experimentation, analysis and improvement. Through the application of the methodology, is obtained information about the significant variables that affect to the response and the model that governs the process. The objective of the methodology is to facilitate the use of statistic applied to the Design of Experiments in the industry, reducing the number of experiments required and facilitating their analysis.

Keywords DoE methodology · Experimental design · Fractional factorial · Robust processes · Process modelling · Training

1 Introduction

Today's manufacturing processes are caught between growing quality needs and reduced variability. In order to meet these demands, manufacturing companies are committed to integrated process monitoring and control, in order to develop flexible processes that can be quickly and efficiently adapted to the variability of the different factors that influence the Manufacturing of products (Rao 2011). Therefore, the use of knowledge-based tools to manage the collected information quickly and efficiently and design robust and adaptable manufacturing processes is a must for survival in the current industrial environment. A methodology that enables the design and

G. Unzueta Aranguren (✉) · J. A. Eguren Egiguren
Dpto. de Organización Industrial, Mondragon Unibertsitatea, C/Loramendi 4, 20500 Arrasate-Mondragon, Spain
e-mail: gunzueta@mondragon.edu

J. A. Eguren Egiguren
e-mail: jaeguren@mondragon.edu

Á. Ortiz et al. (eds.), *Engineering Digital Transformation*, Lecture Notes in Management and Industrial Engineering,
https://doi.org/10.1007/978-3-319-96005-0_35

modelling of robust processes is based on the Design of Experiments (DoE). But the application of this type of tools in industrial companies is limited to the knowledge and training of the people of the organization, being a reality that their use is limited (Tanco et al. 2008). Found that training in these tools is a key element for increasing the competitiveness of industrial organizations (Eguren Egiguren 2012). The DoE is a tool that systematically applies the statistics to the experimentation allowing to develop a mathematical model that predicts how the input variables interact to create output variables or responses of a process (Lye and Science 2005). Through the use of DoE it is possible to: Learn about the researched process; Detect significant factors; Determine how factors interact; Construct a mathematical model for prediction; And Optimize the response.

1.1 Modelling Using Design of Experiment

The Design of Experiments (DoE) emerged in the twentieth century from Sir Ronald Fisher (Box et al. 2002; Fisher 1936). Prior to DoE, the one-factor-for-time (OFAT) strategy was used. According to Montgomery (Montgomery 2008) there have been four stages in the development of DoE.

1. Application of factorial designs and analysis of variance (ANOVA) (Box et al. 2002; Fisher 1936).
2. Development of Response Surface Methodology (RSM) (Box and Wilson 1951).
3. Methodologies developed by Taguchi and Shainin in the 1980s (Taguchi 1987).
4. Six Sigma Methodology and the democratization of statistics due to the massive use of statistical software (Astakhov 2012).

1.2 Classification of Experimental Designs

The most important experimental design techniques can be classified as follows:

- Full Factorial Designs: Full factorial designs include all possible combinations (n^k), taking into account the levels (n) of each factor (k). Main objective: definition of factor significance; and the development of a process model (the fit of the resulting model is limited to the effects of the factors being linear).
- Fractional Factorial Designs: As the number of factors (k) increases, exponentially increases the number of experiments to be performed. To reduce the number of experiments it is possible to select a specific fraction of the complete orthogonal experimental plan, a half, a quarter, etc. Main objective: definition of factor significance; And the development of a model of the process (the fit of the model is more limited than in the full factorial design).

- Plackett-Burman designs: Orthogonal designs at two levels that can be used to study up to k factors, in N experiments, where K = N − 1 and N multiple of 4. Main objective: Sieving, defining the most significant factors.
- Response Surface Design: Once the significant factors are known, RSM examines the relationship between the factors and the response. Main objective: Modeling of nonlinear response (quadratic model), and optimization of response.
- Taguchi Designs: Orthogonal arrangement to design experiments with only a fraction of the complete factorial combinations. Main objective: reduction of process variability; Identify controllable factors and their level to minimize the effect of noise factors on the response.

Usually, different experimental designs are used sequentially to achieve the proposed objectives; Optimization of variables, development of the process model in the performance range, definition of the significance of the variables or reduction of process variability.

2 Modelling Methodology

In order to model the process, the methodology developed by MU (Eguren Egiguren 2012) was divided into 8 phases, of which the first 6 were carried out until the model was defined and validated. The phases described in Fig. 1 are executed sequentially.

CONTENIDO DEL DOE		FASES DEL DOE
FASE 1: Definir	• Equipo de trabajo. • La oportunidad de mejora. • Recopilar información • Elegir la respuesta • Objetivos • Planificación	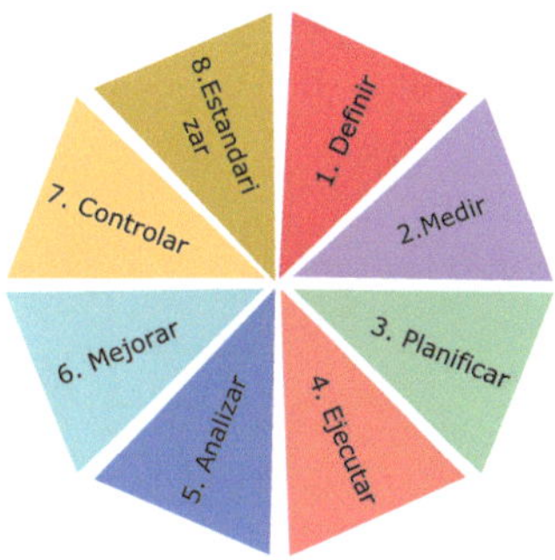
FASE 2: Medir	• Identificar las variables • Clasificar las variables • Validar el sistema de medida • Determinar la variabilidad del proceso y el nº de replicas • Clasificar las variables	
FASE 3: Planificar	• Caracterizar factores • Identificar las características del diseño. • Seleccionar diseño	
FASE 4: Ejecutar	• Preparar experimentos • Realizar experimentos y recopilar datos	
FASE 5: Analizar	• Calcular efectos • Identificar efectos significativos • Construir modelo de predicción • Optimizar • Necesidades de nuevos experimentos	
FASE 6: Mejorar	• Experimentos confirmatorios • Conclusiones • Ejecutar mejoras	
FASE 7: Controlar	• Definir e implantar controles • Validar resultado	
FASE 8: Estandarizar	• Diseñar procedimiento • Formar a los implicados • Planificar expansión • Lecciones aprendidas	

Fig. 1 Modelling methodology

3 Application of the Case

The application of the methodology in this particular case serves as training of people from industrial environments, being a first step for its later application in more complex processes. Through the use of very simple elements and accessible to everyone as are the wooden clips used for the fastening of clothes, a launcher has been built. The aim of experimentation is to identify the influence of the key factors for the purpose to get the maximum throw distance (Fig. 2a).

3.1 Description of the Element to Model

A launcher made of wooden clips will launch an element under certain conditions, and from a constant height (750 mm). The construction of the launcher is performed following the steps defined in Fig. 2b. The process response is the distance traveled to the ground, and the goal is to increase the launch distance.

3.2 Development of Experimentation

The experimental development was carried out through the application of the methodology developed by MU (Eguren Egiguren 2012).

3.2.1 Phase 1: Definition

The objective of experimental development has been to increase launch distance and modelling. In order to know the initial situation, 25 launches have been made,

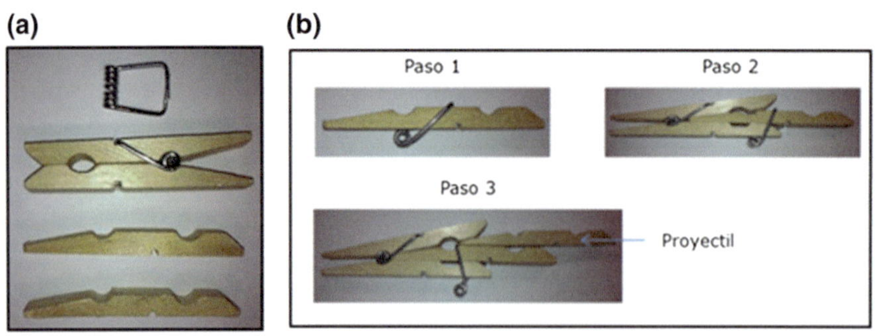

Fig. 2 **a** Elements for the construction of the launcher; **b** steps for mounting the launcher and the projectile

Table 1 Process variables

No	Variable	Type[a]	Clasificatión[b]	Strategy[c]	Expected effect	Range
1	Wear of the elements		Noise	F	–	
2	Load length	D	Controllable	C	+	15–20 mm
3	Quality-life spring		Noise	F	–	
4	Friction	Co	Controllable	C	+	2–4 mm
5	Position of the launcher	D	Noise	F	=	
6	Throw angle	D	Controllable	C	+	0–30°
7	Position of the projectile	D	Controllable	C	=	Up-down

[a]Type: variable continua (Co), variable categórica-Discreta (D)
[b]Variable controllable; variable NOT controllable (noise)
[c]Strategy followed by the variable [Fixed level (F), Controllable level (C)]

obtaining with the initial configuration (Fig. 2b) a mean of 220 mm with a standard deviation of 16.07 mm.

3.2.2 Phase 2: Measurement

After the analysis of the process in its initial conditions, the process variables described in Table 1 have been identified. The strategy followed to manage the noise variables was as follows: variable N°1, wear of the elements of the launcher, has been considered null by the non-significant number of launches made. Variable N°3, the impact has been reduced by changing the spring every 4 throws. Variable N°5, is kept fixed defining a launch procedure. The chosen variables are; Load length, Friction, Throw angle, and Position of the projectile (Fig. 3). With the help of the Minitab software16, taking into account the standard deviation ($\sigma = 16.07$ mm) obtained in the response with the initial conditions (Fig. 2b), and taking as a significant the increase in the response on a 20% (44 in absolute value), with a significance level of 5% and a power of 85%, the number of replicas has been defined in 2.

3.2.3 Phase 3–4: Planning and Execution of the Experiments

In the third phase the experimentation has been planned, the experimentation strategy selected is aimed at reducing the number of experiments required. The aim of experimentation is to increase the response, to identify the appropriate levels of the variables to achieve it, and to model the process. Four controllable variables have been identified (Table 2). It has been assumed that the answer is linear. The Fractional

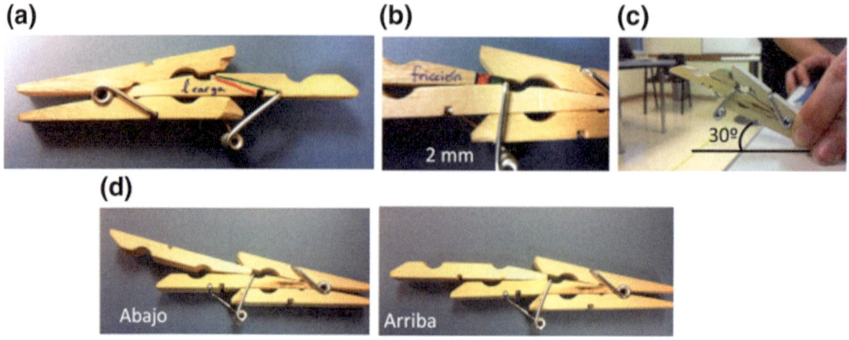

Fig. 3 Controllable variables; **a** load length 20 mm; **b** friction, 2 mm, **c** throw angle, 30°; **d** position of the projectile

Table 2 Controllable variables, experimental factors

No	Variable	Lower level	Upper level
2	Load length	15	20
4	Friction	2	4
6	Throw angle	0°	30°
7	Position of the projectile	Up	Down

Factorial (2^{4-1}) experimental design has been selected, with a resolution IV and two replicates, a total of 16 experiments. Experiments have been randomized to reduce the effect of factors that are not included in the study.

Table 3 summarizes the experimental results (In the table the experiments are not randomized to facilitate their understanding).

3.2.4 Phase 5: Analysis of the Results of the Experimentation

The analysis of the results was done using the software Minitab16. The analysis includes; Normal plot of effects, ANOVA variance analysis, model fit analysis, and effect and interaction graphs. The analysis of variance (ANOVA) (Table 4) shows that the significant effects are A, C and B being its value $p < 0.05$ (level of significance of 5%).

In Fig. 4 the significant effects identified in Table 4 can be graphically observed.

The analysis of the residuals for the response shows that they are normally distributed (Fig. 5a), indicating that the normality assumption is met, and that they are randomly scattered around zero (Fig. 5b).

Once validated the experimentation, the linear model that represents the process has defined. The model generated in Eq. (1) represents the behaviour of the process for the response with a setting of 93.28% (R^2). It is a linear model, the upper level

Table 3 Matrix of experimental design (2^{4-1}, 2 replicas) and results

Orden stand	Load length	Friction	Throw angle	Position of the projectile	Response distance
1	15	2	0	Up	240
2	20	2	0	Down	350
3	15	4	0	Down	280
4	20	4	0	Up	330
5	15	2	30	Down	300
6	20	2	30	Up	440
7	15	4	30	Up	300
8	20	4	30	Down	500
9	15	2	0	Up	230
10	20	2	0	Down	355
11	15	4	0	Down	300
12	20	4	0	Up	380
13	15	2	30	Down	350
14	20	2	30	Up	460
15	15	4	30	Up	390
16	20	4	30	Down	550

Table 4 Analysis of variance (ANOVA)

	Efect	Coef	Error estandar coef.	Valor T	Valor p	
Constante		359.69	8.03	44.77	0	Significant
A—Load length	121.87	60.94	8.03	7.58	0	Significant
B—Friction	38.13	19.06	8.03	2.37	0.045	Significant
C—Angle	103.13	51.56	8.03	6.42	0	Significant
D—Pos. of the projectile	26.88	13.44	8.03	1.67	0.133	
A*B	0.62	0.31	8.03	0.04	0.97	
A*C	30.62	15.31	8.03	1.91	0.093	
A*D	9.37	4.69	8.03	0.58	0.576	

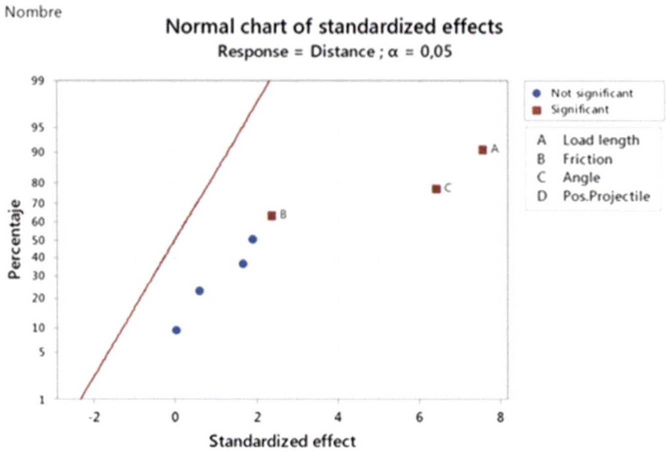

Fig. 4 Normal graph of standardized effects, the response is distance

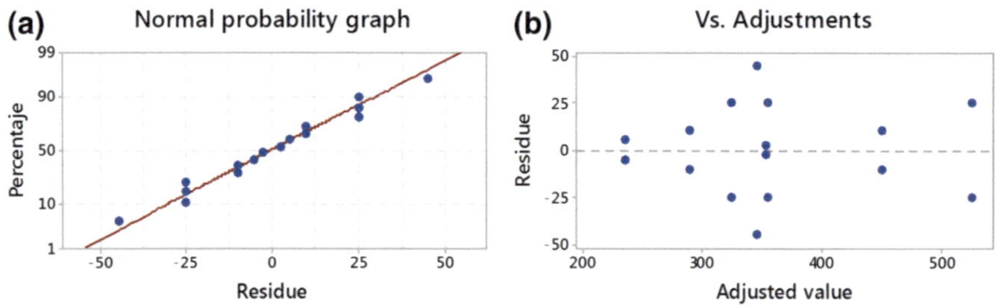

Fig. 5 **a** Normal probability of residuals for "distance" response; **b** residuals versus adjusted values

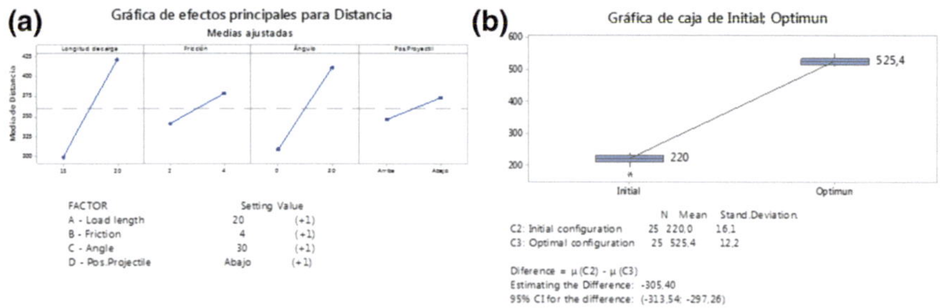

Fig. 6 **a** Effect graph, optimum setting value for increasing launch distance; **b** hypothesis test, initial configuration $\mu = 220$; optimal configuration $\mu = 525.4$ Conclusions

is coded at +1, and the lower level is -1. The value predicted by the model with the optimum configuration (Fig. 6a) is 525 mm. Subsequent experiments (Fig. 6b), experimentally demonstrate the good fit of the model.

$$C9 = 359.69 + 60.94A + 19.06B + 51.56C + 13.44D + 0.31A * B + 15.31A * C + 4.69A * D \quad (1)$$

3.2.5 Phase 6: Improvement, Confirmation Experiments

To confirm the improvement, 25 launches have been made with the optimal configuration [A (+1), B (+1), C (+1), D (+1)] indicated in Fig. 6a, obtaining the average of 525.4 mm. A hypothesis test was performed (Fig. 6b) comparing the results of the initial configuration with those of the optimal configuration. The result of the test shows with a 95% confidence that the improvement is significant.

4 Conclusions

The application of the methodology to the case has served to train the participants in the use of the statistical tools necessary to identify the significant variables and model the behavior of the process. The tools of improvement and experimentation applied in the case have been DoE, fractional factorial design, ANOVA, residue analysis, analysis of effects graphs, and hypothesis test, in addition to classic tools like brainstorming, flowcharts, Ishikawa, etc. The methodology has been validated through the results obtained in the case. The adjustment (93.28%) of the defined model and the improvement achieved in the response of the 138.8% (from 220 to 525.4 mm), has been empirically justified with a 95% confidence level. The knowledge acquired by the participants is valid for their application in complex industrial processes.

References

Astakhov VP (2012) Design of experiment methods in manufacturing: basics and practical applications. In: Statistical and computational techniques in manufacturing, pp 1–54

Box GEP, Wilson KB (1951) On the experimental attainment of optimum conditions. J Roy Stat Soc 13(1):1–45

Box GE, William G, Hunter JS (2002) Estadística para investigadores: introducción al diseño de experimentos, análisis de datos y construcción de modelos/por George EP Box, William G. Hunter y J. Stuart Hunter

Eguren Egiguren JA (2012) Desarrollo de un modelo para abordar proyectos de mejora continua de procesos productivos de forma eficaz y eficiente

Fisher RA (1936) The design of experiments

Lye LM, Science A (2005) Tools and toys for teaching design of experiments, pp 1–9

Montgomery DC (2008) Design and analysis of experiments. Wiley, Hoboken

Rao R (2011) Advanced modeling and optimization of manufacturing processes: international research and development

Taguchi G (1987) System of experimental design; engineering methods to optimize quality and minimize costs
Tanco M et al (2008) Is design of experiments really used? A survey of Basque industries. J Eng Des 19(5):447–460

Formative Assessment for the Analysis and Problem Solving Transversal Competence in Master Degree

María-José Verdecho, P. Gómez-Gasquet, Raúl Rodríguez-Rodríguez and Juan-José Alfaro-Saiz

Abstract The Universitat Politècnica de València has introduced thirteen transversal competences in the curricula of undergraduate and graduate degrees. One of the transversal competences is "Analysis and Problem Solving". This competence had been assessed non-uniformly in the Master Degree in Advanced Engineering Production, Logistics and Supply Chain due to the fact that the assessment lacked a comprehensive and accessible tool to the students that exposed the indicators and levels of achievement of the competence. In addition, it was important to design a tool where all the instructors agreed on the main features so that the tool could be used in both semesters by instructors aiding to support the formative assessment of the competence. This paper presents a methodology and the design of a tool to perform formative assessment of analysis and problem solving competence. It had been implemented in three courses (three pilot applications) and five learning activities to support the formative assessment as well as to update the tool with the feedback collected by each pilot application. The analysis of the pilot applications show the positive results obtained and the recommendation of students for its implementation.

Keywords Transversal competence · Higher education · Analysis and problem solving · Master · Assessment

1 Introduction

The new study plans are structured by a set of generic (or transversal) and specific competences. Both competences should be acquired by students during the degree and, therefore, they should be assessed as they are requirements to obtain the degree. González and Wagenaar (2003) indicate that specific competences are those skills related to the specific domain of knowledge while transversal compe-

M.-J. Verdecho (✉) · P. Gómez-Gasquet · R. Rodríguez-Rodríguez · J.-J. Alfaro-Saiz
Research Centre on Production Management and Engineering (CIGIP), Dep. Business Organisation, Universitat Politècnica de València, C/Camino de Vera S/N, 46022 Valencia, Spain
e-mail: mverdecho@cigip.upv.es

Á. Ortiz et al. (eds.), *Engineering Digital Transformation*, Lecture Notes in Management and Industrial Engineering,
https://doi.org/10.1007/978-3-319-96005-0_36

tences are those skills related to personal development that do not depend on a thematic or specific scope but they appear in all the domains of the professional and academic activity. Transversal competences include, for example, the capability to learn, the capability to communicate, the capability of decision-making, etc. In a changing environment where demands are shifting continuously, these transversal competences are highly important. Specifically, the Universitat Politècnica de València (UPV) defined and introduced thirteen transversal competences within all the curricula of the undergraduate and graduate programs (UPV 2014). Some of the UPV transversal competences are: Analysis and problem solving; Innovation, creativity and entrepreneurship; Effective communication; and Critical thinking.

In 2012, after the completion of the ABET (Accreditation Board for Engineering and Technology) accreditation process of four degrees at the UPV, it was questioned that UPV had defined and implemented systematic procedures to assess the degree of achievement of transversal competences. From this accreditation process, it was concluded that it was necessary to define a general procedure to evaluate the progress and certify the acquisition of transversal competences by students. In order to solve this issue, the UPV has developed a project called "UPV transversal competences" for guiding the general implementation of the transversal competences in the different degrees. Similar approaches to solve this issue had been developed by instructors and Universities such as the works developed by Villa and Poblete (2007), Rodríguez Gómez (2009) and Ibarra Sáiz (2010).

On the other hand, Scriven (1967) introduced the concept of "formative evaluation" and Bloom (1968) incorporated it to student learning. After that, Bloom developed in depth the concept of "formative assessment" in Bloom et al. (1971) and Bloom (1976). There are two types of complementary assessments: summative and formative assessment. Traditional assessment usually relies on summative or final assessment with the main purpose of providing a qualification at the end of a period of time (quarter, semester, etc.) (Morales 2000). However, when the focus is to improve the learning process, the final and summative activities should be complemented with formative and continuous assessment activities (Bloom 1976). The continuous assessment allows providing feedback to students about their achievements and difficulties and motivates them to the achievement of the targets (de Miguel Díaz 2006). In addition, formative assessment provides feedback to the instructor allowing to modify (if necessary) the learning-teaching process (Huhta 2010).

The assessment of the transversal competences in various courses of the Master Degree in Advanced Engineering Production, Logistics and Supply Chain at the Higher Technical School of Industrial Engineering has been performed non-uniformly due to the fact that the transversal competences are evaluated without following a comprehensive tool that exposes to the students the indicators and levels of achievement of the competence. In addition, it was important to design a tool where all the instructors agreed on the main features so that the tool could be used in both semesters by instructors aiding to support the formative assessment of the competence. The idea was to design a methodology and a tool to be used to assess the transversal competence in one course belonging to the first semester and collect by a survey the opinions of students as well as the opinion of the instructors regarding the tool. With this feedback, the tool (if necessary) will be modified to be used in the assessment of the same competence in two other courses belonging to the second semester.

This work emerges from an educational innovation project that aims to develop assessment tools and procedures for the transversal competence "Analysis and Problem Solving" in order to collect evidences of the degree of development of this competence in different moments of the master studies.

This paper is structured as follows. First, the methodology of the research project is showed. Second, the definition of the learning outcome of the competence is presented. Third, the indicators to assess the competence are defined. Fourth, the learning activities are defined. Fifth, the assessment procedures are presented. Sixth, the initial assessment tool (instrument) is introduced. Seventh, the pilot applications are detailed. Finally, the conclusions of the paper are exposed.

2 Methodology

The methodology used in the educational innovation project comprises six phases (Fig. 1):

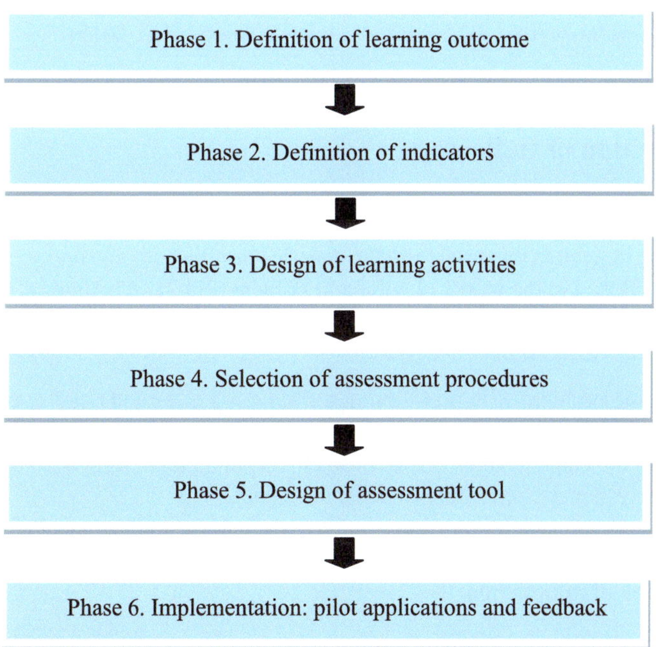

Fig. 1 Research project phases

3 Definition of Learning Outcome

Villa and Poblete (2007) define Analysis and Problem Solving competence as "*identify, analyse and define the significant elements that constitute a problem in order to solve it with criterion and in an effective manner*". This competence aims to improve the confidence of the student in its own thinking; improve the skills and capabilities to learn, understand and apply knowledge and; favour the achievement of a high degree of individual autonomy that allows him/her to continue the learning process. The problems are new situations that demand that people answer with new behaviours. In order to solve a problem, several tasks are to be done demanding more or less difficult reasoning processes.

As a result of the "UPV transversal competences" project, different learning outcomes for every transversal competence depending on the year of the studies were defined. Three levels were established. The first level comprises the first and second year of the degree. The second level comprises the third and fourth year of the degree. Finally, the third level corresponds to master/graduate studies. Thus, the three levels correspond to three levels of increasing demand on the transversal competence as the degree of requirement and development of the generic competence has to increase when moving to a higher level of study. Then, there are three levels of domain for every transversal competence depending on the year of the studies. The learning outcome for the Analysis and Problem Solving competence at master level (level III) is "to *solve problems in individual and/or team manner, in different contexts and different complexity from different approaches*" (Atarés et al. 2015).

4 Definition of Indicators

In order to measure the achievement of the learning outcome, it is necessary to define measurable indicators that transform the generic learning outcome into specific items to achieve. Based in the works of Villa and Poblete (2007) and Atarés et al. (2015), we proposed the following five indicators for assessing the Analysis and Problem Solving competence at master degree:

- Defines the problem with precision describing in a clear and concise manner the most important facts (data) and variables.
- Analyses the causes and effects of problems from a global long-term approach.
- Designs systematic procedure to make decisions.
- Evaluates possible solutions according to its scientific and technical feasibility.
- Applies advanced search criteria information for troubleshooting and evaluates the quality of information.

5 Design of Learning Activities

Once the indicators were established, the next phase is the design of the learning activities. For that purpose, the instructors of the three courses designed the learning activities in which the assessment tool will be used to assess the Analysis and Problem Solving competence. A total of five learning activities were developed for the formative assessment of the competence so that the assessment tool would be applied to five activities. For example, in the course "Technologies and Software Applications for Supply Chain Management" taught in the first semester, the activity to assess corresponds to Unit 4 of the outline which is related to Decision support systems for supply chain contexts. In the activity, the Promethee-Gaia method is applied. The objectives of the assignment were:

- Review scientific literature regarding applications within supply chain contexts.
- Select one the papers in order to model the problem.
- Define the multi-criteria problem and apply the Promethee-Gaia method.
- Develop alternatives-scenarios.
- Use software to solve the problem.
- Use a systematic approach within all the problem definition-resolutions steps.
- Expose conclusions.

The time required to complete the work was about 25 h. The materials needed to perform the activity were the class materials and the software application. The course is scientific oriented and counted with 9 students.

6 Selection of Assessment Procedures

After the design of the activities, it was necessary to establish who will assess the competence: students, instructors or both of them. Two out of three courses used the instructor assessment procedure and one out of three used the student assessment procedure.

7 Design of the Assessment Tool

The next phase was the initial design of the assessment tool. In Verdecho et al. (2016) is presented the assessment tool that was initially developed in the project. Once the pilot case was applied in Phase 6, this initial assessment tool was modified to support the opinions of the students and instructors as presented in next section.

8 Implementation: Pilot Applications and Feedback

Implementation phase consists of three stages in a loop (see Fig. 2). The loop reflects the three pilot applications (one per course). The three stages are: the execution of the pilot in the course, the analysis of the feedback coming from the application and, finally, the introduction of improvements in the assessment tool coming from the feedback. Once the first pilot application finished, the three stages started again in the second application and so on. During the execution of the pilot, students were asked to complete a survey as shown in Table 1. From the survey of the first application, some modifications emerged and the assessment tool was modified. Improvements coming from first application focused on detailing further each indicator in the assessment tool forming two levels of detail.

Regarding the results of degree reached by the students in the analysis and problem solving competence, all the students reached a positive degree of achievement of the competence and there was not a single item with "not pass" grade. The item with lower grade (weighted average is 2.56 out of 4) was the Item Nr. 2 "Analyses the causes and effects of problems from a long-term approach", probably because it was the most complex item for students. Based on the experience of instructors, the initiative was evaluated as satisfactory and, after the improvement suggested, it was tested in the second pilot.

Then, the assessment tool was tested in two learning activities (a case and the course assignment) in the second pilot application during the second semester. The inclusion of two learning activities aided to reinforce the formative assessment objective of the project. From the second application, it was concluded that each indicator of the assessment tool may not have the same weight in the final assessment so that

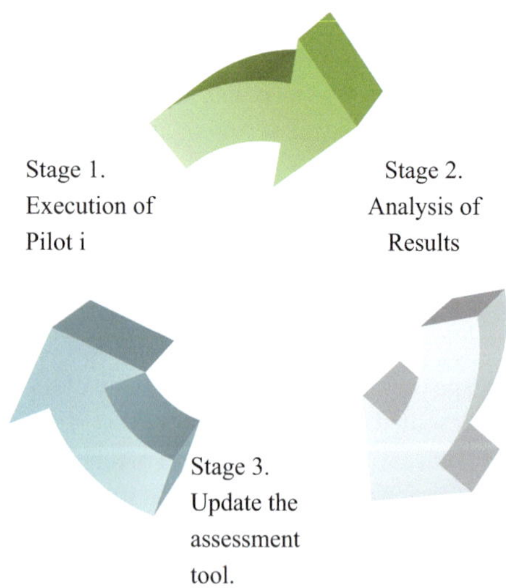

Fig. 2 Implementation phase

Table 1 Student satisfaction survey

	1	2	3	4	5
The assessment procedure is suitable					
The assessment instrument is suitable					
Overall satisfaction with the assessment					
Notes: Comment any aspect you consider relevant to the assessment . Would you implement this instrument to assess similar works? Justify your answer .					

it would help to introduce different weights according to the instructor's experience. The assessment tool was then updated accordingly.

Regarding the results of degree reached by the students in the analysis and problem solving competence in the second application, all the students reached a positive degree of achievement of the competence. In addition, all the indicators of the competence improved from the first learning activity to the second learning activity supporting the main goal of the project: the formative assessment in the competence. This second application was evaluated as highly satisfactory and, after the improvement suggested, it was tested in the third pilot.

Finally, the updated assessment tool was tested in two learning activities in the third pilot application during the second semester as presented in Rodríguez-Rodríguez et al. (2016). In this application, the assessment procedure was a group student assessment. The students achieved in a higher degree of the transversal competence under study in the second activity. This can be due to the own nature of the activity, which was more quantitative than the first one but also to the formative assessment function between two learning activities.

9 Conclusions

This paper presents a methodology developed to assess the "analysis and problem solving" competence in three courses and five learning activities of a master degree in order to support the implementation of formative assessment of that competence.

After testing and updating the assessment tool in a loop of three applications, probably it would still necessary to modify some aspects of it but results show good results in the performance of students as well as positive recommendation from students to implement the tool to assess the competence. In the future, it may be still necessary to test this type of tools in other pilots such as other courses in the same master and/or other master degree's courses. In addition, this initiative may help to proceed similarly in the assessment of other transversal competences.

Acknowledgements This work has been developed within the research project called "Assessment of UPV generic competence "problem analysis and resolution" in master students" (Ref.: PIME-A7-15) funded by the Vice-Rectorate for Studies, Quality and Accreditation at Universitat Politècnica de València.

References

Atarés L, Baviera A, Calvet S, Carballiera J, Escriba C, García E, Llorente R, Muñoz MJ, Palao C, Pastor R, Valero E (2015) Rúbrica UPV CT-3 Análisis y Resolución de Problemas
Bloom BS (1968) Learning for mastery. Eval Comment 1(2):1–12
Bloom BS (1976) Human characteristics and school learning. McGraw-Hill, New York
Bloom BS, Hasting JT, Madaus GF (1971) Handbook on formative and summative evaluation of student learning. McGraw-Hill Book Co, NewYork
de Miguel Díaz M (2006) Modalidades de enseñanza centradas en el desarrollo de competencias. Orientaciones para promover el cambio metodológico en el Espacio Europeo de Educación Superior. Ediciones Universidad de Oviedo
González J, Wagenaar R (2003) Tuning educational structures in Europe. Universidad de Deusto, Bilbao
Huhta A (2010) Diagnostic and formative assessment. In: Spolsky B, Hult F (eds) The handbook of educational linguistics. Blackwell, Oxford, UK, pp 469–482
Ibarra Sáiz MS (2010) INevalCO: INnovación en la EVALuación de COmpetencias Diseño y desa-rrollo de procedimientos e instrumentos para la evaluación de competencias entornos de aprendizaje mixtos/virtuales con la participación de los estudiantes en los títulos de grado. Servicio de Publicaciones de la Universidad de Cádiz, Cádiz
Morales P (2000) Evaluación y aprendizaje de calidad. Universidad Rafael Landívar, Ciudad de Guatemala
Rodríguez Gómez G (2009) EvalHIDA: Evaluación de Competencias con Herramientas de Interacción Dialógica Asíncronas (foros, blogs y wikis). Servicio de Publicaciones de la Universidad de Cádiz, Cádiz
Rodríguez-Rodríguez R, Verdecho MJ, Gómez-Gasquet P, Alfaro-Saiz JJ (2016) Evaluating the "analysis and problem solving" competence at master context with a weighted approach. In: ICERI proceedings, IATED Library
Scriven M (1967) The methodology of evaluation. In: Tyler RW, Gagné RM, Scriven M (eds) Perspectives of curriculum evaluation. AERA monograph series on curriculum evaluation, No. 1. Rand McNally, Chicago, IL, pp 39–83
UPV (2014) Dimensiones competenciales. Marco UPV de definición y evaluación de adquisición de competencias. Vicerrectorado de Estudios, Calidad y Acreditación. http://www.upv.es/contenidos/ICEP/info/DimensionesCompetenciales.pdf. Downloaded 20 Oct 2014
Verdecho MJ, Gómez-Gasquet P, Rodríguez-Rodríguez R, Alfaro-Saiz JJ (2016) A methodology to assess the generic competence "analysis and problem solving" in master degree. In: ICERI proceedings, IATED library

Villa A, Poblete M (2007) Aprendizaje basado en competencias. Una propuesta para la evaluación de las competencias genéricas. Vicerrectorado de Innovación y Calidad de la Universidad de Deusto: Ediciones Mensajero

Part IV
Logistics and Supply Chain

An Overview of Supply Chain Planning That Integrates Financial Issues

Vita Curci, Josefa Mula, Manuel Díaz-Madroñero and Michelle Dassisti

Abstract The main ambition of a supply chain manager is to ensure correct flows of goods and information through all the supply chain nodes to guarantee the right goods in the right place at the right time. To achieve these results, it is very important to consider the flow of items and finances to and from supply chain nodes, which is generated by market demand and productive capacity. In line with a supply chain manager's objective, this paper presents an overview of supply chain modeling methodologies, and analyzes the applicability of current methods and tools to model econometric systems for supply chain planning.

Keywords Supply chain · Production planning · Financial issues · Overview

1 Introduction

Process operations and finances have been traditionally treated as separate problems, and the modeling approaches that support them have been implemented into independent environments (Guillen et al. 2007). In this paper, we review some scientific proposals that account for financial aspects when constructing process operations

J. Mula (✉) · M. Díaz-Madroñero
CIGIP (Research Centre on Production Management and Engineering), Universitat Politècnica de València, Plaza Ferrándiz y Carbonell, 2, 03801 Alcoy, Alicante, Spain
e-mail: fmula@cigip.upv.es

M. Díaz-Madroñero
e-mail: fcodiama@cigip.upv.es

V. Curci · M. Dassisti
Politecnico di Bari, Dip. Meccanica, Management e Matematica. Viale Japigia, 182, 70126 Bari, Italy
e-mail: vitacurci@gmail.com

M. Dassisti
e-mail: michele.dassisti@poliba.it

Á. Ortiz et al. (eds.), *Engineering Digital Transformation*, Lecture Notes in Management and Industrial Engineering,
https://doi.org/10.1007/978-3-319-96005-0_37

models, and integrate both areas operations and financial issues. We specifically focus on supply chain production planning models.

Based on the literature about integrating financial constraints into supply chain models, the objective of this survey is to: (1) provide a focused overview of related references; (2) classify the analyzed papers about the modeling approach, purpose, tools and software, case study, benefits and disadvantages, innovation and application; and (3) identify a research plan for future research in this area.

The rest of the paper is structured as follows: Sect. 2 describes the research methodology and the proposed taxonomy. Section 3 presents an overview and the taxonomy related to the main reviewed references in which the aim and modeling approach of each survey has been dealt with and provides suggestions for further research. Finally, Sect. 5 presents the identified conclusions.

2 Research Methodology

The adopted approach was developed by following a literature survey on supply chain economic models, descriptions and comparisons of different kinds of optimization-based supply chain models. The search process was followed via the Thomson Reuters portal. Suitable articles were searched by introducing the following key words: economic models supply chain, financial models supply chain, economic models of supply chain management, cash flow in supply chain, econometric models in supply chain. The bibliographic references of the studied articles proved most helpful for designating and sustaining this work. Thirty-two references that covered a 20-year period were selected for the research work.

Only two of the studied papers discuss an overview in supply chain literature models (Grossmann 2005; Laínez et al. 2009). Some reviewed articles deal with different supply chain models related to not only a range of distinct economics metrics, but also to dissimilar kinds of firms; e.g., Lee et al. (1996), Romero et al. (2003), Badell et al. (2004), Georgiadis et al. (2005), Zhang (2006), Guillén et al. (2006), Xu (2006), Guillén et al. (2007), Badell et al. (2007), Laınez et al. (2007), Gong (2007), Smith and Martinez-Flores (2007), Guiffrida et al. (2007), Shin et al. (2007), Cruz et al. (2007), Bertel et al. (2008), Comelli et al. (2008), Tian-jian (2009), Naraharisetti et al. (2008), Xue and Zeng (2009), I-Ping and Chiuh-Cheng (2009), Sodhi and Tang (2009), Puigianer and Guillén-Gosalbes (2014), Gupta and Dutta (2011), Mahata and Mahata (2011), Chung and Lin (2011), Longidinis and Georgiadis (2011), Jeong and Han (2012), Ramezani et al. (2014) and Jin et al. (2017).

One group of eight journals represented 74.07% of all the references included in this work, these being: Computers and Chemical Engineering, Industrial & Engineering Chemistry Research, Computers & Industrial Engineering, European Journal of Operational Research, Mathematical and Computer Modelling, Industrial Engineering and Engineering Management, International Journal of Production Economics and International Journal of Production Research.

2.1 Taxonomy

The proposed taxonomy considers the following criteria: aim, modeling approach, tools, kind of presented case study, benefits, and disadvantages. All these parameters are briefly described below:

Aim: it defines the general purpose of each reference.
Modeling approach: several kinds of modeling approaches can be distinguished: linear programming, mixed integer/integer linear programming, nonlinear programming, multiobjective programming, stochastic programming, fuzzy mathematical programming, heuristics algorithms and meta-heuristics.
Tools: this measure represents the tool that the references employed; e.g., mathematical calculations software like GAMS, CPLEX, Matlab, AMPL, Excel, and other software.
Kind of presented case study: we differentiate two kinds of possible case studies: the treated case study conducted either from a real industrial situation or based on assumed data.
Benefit and disadvantages: this criterion represents the benefits that each reference has in the research field, and thus its disadvantages, e.g., limitations.

3 Overview of Supply Chain Models That Integrate Financial Issues

We identified that until the present day, scheduling/planning and budgeting modeling have been treated as separate problems, and have been traditionally implemented into independent environments. This is the main reason why financial and operative matters are still not integrated into support enterprise decision making. A tool for supply chain financial simulation allows different planning alternatives to come into play by taking into account all the available information, and providing absolute transparency of the limitations and interactions that occur at the plant for each tested alternative (Guillén et al. 2006).

More than half the analyzed supply chain models are able to improve financial integration, but many authors report that this is difficult to achieve. Therefore, each work describes the present industrial context and underlines the enterprises that deal with outside and inside uncertainties, with fierce and strict competition. Consequently, flexibility is one of the tactical aims of many firms and of researchers' studies. Many works provide models that help to improve production scheduling in order to increase the customer satisfaction level. The customer satisfaction level is one of the variables with which a company can compete and evaluate a supply chain framework.

Nowadays traditional performance measures, such as profitability, are less relevant for measuring supply chain planning as they tend to focus individually and fail to consider chain-wide areas for performance improvement. It has been proved in practice than even when profits go down for a while, this does not necessarily mean that the supply chain is upside down.

Most of the analyzed references use the linear programming modeling approach, particularly mixed integer linear programming models, wand this approach represents approximately 40% of all the articles (Table 1). Then there is linear programming models, which 30% of the papers present. Nonlinear programming is used only in two references (Smith and Martinez-Flores 2007; Cruz et al. 2007). Only three references refer to the multiobjective programming-based modeling approach. Two of these references employ multiobjective nonlinear programming, these being the works of Comelli et al. (2008) and I-Ping and Chiuh-Cheng (2009). The third prefers linear modeling and is the survey of Guiffrida et al. (2007). Researchers are able to make an efficient supply chain model by dealing with variables approximately. This is possible thanks to the implementation of either fuzzy programming, as in the case of Mahata and Mahata (2011), or stochastic programming, as in Sodhi and Tang (2009) or Longidinis and Georgiadis (2011).

These have been implemented through different software types, but the most widely used ones are GAMS, CPLEX and MATLAB. Visual Basic has been implemented only into multiobjective nonlinear programming models. Thus we can state that other employed softwares, like Xpress, Lingo and Fortran, have been used in the same proportion.

Each reference has engaged a case study to explain the proposed model. As it is difficult to adapt real situations to the formulated supply chain models, most references have prepared assumed examples. Indeed 62% of the reviewed works have dealt with the proposed items using assumed numerical case studies, while the remaining 38% have employed case studies and examples by collecting data from real industrial situations.

From the previous literature review, we identified the following further research guidelines of the analyzed references.

On the one hand, one further research suggestion could involve developing tools capable of facing industrialized problems: e.g., Lee et al. (1996), Romero et al. (2003), Georgiadis et al. (2005), Xu (2006), Badell et al. (2007), Comelli et al. (2008), I-Ping and Chiuh-Cheng (2009) and Sodhi and Tang (2009).

Bertel et al. (2008) presented a framework that was unable to fulfill 100% demand satisfaction. Therefore, this feature moves managers far from the highest efficient supply chain administration. Further research could be conducted by increasing the demand satisfaction value by integrating both ABC and advanced production scheduling to optimize the financial flow and physical flow together in operational, tactical and strategic planning.

On the other hand, further research makes the same model changing assumptions. The following references belong to this second set: Badell et al. (2004), Guillén et al. (2006), Zhang (2006), Guillén et al. (2007), Guiffrida et al. (2007), Smith and Martinez-Flores (2007), Shin et al. (2007), Laınez et al. (2007), Naraharisetti et al. (2008), Tian-jian (2009), Xue and Zeng (2009), Mahata and Mahata (2011), Gupta and Dutta (2011) and Chung and Lin (2011).

Table 1 Modeling approach of some of the reviewed works

Authors	LP	ILP	INLP	NLP	MOLP	MONLP	FMP	SP	HEU
Lee et al. (1996)		●							
Romero et al. (2003)	●								
Badell et al. (2004)	●								
Zhang (2006)	●								
Guillén et al. (2006)		●							
Georgiadis et al. (2005)		●							
Badell et al. (2007)		●							
Xu (2006)	●								
Bertel et al. (2008)		●							
Guillén et al. (2007)		●							
Smith and Martinez-Flores (2007)				●					
Cruz et al. (2007)			●						
Lainez et al. (2007)		●							
Gong (2007)		●							
Shin et al. (2007)		●							
Guiffrida et al. (2007)	●				●				
Naraharisetti et al. (2008)		●							
Tiajian (2008)	●								

(continued)

Table 1 (continued)

Authors	LP	ILP	INLP	NLP	MOLP	MONLP	FMP	SP	HEU
Comelli et al. (2008)						●			
Xue and Zeng (2009)									●
I-Ping and Chiuh-Cheng (2009)						●			
Laínez et al. (2009)		●							
Sodhi and Tang (2009)								●	
Puigjaner and Guillén-Gosalbes (2014)		●						●	
Mahata and Mahata (2011)							●		
Gupta and Dutta (2011)		●							
Chung and Lin (2011)	●								
Longidinis and Georgiadis (2011)		●							
Ramezani et al. (2014)		●							
Jin et al. (2017)		●							

Gong (2007) presents a framework that can be improved in a variety of ways: first, market demand uncertainty is omitted in these models; second, the model can be extended to consider new product flexibility, which can be achieved by appropriately defining demand distributions; third, models can be extended to judge the feasibility of flexibility investment; finally, the uncertainty of logistics factors, such as distribution and transportation, is not included in the model. The model does not consider the concurrent variation of more factors. Hence when the model analyzes the relationship between supply chain flexibility and a factor, the other flexibility factors remain unchanged. It would be interesting to study their interaction.

Laínez et al. (2009) formulates a framework whose future challenges could systematically incorporate supplier/buyer options contracts within the supply chain problems framework. Then making significant progress to solve multistage stochastic optimization problems with large scenario trees would be required to address practical problems. Finally, the scope of enterprise-wide decision problems could further expand by incorporating the full life cycle consideration.

Regarding benefits and disadvantages, the majority of the studied works have developed complex supply chain frameworks that can integrate financial constraints into operational management; despite this, the same authors state that these frameworks need very advanced software, which has not been available until now, or are unable to sustain such modeling even if it does exist (Romero et al. 2003; Gong 2007; Tian-jian 2009; Sodhi and Tang 2009; Mahata and Mahata 2011; Gupta and Dutta 2011). We found only one reference that implements benefit into inventory scheduling (Lee et al. 1996) as all the others references implement benefit into production scheduling.

4 Conclusions

This paper has presented an overview of selected works that pursues the integration of supply chain planning models with financial issues, which have been traditionally addressed separately. Thirty-two papers were analyzed and compared in detail. A taxonomy is proposed that contemplates the aim, modeling approach, tool, case study and benefits and disadvantages of the different proposals.

We have generally concluded that mixed integer linear programming is the most widely used modeling approach. The mathematical programming tools most frequently used by the reviewed works are GAMS, CPLEX and MATLAB. Numerical examples are used mainly to validate proposals. Finally, the previous section has provided more detailed information on the identified benefits, disadvantages and further research from the reviewed works.

Acknowledgment This research has been carried out within the framework of the project funded by the Spanish Ministry of Economy and Competitiveness entitled 'Operations design and Management of Global Supply Chains' (GLOBOP) (Ref. DPI2012-38061-C02-01).

References

Badell M, Romero J, Huertas R et al (2004) Planning, scheduling and budgeting value-added chains. Comput Chem Eng 28:45–61
Badell M, Fernandez E, Guillen G et al (2007) Empowering financial tradeoff with joint financial and supply chain planning models. Math Comput Model 46:12–23
Bertel S, Fenies P, Roux O (2008) Optimal cash flow and operational planning in a company supply chain. Int J Comput Int Manuf 21:440–454
Chung KJ, Lin SD (2011) The inventory model for trade credit in economic ordering policies of deteriorating items in a supply chain system. Appl Math Model 35:3111–3115
Comelli M, Fenies P, Tchernev N (2008) A combined financial and physical flows evaluation for logistic process and tactical production planning: application in a company supply chain. Int J Prod Econ 112:77–95
Cruz JM, Nagurney A, Wakolbinger T (2007) Financial engineering of the integration of global supply chain networks and social networks with risk management. Nav Res Log 53:674–696
Georgiadis MC, Aaron AL, Tsiakisc P et al (2005) Optimisation based scheduling: A discrete manufacturing case study. Comput Ind Eng 49:118–145
Gong Z (2007) An economic evaluation model of supply chain flexibility. Eur J Oper Res 184:745–758
Grossmann I (2005) Enterprise wide optimization: a new frontier in process systems engineering. AIChE J 51:1846–1857
Guiffrida AL, Jaber MY, Rzepka RA (2007) An economic model for justifying the reduction o deliver. INFOR Inform Syst Oper Res 46:147–153
Guillen G, Badell M, Puigjaner L (2007) A holistic framework for short-term supply chain management integrating production and corporate financial planning. Int J Prod Econ 106:288–306
Guillen G, Badell M, Espuna A et al (2006) Simultaneous optimization of process operations and financial decisions to enhance the integrated planning/scheduling of chemical supply chains. Comput Chem Eng 30:421–436
Gupta S, Dutta K (2011) Modeling of financial supply chain. Eur J Oper Res 211:47–56
I-Ping H, Chiuh-Cheng C (2009) A biobjective economic model for a two-level supply chain production process. In: Proceedings of international conference on computers & industrial engineering, CIE 2009, pp 881–885
Jeong S, Han J (2012) Integrated supply chain and finance model for value maximization of a firm with multi-divisions. Information 15(12C):5985–5998
Jin SH, Jeong SJ, Kim KS (2017) A linkage model of supply chain operation and financial performance for economic sustainability of firm. Sustainability 9(1):139
Laınez JM, Guillén-Gosàlbez G, Badell M et al (2007) Enhancing corporate value in the optimal design of chemical supply chains. Ind Eng Chem Res 46:7739–7757
Laínez JM, Puigjaner L, Reklaitis GV (2009) Financial and financial engineering considerations in supply chain and product development pipeline management. Comput Chem Eng 33:1999–2011
Lee H, Pinto JM, Grossmann IE et al (1996) Mixed integer linear programming model for refinery short term scheduling of crude oil unloading with inventory management. Ind Eng Chem Res 35:1630–1641
Longidinis P, Georgiadis MC (2011) Integration of financial statement analysis in the optimal design of supply chain networks under demand uncertainty. Int J Prod Econ 129:262–276
Mahata GC, Mahata P (2011) Analysis of a fuzzy economic order quantity model for deteriorating items under retailer partial trade credit financing in a supply chain. Math Comput Model 53:1621–1636
Naraharisetti PK, Karimi IA, Srinivasan R (2008) Supply chain redesign through optimal asset management and capital budgeting. Comput Chem Eng 32:3153–3169
Puigjaner L, Guillén-Gosálbez G (2014) Bridging the gap between production, finances, and risk in supply chain optimization. Process Syst Eng 3–7:1–44

Ramezani M, Kimiagari AM, Karimi B (2014) Closed-loop supply chain network design: a financial approach. Appl Math Model 38:4099–4119
Romero J, Badell M, Bagajewicz M et al (2003) Integrating budgeting models into scheduling and planning models for the chemical batch industry. Ind Eng Chem Res 42:6125–6134
Shin T, Chen S, MacCarthy BL, Muyldermans L et al (2007) Supply chain grounded on information theory: A hierarchical economic information filtering model of supplier selection. In: IEEE proceedings on industrial engineering and engineering management, pp 1471–1475
Sodhi MS, Tang CS (2009) Modeling supply-chain planning under demand uncertainty using stochastic programming: a survey motivated by asset–liability management. Int J Prod Econ 121:728–738
Smith NR, Martinez-Flores JL (2007) Discrepancies in solutions between traditional and net present value formulations of finite horizon, discrete time economic lot size problems. Int J Prod Res 45:5731–5741
Tian-jian Y (2009) Financial performance analysis of e-collaboration supply chain under retailer's stock disruption. In Proc Int Conf Inform Technol Comput Sci ITCS 2:608–612
Xu L (2006) An integrated supply chain planning model for manufacture firm based on logistics chain. IEEE Proc Int Conf Ser Syst Ser Managet 1:589–593
Xue X, Zeng Z (2009) The modeling framework for collaborative economic organization: cluster supply chain. IEEE Int Conf Intell Comput Intell Syst ICIS 2:72–76
Zhang D (2006) A network economic model for supply chain versus supply chain competition. Omega 34:283–295

Adaptive and Hybrid Forecasting Models—A Review

Carlos Hernán Fajardo-Toro, Josefa Mula and Raul Poler

Abstract Nowadays, good supply chain management is most important to guarantee a competitive advantage and to accomplish the value promise offered to the company's clients. To this end, it is important to reduce uncertainty associated with demand, and it is important that demand forecast is as accurate as possible. To achieve this, it is necessary to know the features of the demand to be forecast and, based on this, to build or choose the best and the most accurate model or technique, which is based normally on that with fewer errors. Many statistical techniques exist, but for some 20 years, many heuristic algorithms have been developed that allow to absorb the variance associated with demand, and to reduce forecasting errors with better results than those obtained by statistical methods. These methods are normally adaptive and allow to hybridize techniques to construct different models. This document reviews these adaptive techniques, such as neural networks (NN) and hybrid methods, and in particular models based on case-based reasoning (CBR).

Keywords Forecasting · Time series · Case-based reasoning
Neural networks · Adaptive methods

C. H. Fajardo-Toro
Facultad de Administración, Finanzas y Ciencias Económicas, Universidad Ean, Calle 79 # 11-45, Bogotá, Colombia
e-mail: cfajardo.toro@gmail.com

J. Mula (✉) · R. Poler
CIGIP (Centro de Investigación Gestión e Ingeniería de Producción), Universitat Politècnica de València, Escuela Politécnica Superior de Alcoy, Plaza Ferrándiz y Carbonell 2, 03801 Alcoy (Alicante), Spain
e-mail: fmula@cigip.upv.es

R. Poler
e-mail: rpoler@cigip.upv.es

Á. Ortiz et al. (eds.), *Engineering Digital Transformation*, Lecture Notes in Management and Industrial Engineering,
https://doi.org/10.1007/978-3-319-96005-0_38

1 Introduction

Companies permanently attempt to reduce the uncertainty in their decisions. This decision-making process is now more complex because of markets' variability which, in many cases, cuts products' life cycling, and has a direct effect on inventories and, therefore, on costs, which could contribute to a less competitive advantage. According to this scenario, it is easy to conclude that many factors exist and helps to increase uncertainty, e.g., market demand, capacity data and cost information (Mula et al. 2007).

In order to reduce uncertainty in decision-process making, adopting different kinds of forecasting methods is important. These methods can be classified as qualitative and quantitative. For quantitative methods, different statistical models have been developed (Abdollahzade et al. 2015; Gooijer and Hyndman 2005). However, with the advances and improvements of computational techniques, artificial intelligence (AI) techniques have occupied a relevant spot in the developed models (Acosta-Cervantes et al. 2013; Aladag et al. 2009; Andrawis et al. 2011; Du et al. 2014). The advantage that employing AI techniques allows is that they adapt easily to manage stochastic non-linear behavior, and also to problems in which the collinearity of variables is present (Behnamian and Ghomi 2010; Wang et al. 2013; Yu et al. 2009; Wang et al. 2010). The AI techniques used to obtain forecasts include neural networks (NN) (Ettouney et al. 2009; Gupta and Kashyap 2015; Pan et al. 2013; Venugopal and Baets 1994; Adhikari 2015; Kourentzes et al. 2014; Mohammadi et al. 2014; Yolcu et al. 2013), support vector machines (Wang et al. 2003; Lydia et al. 2016; Kim 2003), case-based reasoning (CBR) (Chang et al. 2008; Chang and Lai 2005; Lee 1997; Li et al. 2011; Nyberg and Saikkonen 2014; Zhang et al. 2013), and fuzzy logic-based techniques and optimisation heuristics, like genetic algorithms (Fan et al. 2011; Li et al. 2011; Aksoy et al. 2014; Huarng et al. 2012; Mula et al. 2007; Peidro et al. 2009).

This work centers on reviewing hybrid models. For a model to be considered hybrid, the conjunction or combination of two techniques or more is necessary, which need to be statistical, and based on AI, algorithms, or on both. CBR models can be mostly considered hybrid given the characteristics that this methodology has (Kamp et al. 1998; Watson 1998). The CBR has four stages (Cheung et al. 2012; Choy et al. 2002b; Baruque et al. 2010; Fdez-Riverola and Corchado 2003): (i) case retrieving, where similar cases are retrieved from a knowledge database. To do this, statistical techniques can be used, such as clustering and classification AI techniques, like SOM (self-organizing maps) networks; (ii) reusing the cases selected to provide a solution to the problem to be solved. In this stage, and depending on the nature of the problem, it is possible to use neural networks, fuzzy logic-based algorithms or genetic algorithms; (iii) case reviews, which consists in validating the given error, but with the solution proposed in the previous step for real cases. In this stage, validation can be as simple as a minimum acceptable percentage, as well as a fuzzy evaluation of the estimated parameters; (iv) case retaining, where the system learns from errors and decides which element is to be taken to adapt the results and improve them. This document presents, on the one hand, hybrid models that have not been constructed based on the CBR methodology and, on the other hand, some CBR-based models found in the literature, especially in the Emerald and Science Direct Editors.

2 Hybrid Models

A model can be considered hybrid if it has and combines two statistical and/or AI techniques, or more, to obtain a result. Wang et al. (2010) developed a model based on chaotic time series and a diagonal recurrent neural network (DRNN) to obtain short-term daily energy power utility forecasts. To obtain better results, and given the characteristics of the NN, it was necessary to define and optimize the weights on the layers of networks. These authors used a genetic algorithm to achieve this. Wang et al. (2003) attempted to construct a model to help forecast air pollution. To achieve this, they used an adaptive radial basis function (ARBF) network and improved the support vector machine (SVM). They also constructed a hybrid model that used a principal component analysis (PCA) as the technique to allow the ARBF to improve its learning process. Pan et al. (2013) developed a hybrid model that combined empirical mode decomposition (EMD), which is an important part of the Hilbert–Huang transform (HHT), with a multi-layer perception MLP-NN, which they called EMD-NN. They also compared the results with an MLP model and an SARIMA model, and found that EMD-NN was more stable than the other techniques and in volatile situations. Many works have built models based on fuzzy logic techniques to generate adequate forecasts (Mula et al. 2006; Peidro et al. 2009). Many are hybrid models because they use this fuzzy logic in conjunction with statistical or AI techniques. Aksoy et al. (2014) used neuro-fuzzy techniques to estimate the demand of apparels. Huarng et al. (2012) applied fuzzification to a time series, and then fed a back propagation (BP) NN to obtain better results than with statistical techniques. Talarposhti et al. (2016) built a model based on an exponential fuzzy time series and learning automata-based optimisation for stock forecasting. Other fuzzy logic-based hybrid works are those by Deng et al. (2015), who developed a model that first performs Gaussian cloud transformation and then applies a fuzzy time series model to make forecasts. The combination of fuzzy algorithms with other techniques used for stock market forecasts (Su and Cheng 2016; Wei 2016; Chen and Chen 2015) is frequently find. Other fuzzy logic-based works have been done by Cai et al. (2013), Egrioglu et al. (2013), (2009), Lee and Hong (2015), Peng et al. (2015), Sadaei et al. (2014), Singh and Borah (2013) and Wang and Xiong (2014), who have developed a conjunction of time series, fuzzy logic and optimisation or adaptive techniques as neural networks. Many models have combined time series to statistical forecasting techniques with different AI algorithms. Ruiz-Aguilar et al. (2014) used the results obtained with an SARIMA method to feed a BP-NN. Khandelwal et al. (2015) also constructed an ARIMA-NN model, but used a decomposition wavelet transform (DWT) to feed the ARIMA-NN model. Other such works have been developed by Babu and Reddy (2014), Barak and Sadegh (2016), Egrioglu et al. (2009), Ismail et al. (2011), Li and

Hu (2012), Maia and de Carvalho (2011), Mohammadi et al. (2014) and Peng et al. (2015).

3 Case Based-Reasoning Models

As explained above, the models developed according to the case-based reasoning methodology can be considered hybrid in nature. The range of possibilities with CBR is very wide. It can be used to develop a system to help integrate or optimise supply chains (Cheung et al. 2012), improve relations with suppliers' costs and management (Choy et al. 2002a, b), manage warehouses (Chow et al. 2006; Choy et al. 2009; Lao et al. 2012; Poon et al. 2009) and manage costs (Raphael et al. 2007).

In relation with forecasting models, some CBR applications and works are shown below. Barrientos and Vargas (1998) developed a framework that combines Bayesian neural networks with a CBR structure to forecast ozone levels in Mexico City. Baruque et al. 2010 developed a CBR model that forecasts the zones in which oil slicks can be found by combining SOM and GRBF networks. Chang and Lai (2005) hybridised classification techniques, e.g., SOM maps and a CBR, to forecast the sales of a newly released book. They compared SOM/CBR with K-mean/CBR. The model basically classified some given features, and was based on a data vector, to make forecasts. Chang et al. (2006) used the technique to forecast returning books sales by developing the CBR with a genetic algorithm. Chang et al. (2008) also developed a fuzzy logic-based CBR model because of the difficulty to efficiently collect information. Chou et al. (2015) constructed a framework to forecast project award prices. This framework can be considered hybrid, not only by using the CBR methodology, but also by employing genetic algorithms with non-linear regressions, a neural network and a CBR. Chun and Park (2006) developed what they called RCBR, or regression CBR, to obtain financial forecasts. Basically, they constructed a CBR in which a regression analysis helped reduce the weight vector, which helped cut errors. Fan et al. (2011) proposed a model that sought to obtain medical data classifications to help forecast diseases. They integrated a classification technique with a fuzzy decision tree. For classifications, they chose case-based weighted data clustering, which feeds a decision tree based on fuzzy logic algorithms, and which forecast breast cancer according to that information. Fdez-Riverola and Corchado (2003) built a model to forecast red tides on the Galician coast (Spain). The model used conceptual graphs (CGs), networks to extract data, and the radial basis function (RBF) network to produce the first solution, fuzzy sets to confirm the solution, and all four techniques for the learning process. Green and Armstrong (2007) developed a CBR that starts out with the hypothesis that the analogies which people use when attempting to forecast something reflect their beliefs, based on the fact that the model makes a classification and forecasts a specific problem. Lertpalangsunti and Chan (1998) built a framework to estimate electricity demands, and Lee (1997) proposed a model that used CBR, and a multilayer neural network and discriminate analysis. Li et al. (2011) proposed a CBR model by combining techniques, which was inspired by the order performance technique for its similarity to the ideal solution (TOPSIS), to forecast business failure. They also approached the problem in another work, in which they developed Gaussian CBR, where they used a Gaussian function to obtain the case selection in the retrieval stage (Li and Sun 2009). Liu and Yu (2009) constructed a CBR model where they hybridized a retrieval method with fuzzy techniques and importance-performance analysis (IPA). Platon et al. (2015) combined PCA, NN and CBR to forecast buildings' hourly energy consumptions. Hybridisation of techniques has been demonstrated to be most useful for environment-related problems, such as forecasting cyclones (San et al. 2005), daily wind forecasts (Shukur and Lee 2015) and river flow forecasts (Toro et al. 2013).

4 Conclusions

As this document reveals, obtaining accurate forecasts is an important element in the decision-making process as it is necessary to reduce the uncertainty associated with the factors and uncontrollable variables that affect decisions. To achieve accuracy in forecasts and estimations, constructing and selecting the most adequate model is of utmost importance. This model has to be able to absorb the stochastic and chaotic nature of the behavior of the variables which govern the problem on which decisions are made. Given the complexity of problems, especially their stochastic behavior, hybrid models are a useful alternative to obtain a good approach to reality. This is because many problems cannot be solved with algorithmic solutions, but heuristic methods must be used. Hybrid models allow to well absorb variability. This is especially true as they are capable of adapting to the characteristics of the problem by assuming that each problem is unique and requires a unique model. Further research needs to be done to develop new adaptive and hybrid forecasting methods, and to apply them to real-world supply chain management problems.

References

Abdollahzade M, Miranian A, Hassani H et al (2015) A new hybrid enhanced local linear neuro-fuzzy model based on the optimized singular spectrum analysis and its application for nonlinear and chaotic time series forecasting. Inf Sci 295:107–125

Acosta-Cervantes MC, Villarreal-Marroquín MG, Cabrera-Ríos M (2013) Estudio de validación de un método para seleccionar técnicas de pronóstico de series de tiempo mediante redes neuronales artificiales. Ing Invest Tecnología 14:53–63

Adhikari R (2015) A neural network based linear ensemble framework for time series forecasting. Neurocomputing 157:231–242

Aksoy A, Öztürk N, Sucky E (2014) Demand forecasting for apparel manufacturers by using neuro-fuzzy techniques. J Model Manag 9:18–35

Aladag CH, Basaran MA, Egrioglu E et al (2009) Forecasting in high order fuzzy times series by using neural networks to define fuzzy relations. Expert Syst Appl 3:4228–4231

Andrawis RR, Atiya AF, El-Shishiny H (2011) Forecast combinations of computational intelligence and linear models for the NN5 time series forecasting competition. Int J Forecasting 27:672–688

Babu CN, Reddy BE (2014) A moving-average filter based hybrid ARIMA–ANN model for forecasting time series data. Appl Soft Comput 23:27–38

Barak S, Sadegh SS (2016) Forecasting energy consumption using ensemble ARIMA–ANFIS hybrid algorithm. Int J Elec Power 82:92–104

Barrientos MA, Vargas JE (1998) A framework for the analysis of dynamic processes based on Bayesian networks and case-based reasoning. Expert Syst Appl 15:287–294

Baruque B, Corchado E, Mata A et al (2010) A forecasting solution to the oil spill problem based on a hybrid intelligent system. Inform Sci 180:2029–2043

Behnamian J, Fatemi Ghomi SMT (2010) Development of a PSO-SA hybrid metaheuristic for a new comprehensive regression model to time-series forecasting. Expert Syst Appl 37:974–984

Cai Q, Zhang D, Wu B, Leung SCH (2013) A novel stock forecasting model based on fuzzy time series and genetic algorithm. Procedia Comput Sci 18:1155–1162

Chang P, Lai C (2005) A hybrid system combining self-organizing maps with case-based reasoning in wholesaler's new-release book forecasting. Expert Syst Appl 29:183–192

Chang P, Lai C, Lai KR (2006) A hybrid system by evolving case-based reasoning with genetic algorithm in wholesaler's returning book forecasting. Decis Support Syst 42:1715–1729

Chang PC, Liu CH, Lai RK (2008) A fuzzy case-based reasoning model for sales forecasting in print circuit board industries. Expert Syst Appl 34:2049–2058

Chen MY, Chen BT (2015) A hybrid fuzzy time series model based on granular computing for stock price forecasting. Inf Sci 294:227–241

Cheung CF, Cheung CM, Kwok SK (2012) A knowledge-based customization system for supply chain integration. Expert Syst Appl 39:3906–3924

Chou J, Lin C, Pham A et al (2015) Automation in Construction Optimized artificial intelligence models for predicting project award price. Automat Constr 54:106–115

Chow HKH, Choy K, Lee WB et al (2006) Design of a RFID case-based resource management system for warehouse operations. Expert Syst Appl 30:561–576

Choy KL, Chow KH, Moon KL et al (2009) A RFID-case-based sample management system for fashion product development. Eng Appl Artif Intel 22:882–896

Choy KL, Lee WB, Lo V (n.d.) Design of an intelligent supplier relationship management system: a hybrid case based neural network approach. Expert Syst Appl 24:225–237

Choy KL, Lee WB, Lo V (2002a) An intelligent supplier management tool for benchmarking suppliers in outsource manufacturing. Expert Syst Appl 22:213–224

Choy KL, Lee WB, Lo V (2002b) Development of a case based intelligent customer—supplier relationship management system. Expert Syst Appl 22:281–297

Chun S, Park Y (2006) A new hybrid data mining technique using a regression case based reasoning: Application Financ. Forecasting 31:329–336

Deng W, Wang G, Zhang X (2015) A novel hybrid water quality time series prediction method based on cloud model and fuzzy forecasting. Chemometr Intell Lab 149:39–49

Du W, Leung SYS, Kwong CK (2014) Time series forecasting by neural networks: A knee point-based multiobjective evolutionary algorithm approach. Expert Syst Appl 41:8049–8061

Egrioglu E, Aladag CH, Yolcu U (2013) Fuzzy time series forecasting with a novel hybrid approach combining fuzzy c-means and neural networks. Expert Syst Appl 40:854–857

Egrioglu E, Aladag CH, Yolcu U et al (2009) A new hybrid approach based on SARIMA and partial high order bivariate fuzzy time series forecasting model. Expert Syst Appl 36:7424–7434

Ettouney RS, Mjalli FS, Zaki JG et al (2009) Forecasting of ozone pollution using artificial neural networks. Manag Environ Qual Int J 20:668–683

Fan C, Chang P, Lin J et al (2011) A hybrid model combining case-based reasoning and fuzzy decision tree for medical data classification. Appl Soft Comput 11:632–644

Fdez-Riverola F, Corchado JM (2003) CBR based system for forecasting red tides. Knowl-Based Syst 16:321–328

Gooijer JG, De Hyndman RJ (2005) 25 Years of IIF Time Series Forecasting: a Selective Review. Working paper 12/05 Department of Econometrics & Business Statistics, Monash University, Australia, 1–39
Green KC, Armstrong JS (2007) Structured analogies for forecasting. Int. J. Forecasting 23:365–376
Gupta S, Kashyap S (2015) Forecasting inflation in G-7 countries: an application of artificial neural network. Foresight 17:63–73
Huarng K, Hui-Kuang YuT, Moutinho L et al (2012) Forecasting tourism demand by fuzzy time series models. Int J Cult Tourism Hospitality Res 6:377–388
Ismail S, Shabri A, Samsudin R (2011) A hybrid model of self-organizing maps (SOM) and least square support vector machine (LSSVM) for time-series forecasting. Expert Syst Appl 38:10574–10578
Kamp G, Lange S, Globig C (1998) Case-based reasoning technology: related areas case-based reasoning technology from foundations to application. Springer, Berlin, pp 327–351
Khandelwal I, Adhikari R, Verma G (2015) Time series forecasting using hybrid ARIMA and ANN models based on DWT decomposition. Procedia Comput Sci 48:173–179
Kim K (2003) Financial time series forecasting using support vector machines. Neurocomputing 55:307–319
Kourentzes N, Barrow DK, Crone SF (2014) Neural network ensemble operators for time series forecasting. Expert Syst Appl 41:4235–4244
Lao SI, Choy KL, Ho GTS et al (2012) A real-time food safety management system for receiving operations in distribution centers. Expert Syst Appl 39:2532–2548
Lee H (1997) Bankruptcy prediction using case-based reasoning, neural networks and discriminant analysis. Expert Syst Appl 13:97–108
Lee WJ, Hong J (2015) A hybrid dynamic and fuzzy time series model for mid-term power load forecasting. Int J Elect Power 64:1057–1062
Lertpalangsunti N, Chan CW (1998) An architectural framework for the construction of hybrid intelligent forecasting systems: application for electricity demand prediction. Eng Appl Artif Intel 11:549–565
Li B, Wang H, Yang J et al (2011a) A belief-rule-based inventory control method under nonstationary and uncertain demand. Expert Syst Appl 38:14997–15008
Li C, Hu JW (2012) A new ARIMA-based neuro-fuzzy approach and swarm intelligence for time series forecasting. Eng Appl Artif Intel 25:295–308
Li H, Adeli H, Sun J et al (2011b) Hybridizing principles of TOPSIS with case-based reasoning for business failure prediction. Comput Oper Res 38:409–419
Li H, Sun J (2009) Gaussian case-based reasoning for business failure prediction with empirical data in China. Inform Sci 179:89–108
Liu KF, Yu C (2009) Environmental modelling and software integrating case-based and fuzzy reasoning to qualitatively predict risk in an environmental impact assessment review. Environ Modell Softw 24:1241–1251
Lydia M, Suresh Kumar S, Immanuel Selvakumar A et al (2016) Linear and non-linear autoregressive models for short-term wind speed forecasting. Convers Manage 112:115–124
Maia ALS, de Carvalho FAT (2011) Holt's exponential smoothing and neural network models for forecasting interval-valued time series. Int J Forecasting 27:740–759
Mirzaei Talarposhti F, Javedani Sadaei H, Enayatifar R et al (2016) Stock market forecasting by using a hybrid model of exponential fuzzy time series. Int J Approx Reason 70:79–98
Mohammadi R, Fatemi Ghomi SMT, Zeinali F (2014) A new hybrid evolutionary based RBF networks method for forecasting time series: a case study of forecasting emergency supply demand time series. Eng Appl Artif Intel 36:204–214
Mula J, Poler R, García-Sabater JP et al (2006) Models for production planning under uncertainty: a review. Int J Prod Econ 103:271–285
Mula J, Poler R, Garcia-Sabater JP (2007) Material requirement planning with fuzzy constraints and fuzzy coefficients. Fuzzy Set. Syst. 158:783–793

Nyberg H, Saikkonen P (2014) Forecasting with a noncausal VAR model. Comput Stat Data An 76:536–555
Pan Y, Pohlen T, Manago S (2013) Hybrid Neural Network Model in Forecasting Aggregate U.S. Retail Sales. In: Advances in Bussiness Management Forecasting Vol. 9. Emerald, pp. 153–170
Peidro D, Mula J, Poler R et al (2009) Fuzzy optimization for supply chain planning under supply, demand and process uncertainties. Fuzzy Set Syst 160:2640–2657
Peng HW, Wu SF, Wei CC et al (2015) Time series forecasting with a neuro-fuzzy modeling scheme. Appl Soft Comput 32:481–493
Platon R, Dehkordi VR, Martel J (2015) Hourly prediction of a building's electricity consumption using case-based reasoning, artificial neural networks and principal component analysis. Energ Buildings 92:10–18
Poon TC, Choy KL, Chow HKH et al (2009) A RFID case-based logistics resource management system for managing order-picking operations in warehouses. Expert Syst Appl 36:8277–8301
Raphael B, Domer B, Saitta S et al (2007) Incremental development of CBR strategies for computing project cost probabilities. Adv Eng Inform 21:311–321
Ruiz-Aguilar JJ, Turias IJ, Jiménez-Come MJ (2014) Hybrid approaches based on SARIMA and artificial neural networks for inspection time series forecasting. Transport E Log 67:1–13
Sadaei HJ, Enayatifar R, Abdullah AH et al (2014) Short-term load forecasting using a hybrid model with a refined exponentially weighted fuzzy time series and an improved harmony search. Int J Elec Power 62:118–129
San J, Burstein F, Sharp A (2005) A case-based fuzzy multicriteria decision support model for tropical cyclone forecasting. Eur J Oper Res 160:308–324
Shukur OB, Lee MH (2015) Daily wind speed forecasting through hybrid KF-ANN model based on ARIMA. Ren Energy 76:637–647
Singh P, Borah B (2013) High-order fuzzy-neuro expert system for time series forecasting. Knowl-Based Syst 46:12–21
Su CH, Cheng CH (2016) A hybrid fuzzy time series model based on ANFIS and integrated nonlinear feature selection method for forecasting stock. Neurocomputing 205:264–273
Toro CHF, Gómez Meire S, Gálvez JF et al (2013) A hybrid artificial intelligence model for river flow forecasting. Appl Soft Computing 13:3449–3458
Venugopal V, Baets W (1994) Neural networks and statistical techniques in marketing research. Market Intell Plan 12:30–38
Wang JG, Zhou JZ, Peng B (2010) A hybrid neural genetic method for load forecasting based on phase space reconstruction. Kybernetes 39:1291–1297
Wang J, Xiong S (2014) A hybrid forecasting model based on outlier detection and fuzzy time series—a case study on Hainan wind farm of China. Energy 76:526–541
Wang W, Xu Z, Lu JW (2003) Three improved neural network models for air quality forecasting. Eng Comput 20:192–210
Wang Y, Wang S, Zhang N (2013) A novel wind speed forecasting method based on ensemble empirical mode decomposition and GA-BP neural network. IEEE Power Energy Soc General Meet 94:629–636
Watson I (1998) CBR is a methodology not a technology. In: Miles R, Moulton M and Bramer M. (eds), Research and development in expert systems. Springer-Verlag, pp 213–223
Wei LY (2016) A hybrid ANFIS model based on empirical mode decomposition for stock time series forecasting. Appl Soft Computing 42:368–376
Yolcu U, Egrioglu E, Aladag CH (2013) A new linear & nonlinear artificial neural network model for time series forecasting. Decis Support Syst 54:1340–1347
Yu L, Wang S, Lai KK (2009) A neural-network-based nonlinear metamodeling approach to financial time series forecasting. Appl Soft Comput 9:563–574
Zhang B, Han X, Zhou Z et al (2013) Construction of a new BRB based model for time series forecasting. Appl Soft Comput 13:4548–4556

Additive Manufacturing and Supply Chain: A Review and Bibliometric Analysis

Jairo Nuñez, Ángel Ortiz, Manuel Arturo Jiménez Ramírez, Jairo Alexander González Bueno and Marianela Luzardo Briceño

Abstract The additive manufacturing studies has been increased in the last years, positioning in the scientific community. This paper analyze the literature about the relation between Additive Manufacturing and Supply Chain, with the final purpose of giving an exhaustive vision and to present and initial parameter in future researches. Using a bibliometric analysis 497 published articles in the last 25 years were evaluated, 78 articles of which were selected that are related in the AM and SC topic, identifying the publishing tendency, the principles authors and the impact of the articles and the magazines. Also, there are knowledge networks and geographic location of the most influential countries.

Keywords Bibliometric analysis · Additive manufacturing · Supply chain

1 Introduction

The Supply Chain (SC) enterprises looks to improve their competitively and productivity through the innovation in their processes, with the development of Additive Manufacturing in the 80s years (Hull 2013), a new fabrication method were born, the prototype and products generation from a printer machine, It evolves from the subtractive methods which eliminated the material to the generation of objects that adds the material layer by layer (Thomas 2016), the improving materials set available for 3D print, as polymers, ceramics, metals, aluminum, copper (Vaezi et al. 2013), steel, chrome, cobalt (Singh et al. 2017) has allowed and endless applications in different industries as: Biomedical (Emelogu et al. 2016), automotive (Savastano et al. 2016), Aviation (Wagner and Walton 2016), Medical (Hieu et al. 2003), dental (Atzeni and Salmi 2015) and many others more.

The implementation of new manufacturing technologies suppose a fundamental change in the production infrastructure and in the organizational processes (SMLC

J. Nuñez (✉) · Á. Ortiz · M. A. J. Ramírez · J. A. González Bueno · M. L. Briceño
Pontificia Bolivariana University, Medellín, Antioquia, Colombia
e-mail: jairo.nunez@upb.edu.co

Á. Ortiz et al. (eds.), *Engineering Digital Transformation*, Lecture Notes in Management and Industrial Engineering,
https://doi.org/10.1007/978-3-319-96005-0_39

2009), the enterprises will depend of the university education, the workforce will be distributed again in all the supply chain, advanced technology suppliers, innovation and start-ups companies (Davis et al. 2012) for the product consumption, the dynamics of the demand has change the reasoning about the manufacturing processes, shorter times of launching, quicker productions processes, flexible systems, mass personalization, simplified work networks and highly simplified collaborative systems has created the need of modification the information flows, financiers and the materials trough the SC.

The results obtained of this paper is the valuation of the scientific activity visualizing the real state, through bibliometric indicators as the distribution of publications in time, the identification of representative authors, the investigation networks between authors, the countries that have more density of publications and the themes of interest. Finally, from the revised literature, there is a vision of the AM process presented and for the variables that affect in the SC.

2 Methodology

The structure of the methodology is presented in two phases: Phase I Information needs: (i) Literature review. (ii) Identification of the key terms of research. Phase II Bibliometric: (iii) Information research based on data: belonging to El Sevier editorial, which offers a multi-disciplinary thematic allowing to encompass any type of study, it is also a scientific information revised by academic pairs. (iv) Refine initial results: Process and analyse the found information for the election of relevant research from the title and abstract relevance. (v) Statistical data: Value of the data found through the text mining using the Vantage Point software and identifying the distribution of the publications in time, the identification of representative authors, the knowledge networks between authors, the countries with higher density of publication and the themes of interest.

3 Results

Phase I Information Needs

(i) Literature review

In accordance with the introduction of the present paper, the advanced in additive manufacturing (AM) and the tools according Masood (2014), defines the additive manufacturing as a new generation of manufacturing processes that has acquired recognition from its appearance at the end of the 80s decade. Where each piece is manufactured layer by layer from a computer aided design (CAD). In contrast, the traditional method where the material it is removed from a solid pieces with tools or a control numeric computerized (CNC) it requires a manufacturing processes planning (Chen et al. 2015). The spectrum of applications of AM is comprised by several industries as automotive, aerospace, medicine, engineering, the biological systems, the food and supply chains (Gao et al. 2015). The main benefits of AM compared with the traditional methods are the construction of pieces with a complex geometries; the use of a single tool to manufacture components from start to the end, a decrease in production time and cost.

AM machines use the stereolithographic (STL) (Stroud and Xirouchakis 2000) as a standard input file for construct objects in different additive manufacturing processes. Recently, the additive manufacturing file format was introduced (AMF) (ASTM 2011) although the current commercial AM machines still using STL files (Zha and Anand 2015). The STL file format is compatible with many software packages, it consists of triangular facet data which represents the 3D surface geometry of a CAD model, the smaller this triangles are, the higher the resolution will be, the size of the triangles are directly provide with the weight of the file, so it is advisable to reach a balance between the resolution and the file weight (R3a1D 2017). Based in the application area, the techniques used for the modelling fall into two categories: complex surface modelling and solid modelling. A system for solid modelling maintains two principles to describe the model, geometric and topographic data, the first one indicate the parameters that defined the shape and the second one relates each one of the geometric components (Yan and Gu 1996). After that, the CAD file is entered to the AM software, the model is sliced in thousands of layers forming a prototype or a piece of product.

The modern techniques of the AM process are based on four patents: Photo polymerization tubs (United States patent nº 4575330, 1986), powder melting (United States Patent nº 4863538, 1989), materials extrusion (United States Patent nº 5121329, 1992) and binder jet (United States patent nº 5204055, 1993) (Gao et al. 2015) that according to their characteristics allows to manufacture objects in different materials. The classifications of the AM manufacturing processes are: Stereolithographic (SLA), Melt deposition modelling (FDM), laminated manufacturing (LOM), Laser selective sintering (SLS or SLM), Electron beam fusion (EBM) and 3D printing (3DP). (Li et al. 2017; ASTM 2013; Wagner and Walton 2016; Kruth et al. 1998; Chen et al. 2015).

The SC it is affected by the implementation of the AM, in which generates: more (+) personalization, less (−) residues, (−) energy consumption, (−) object weight, (−) cycle time, (−) associated tools, (−) transport cost, (−) environmental impact, (−) inventory, (−) number of people, (+) capital ratio, (+) geometric & design, (+) flexibility, (−) lots, (−) production, (−) waste, (−) chain links, (+) relationship with the consumer, (+) greater capital investment, (−) material, (+) collaborative relationships, (−) product cost. (Wagner and Walton 2016; Li et al. 2016; De la Torre et al. 2016; Masood 2014).

(ii) Identification of key terms of research.

The terms identification is proposed starting from a articles reading associated to the subject of the study, defining the semantic field of the AM and on the other

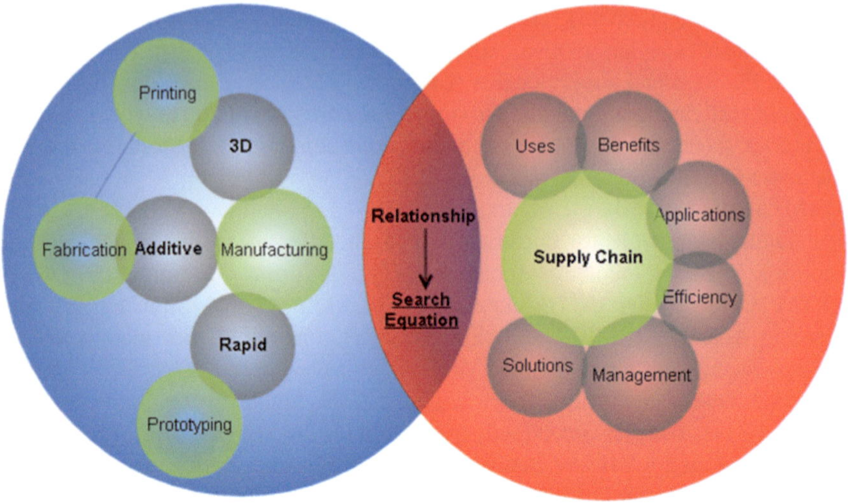

Fig. 1 Venn diagram of intersection AM and SC emphasis, key terms

Table 1 Traceability of scientific coverage applying the search equation

Search date	Search results	Documents analysed	Chosen documents
13/10/2016	464	464	88
13/12/2016	486	22	15
13/02/2017	497	11	6

Table 2 Classification of scientific documentation

AM and SC	AM and other topics	Conferences	Others (discarded)
78	31	48	329

hand, the management, applications, benefits, solutions, uses and efficiency of SC. As shown in the Venn diagram Fig. 1.

(iii) Searching of information in data base

The documents that was collected were used in the data based named Scopus. The reason to limit the investigation is that Scopus is the largest database of abstracts and citations in the pair reviewed literature, over 22.748 journals reviewed (Elsevier 2017). The traceability of scientific coverage is reflected in Table 1. Where the date of search, the search result, documents analysed and chosen are discriminated.

(iv) Refine the initials results

The vantage point and the Vos viewer software were used to improve research results, 11 duplicated titles were eliminated, the author's names and the affiliations were standardized and selection groups were created as shown in Table 2.

Table 3 Most relevant articles

Title	Year	Source	No cites
dditive manufacturing and its societal impact: A literature review	2013	International Journal of Advanced Manufacturing Technology	83
Additive manufacturing: A framework for implementation	2014	International Journal of Production Economics	37
Rapid manufacturing in the spare parts supply chain: Alternative approaches to capacity deployment	2010	Journal of Manufacturing Technology Management	37
Rapid manufacturing: Impact on supply chain methodologies and practice	2007	International Journal of Services and Operations Management	27

AM and SC: The selection criteria is from the moment that the research reflects the impact of AM in the chain, including specific industries and/or impact on the information, materials and financial flow of the SC.

AM and other topics: There are issues that are not taking into account for the results analysis because it refers to AM, but do not relate to SC issues. Some of these discarded topics are: print quality, AM certificates, Am design, AM electrical part, semantic and micro components.

Conferences: The documents selection has not taken into account the results that have the title "conferences", because it cannot analyse the fields as authors, journals and representative countries.

Other topics: These are the documents that were discarded because they do not refers to the relationship between AM and SC, it talks about topics as EFID, SMEs, sustainability, and additive decomposition among others.

(iv) Statistics of the initial data

The number of analysed documents corresponds to the ratio of AM and SC, a total of 78 documents in this investigation field, it is not discriminated the time period, reflecting that despite the fact that 30 years of AM there are studies related with SC initiated from 2002, with a dramatic increase of the publications in the last 3 years, 2016 with 33 publications, 2015 with 18 publications and 2014 with 11 publications.

The most relevant documents were selected from the articles that have more than 20 citations according to the CSV file exported from the Scopus database as shown in Table 3.

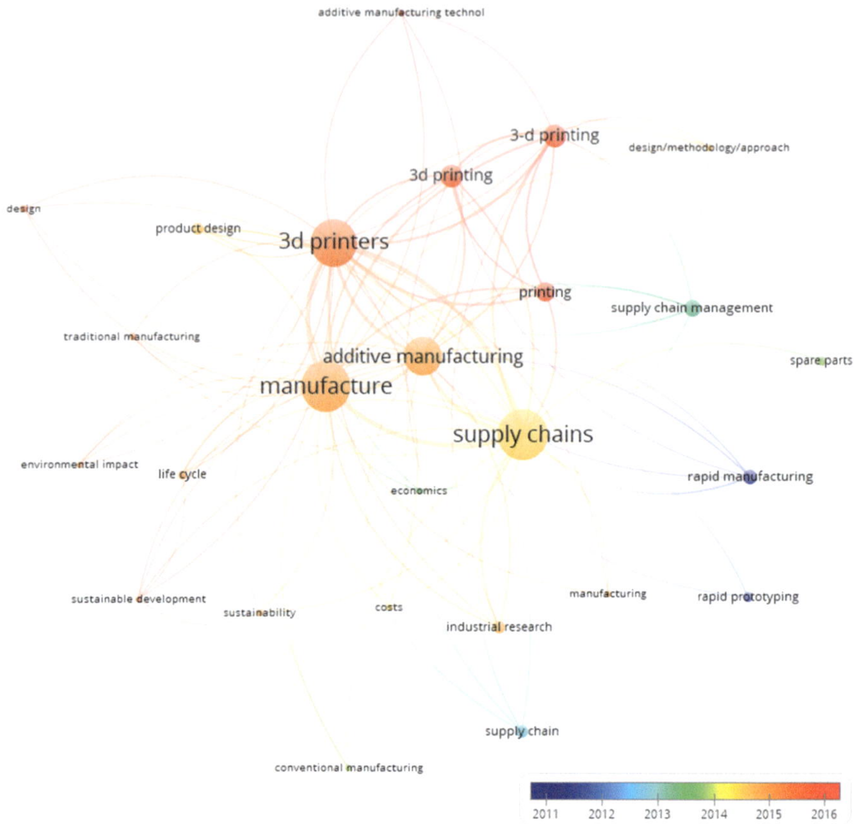

Fig. 2 Key words analyse

Table 4 Authors representatives

Authors	Country	University	Records
Holmstrom, J	Finland	Aalto University	3
Partanen, J	Finland	Aalto University	3
Truck, C.	United Kingdom	Loughborough University	3

The analysis of the key words mainly refers to the evolution of the term referred to AM, as it is shown in the Fig. 2, in the beginning the authors referred their documents as rapid manufacturing and rapid prototype, moving to additive manufacturing and finally a 3d printers. Additional, topics related to SC and AM can be identified as life cycle, sustainability, costs, spare parts and environmental impact.

The results do not present a relevant author, there were identified 210 authors, of which 3 authors' presents the participation in three investigations as shown in the Table 4.

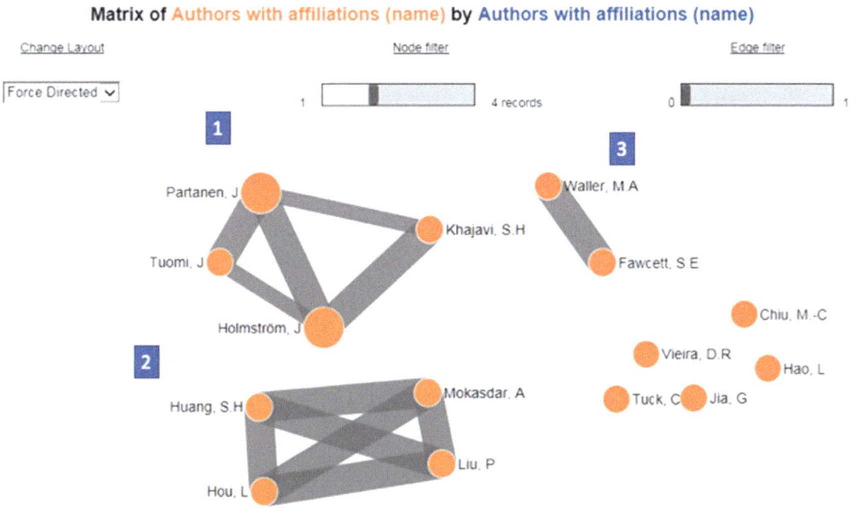

Fig. 3 Knowledge networks

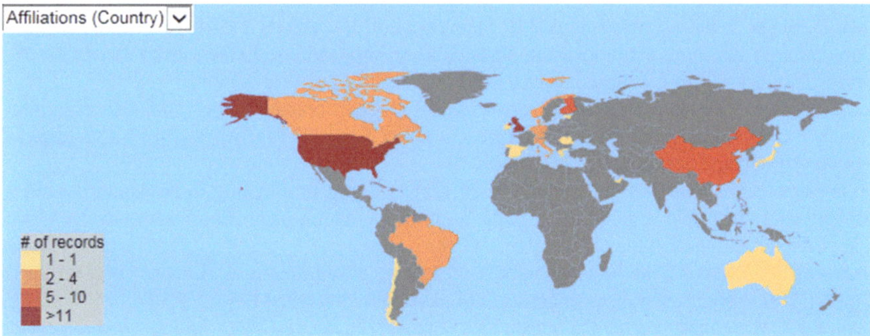

Fig. 4 Geographic location

The selected papers presents three knowledge networks on authors' part, the first network presents the authors largest number of publications, the second is the one with highest impact by citing writing articles (Fig. 3).

The main countries found are: In the first place, United States with a total of 22 articles, in the second place, United Kingdom with 16 articles and in third place China with 9 articles, Highlight that Central America and Africa do not present any record (Fig. 4).

4 Discussion

The Thesaurus and search terms may vary to broaden the information search and if another data base are used, the literature in AM is improving exponentially, but the literature that is associated to SC, has had a greater interest in the last two years, although its appearance its very recent from 2012, given that until now the companies are appropriating the technologies, in general, there is a large period of time needed to defined the new topics and to know which authors and institutions can be really representative. Finally, this work may be interested as a starting point for investigators who wants to carry out investigations in issues associated with AM and SC.

References

ASTM (2013) Standard terminology for additive manufacturing technologies

Atzeni E, Salmi A (2015) Evaluation of additive manufacturing (AM) techniques for the production of metal–ceramic dental restorations 20. J Manuf Process Part 1 20:40–45. https://doi.org/10.1016/j.jmapro.2015.09.010

Chen D, Heyer S, Ibbotson S, Salonitis K, Steingrímsson JG, Thiede S (2015) Direct digital manufacturing: definition, evolution, and sustainability implications. J Clean Prod 107:615–625. https://doi.org/10.1016/j.jclepro.2015.05.009

Davis J, Edgar T, Porter J, Bernaden J, Sarli M (2012) Smart manufacturing, manufacturing intelligence and demand-dynamic performance. Comput Chem Eng FOCAPO 47:145–156. https://doi.org/10.1016/j.compchemeng.2012.06.037

De la Torre N, Espinosa MM, Domínguez M (2016) Rapid prototyping in humanitarian aid to manufacture last mile vehicles spare parts: an implementation plan. Hum Factors Ergon Manuf Serv Ind 26:533–540. https://doi.org/10.1002/hfm.20672

Elsevier (2017) Scopusl the largest database of peer-reviewed literaturel Elsevier [WWW Document]. URL https://www.elsevier.com/solutions/scopus. Accessed Feb 22 2017

Emelogu A, Marufuzzaman M, Thompson SM, Shamsaei N, Bian L (2016) Additive manufacturing of biomedical implants: a feasibility assessment via supply-chain cost analysis. Addit Manuf 11:97–113. https://doi.org/10.1016/j.addma.2016.04.006

Gao W, Zhang Y, Ramanujan D, Ramani K, Chen Y, Williams CB, Wang CCL, Shin YC, Zhang S, Zavattieri PD (2015) The status, challenges, and future of additive manufacturing in engineering. Comput-Aided Des 69:65–89. https://doi.org/10.1016/j.cad.2015.04.001

Hieu LC, Bohez E, Vander Sloten J, Phien HN, Vatcharaporn E, Binh PH, An PV, Oris P (2003) Design for medical rapid prototyping of cranioplasty implants. Rapid Prototyp J 9:175–186. https://doi.org/10.1108/13552540310477481

Hull C (2013) Q & A con Chuck Hull. Co-Fundador. 3D Systems

Kruth J-P, Leu MC, Nakagawa T (1998) Progress in Additive Manufacturing and Rapid Prototyping. CIRP Ann Manuf Technol 47:525–540. https://doi.org/10.1016/S0007-8506(07)63240-5

Li Q, Kucukkoc I, Zhang DZ (2017) Production planning in additive manufacturing and 3D printing (Article). https://doi.org/10.1016/j.cor.2017.01.013

Masood SH (2014) 10.01—Introduction to advances in additive manufacturing and tooling, In: Hashmi S, Batalha GF, Tyne CJV, Yilbas B (eds), Comprehensive materials processing. Elsevier, Oxford, pp 1–2. https://doi.org/10.1016/b978-0-08-096532-1.01016-5

R3a1D (2017) ¿Qué es un fichero STL? - R3ALD [WWW Document]. URL http://www.r3ald.com/que-es-un-fichero-stl. Accessed Feb 22 2017

Savastano M, Amendola C, Fabrizio D, Massaroni E et al (2016) 3-D printing in the spare parts supply chain: an explorative study in the automotive industry, In: Digitally supported innovation. Springer, pp 153–170

Singh S, Ramakrishna S, Singh R (2017) Material issues in additive manufacturing: a review. J Manuf Process 25:185–200. https://doi.org/10.1016/j.jmapro.2016.11.006

SMLC (2009) Smart process manufacturing: A technology and operations roadmap. [WWW Document]. Smart Manuf. Leadersh. Coalit. URL www.smart-process-manufacturing.ucla.edu

Stroud I, Xirouchakis PC (2000) STL and extensions. Adv Eng Softw 31:83–95. https://doi.org/10.1016/S0965-9978(99)00046-0

Thomas D (2016) Costs, benefits, and adoption of additive manufacturing: a supply chain perspective. Int J Adv Manuf Technol 85:1857–1876. https://doi.org/10.1007/s00170-015-7973-6

Vaezi M, Chianrabutra S, Mellor B, Yang S (2013) Multiple material additive manufacturing—Part 1: a review. Virtual Phys Prototyp 8:19–50. https://doi.org/10.1080/17452759.2013.778175

Wagner SM, Walton RO (2016) Additive manufacturing's impact and future in the aviation industry. Prod Plan Control 27:1124–1130. https://doi.org/10.1080/09537287.2016.1199824

Yan X, Gu P (1996) A review of rapid prototyping technologies and systems. Comput-Aided Des 28:307–318

Zha W, Anand S (2015) Geometric approaches to input file modification for part quality improvement in additive manufacturing. J Manuf Process Add Manuf Part 20 3:465–477. https://doi.org/10.1016/j.jmapro.2015.06.021

Towards a Sustainable Agri-food Supply Chain Model. The Case of LEAF

David Pérez Perales, Faustino Alarcón Valero, Caroline Drummond and Ángel Ortiz

Abstract The Agri-food Supply Chain (ASC) has received increasing attention lately due to environmental and social impacts. This is likely to lead to tighter regulations and closer control of these supply chains, where the traditional ASC practices will be reviewed and changed. One of the key solutions may be the necessity of establishing procedures accounting for the sustainability of the whole ASC. This is directly related to the increasing presence of different ASC third parties offering this type of services. In this paper, different aspects about ASC sustainability are reviewed from the literature, with the aim of finding out which are the most addressed and remarkable ones. These ones will be used as a benchmark against an example of a real sustainable ASC management model implemented by the UK established company (LEAF-Linking Environment And Farming).

Keywords Agri-food supply chain · Sustainability models · LEAF company

1 Introduction

Agri-food industry is one of the most important sectors in the world, and remains one of the largest manufacturing sectors in many developed and developing countries. Although this industry has become very efficient, it is still dependent on significant

D. Pérez Perales (✉) · F. Alarcón Valero · Á. Ortiz
Research Centre on Production Management and Engineering (CIGIP), Universitat Politècnica de València, Camino de Vera, S/N, 46022 Valencia, Spain
e-mail: dapepe@omp.upv.es

F. Alarcón Valero
e-mail: faualva@omp.upv.es

Á. Ortiz
e-mail: aortiz@cigip.upv.es

C. Drummond
LEAF, Stoneleigh Park, Warwickshire CV8 2LG, UK
e-mail: caroline.Drummond@leafuk.org

Á. Ortiz et al. (eds.), *Engineering Digital Transformation*, Lecture Notes in Management and Industrial Engineering,
https://doi.org/10.1007/978-3-319-96005-0_40

amounts of natural resources and faces growing demands in demonstrating responsibility. Increasingly there is a spotlight on the sustainable food supply to reduce its impact environmentally and socially.

Improving sustainability requires a deep understanding of the relationships between food consumption behaviours, retailing, processing and distribution activities, as well as agricultural production practices. The links between social needs, the economic processes involved in meeting them, and the associated environmental consequences must be properly determined (Heller and Keioleian 2003).

The opening of domestic markets to international competition throughout the world will undoubtedly result in shifting the focus from a single echelon, such as the farmer, to the sustainability of the overall supply chain. In order to meet these new challenges, it is necessary to take a critical look at the current supply chain practices to determine the best strategies to accommodate the new global conditions. This is increasingly important as standards are raised whether for food safety, environmental and social responsibility, animal welfare and quality.

Therefore, more than ever, sustainability must be seen in terms of the wider agri-food system, especially the vertical linkages in the food supply chain beyond, but including, the farm. Emphasis on sustainable food chains is critical since the food chain, as a whole, is the ultimate framework to demonstrate sustainability (Ilbery and Maye 2005).

This focus on achieving "the overall ASC sustainability" has become an increasingly important business imperative as "stakeholders are demanding more transparency and companies themselves are under increasing competitive and regulatory pressures to demonstrate a commitment to sustainable practices (Jones et al. 2014). In this sense new businesses (acting as ASC independent external organizations) are becoming more relevant offering services such as promoting, enabling, certifying and endorsing sustainable practices to the different stakeholders making up the ASC.

From the analysis of the literature this paper aims to clearly define sustainability and addressing what are the most important issues for a successful and sustainable ASC management. These issues will be subsequently be used as a benchmark of a real sustainable ASC management model implemented by the UK established company (LEAF- Linking Environment And Farming).

The paper is structured as follows: in Sect. 2 a brief literature review about the concept of "sustainability" in the ASC is conducted, while in Sect. 3 the most important aspects that must be addressed for sustainable ASC management are analyzed. In Sect. 4 a real case-study is described. Finally, in Sect. 5 some conclusions are drawn.

2 Sustainability in the Agri-food Supply Chain

The term agri-food supply chains (ASC) has been coined to describe the activities from production to distribution that bring agricultural or horticultural products from the farm to the fork. Therefore, ASC are formed by the organizations responsible for production (farmers), distribution, processing, and marketing of agricultural products to the final consumers (Prima et al. 2016).

The supply chain of agri-foods, as any other supply chain, is a complex network of different actors working together in different processes and activities in order to bring products and services to the market, with the purpose of satisfying customers' demands. What mainly differentiates ASC from other supply chains is the limited products shelf-life and the importance that consumers give to aspects such as quality and health. Other relevant characteristics of agri-foods include coping with high levels of uncertainty due to the weather's unexpected variations and products demand and price variability and volatility. Furthermore, trade, tariffs and political agreements and complexity also impact the ASC. These issues together with the increasing awareness in environmental and social issues make the underlying supply chain more complex and harder to manage than other supply chains.

Environmental and social issues are why ASC's sustainability has become such a popular term in the last decade. But, what does sustainability really mean? Within our review the literature states that three broad dimensions characterize the term "sustainable": economic, environmental and social:

1. Economic sustainability: economic aspects are critical drivers in food business processes which implement sustainability. These dimensions refer to an allocation of resources in an appropriate manner to achieve efficiency and competitiveness to enhance contribution to the society. The implementation of sustainability may lead to an incremental cost while adjusting internal and external facilities to create the advantages for all food business partners. It is noticeable to remark that sustainability in the food industry must find the balance between quality and cost-effectiveness (Li et al. 2014), while also ensuring a fair return to each part of the ASC.
2. Environmental sustainability: decision making does not only consider the economic aspects, but also environmental (or ecological). These later ones basically include prevention of pollution and waste and efficient use of scarce resources such as land, energy and water (Gerbens-Leenes et al. 2003), but also reducing Greenhouse gas emissions and the enhancement of habitats and biodiversity.
3. Social sustainability: social aspects are mainly related to health and safety, employee motivation, turnover and recruitment costs, working conditions, organizational reputation and supplier relationships (Bendul et al. 2016). Other areas are also the importance of fair trading conditions and also increasingly public engagement within an increasingly urban society.

3 Key Issues for Sustainable ASC Management

In this third section a brief review from existing literature about the most addressed and remarkable issues for sustainable ASC management is conducted.

Integration along the whole ASC to achieve sustainability is complex. All the actors must prioritize their financial benefits but at the same time considering the increasing demand on social and environmental aspects. Sustainable ASC management is important to be applied as it will influence not only internal organizations but also external relationship with other parties. (Li et al. 2014).

In this sense, three main ideas from the state of the art have been concluded.

Firstly, that *there are just a few studies focusing on integrated collaboration to achieve sustainable ASC* (Mota et al. 2015). Additionally, most of the studies only consider the trade-off between economic and environmental aspects, neglecting social ones. These studies only focus on environmental (known as green ASC by some authors) and economic aspects attempting to turn environmental impact into economic value in their models. Just a few articles focus on social aspects, mainly related with health and safety, employee motivation, turnover and recruitment costs, working conditions, organizational reputation and supplier relationships. Besides they are mostly tackled in a operational manner, leaving apart their necessary strategic perspective. However, some research considered the 3 sustainability dimensions in their proposed model such as Bourlakis et al. (2014).

Secondly, *many studies assume that all sustainability measurement indicators are independent* (Gerbens-Leenes et al. 2003). The enormous number of indicators found in the literature generates too much data that often provide no additional knowledge on the sustainability of a system. Moreover, although research has addressed many aspects of sustainability, it has often ignored the interactions.

Finally, *there are a significant amount of studies just covering certain ASC stages* (Prima et al. 2016). Whole supply chain sustainability is not an easy task since it requires the joint collaboration of all its stakeholders (a high degree of vertical integration must exist). An ideal integration scheme would involve sharing information and infrastructure, skills and knowledge among all of them. Unfortunately, there is great complexity in its real-life application, and particularly in the agri-food industry due to its special characteristics (as aforementioned in Sect. 2). Moreover, the existence of global regulations, global trading, and emerging consumer preferences also brings more risk and uncertainty to the collaboration system and makes that this collaboration for the whole ASC sustainability only operate effectively among certain stages.

4 The Case of LEAF: A Sustainable ASC Management Model

LEAF is a leading global charity and membership organization (UK established) whose business model consist of promoting, enabling, assuring and integrating sustainable practices throughout the different stages of the ASC (LEAF 2016).

As aforementioned, the size and complexity of sustainability issues make an integrated sustainable ASC management model to be a challenge. This is why LEAF has such a crucial role to play. It helps to integrate the whole ASC actors, from 'farm to folk', to deliver a shared vision of sustainability.

4.1 What Does Sustainability Mean for LEAF?

LEAF has a broad view about the "sustainability" term as it implies three basic pillars. First, the economic performance, promoting high productivity with low environmental impact, tangible financial benefits delivered by increased attention to detail, being prepared for future challenges and keeping ahead of legislation. Secondly, the environmental quality, maintaining and enhancing the wildlife value and character of the countryside and landscape, reducing the risk of pollution and environmental degradation, monitoring and demonstrating improvements in the quality of soil, water, air, wildlife habitats and landscape. Finally, the social health, reaching out and connecting with suppliers, customers and the wider agricultural industry, building public understanding, knowledge and trust in farming.

4.2 LEAF ASC

The LEAF ASC is made up of "LEAF Marque" producers, packers, processors, distributors and retailers who are members of LEAF. It is a sustainability assurance certificate system with third independent verification, given to those that meet the LEAF Marque standard requirements (33% of UK fruit and vegetables are LEAF Marque certified and the standard is currently applied on farms in some 36 countries around the world). The LEAF Marque Standard (LEAF 2017) is a list of questions about how it is farmed and how it is managed. For each question there is a control point that must be required to meet in order to qualify for LEAF Marque certification. The questions must be applied to the whole farm.

The current LEAF Marque supply chain directory and other industry services are available to genuine buyers and sellers who are committed to buying food grown with care to the environment that is LEAF Marque certified. The supply chain directory gives the opportunity for buyers to purchase LEAF Marque certified product and promote this to their customers.

4.3 LEAF Sustainable ASC Management Model

LEAF main customers are the farms. Its sustainable ASC management model is primarily based on promoting, enabling and assuring that farmers produce in a sustainable manner. This is achieved by the adoption of Integrated Farm Management (IFM). Those ones meeting the specific requirements within the LEAF Marque stan-

Fig. 1 LEAF integrated farm management (IFM) model

dard, are given the sustainable assurance certificate following independent, third party inspection from an approved certification body, thus recognizing and rewarding these farming practices. LEAF management model focuses not only upstream on producers (in spite of being the most important ones) but also downstream on the different stakeholders throughout the ASC such as processors, retailers and finally the consumers. Some others stakeholders are also considered such as producers groups, government bodies, the scientific community, machinery manufacturers, environmental groups and all other interested businesses. This holistic view by focusing in the different stakeholders, as well as their integration and traceability, aims to achieve the sustainability of the ASC as a whole.

In order to better understand how LEAF interacts with its different customers and which type of services are offered to them is important to know what is the core of its "sustainable ASC management model". Such a core is based on its "Integrated Farm Management (IFM)" model.

4.3.1 IFM Model

IFM is a whole farm business approach that delivers more sustainable farming. IFM uses the best of modern technology and traditional methods to deliver prosperous farming that enriches the environment and engages local communities. LEAF's IFM is made up of 9 sections, shown in Fig. 1, addressing the entire farm business. An appreciation of the importance of each section and the integration between them is essential for its effective implementation.

4.3.2 LEAF ASC Services

Once the LEAF's IFM core business model has been described, in Table 1 a wide range of services offered for all its stakeholders are shown.

4.3.3 Linking IFM and ASC Management: LEAF Marque Chain of Custody

LEAF Marque Chain of Custody system is a *key aspect* in the "LEAF sustainable ASC management model". It is a mechanism for tracking LEAF Marque certified product from certified farms to the final consumer product. It is aimed to ensure that the credibility, transparency and integrity of all LEAF Marque supply chains are maintained which will help everybody to trust the LEAF Marque logo and any claims made. The acquisition of a LEAF Marque Chain of Custody certificate demonstrates the sustainability commitments of businesses. It provides companies with a commercial advantage as it allows them to use the LEAF Marque logo on products, with the aforementioned advantages.

It is an essential part of the LEAF Marque assurance system, it will ensure that the use of the LEAF Marque logo and marketing claims about products originating from LEAF Marque certified farms are credible and verifiable throughout the whole supply chain. It is used to approve and verify sites all along the food and agricultural products value chain. The system works at product level and tracks movement from site to site.

5 Conclusions

This paper highlights the increasing awareness about sustainability. Although economic issues still prevails, environmental and social ones are becoming not just a competitive advantage but also an imperative. Additionally, it is becoming an imperative to set procedures for the sustainability of the whole ASC. In this sense, the research outlines that it is not an easy task since a strong collaboration among all the stakeholders is required, and not only dyadic relationships. Besides, sustainability must be seen in its broadest view, considering not only economic and environmental issues, but also social ones and assuming their interdependences.

Table 1 LEAF ASC Services

LEAF ASC services	Service description	Customer target
Integrated Farm Management: A guide	Booklet for members which goes through the key principles of each section of IFM and the benefits of following an integrated approach	Farmers
LEAF Sustainable Farming Review	Self-assessment on-line management tool, based on IFM, to help farmer members monitor their performance, identify strengths/weaknesses and set targets for future improvements	Farmers
LEAF Marque	Assurance scheme recognizing produce that have been grown lo LEAF's IFM principles. Investment in the LEAF Marque certification enables to demonstrate the environmental commitment and provides a genuine commercial advantage in the demanding premium and assured food market	Farmers, processors, retailers
Integrated Farm Management Bulletin	Electronic update sent to LEAF members several times a year and includes articles from LEAF's Demonstration Farmers, LEAF Innovation Centres and corporate members about cutting edge research and practices going on in IFM at the moment	Stakeholders except final consumers
Information Centre	LEAF's Online library of IFM information	Farmers
LEAF Demonstration Farms	There are 38 working farmers throughout the UK committed to the sustainable farming practices of IFM. They demonstrate best practice IFM through farm walks, talks and demonstrations	Stakeholders
LEAF Innovation Centres	There are 9 research sites throughout the UK developing new approaches to push forward the boundaries of IFM	Stakeholders except final consumers
Open Farm Sunday and Open Farm School Days	Events held in farms and managed by LEAF to engage the public in sustainable food/farming	Stakeholders, mainly consumers

Finally, all the previous outlines have been used as a benchmark to describe the case of LEAF and how it manages the interactions demanded in the delivery of more sustainable systems promoting, enabling, assuring and integrating sustainable practices throughout the ASC.

Acknowledgements Authors of this publication acknowledge the contribution of the Project 691249, RUC-APS: Enhancing and implementing Knowledge based ICT solutions within high Risk and Uncertain Conditions for Agriculture Production Systems (www.ruc-aps.eu), funded by the European Union under its funding scheme H2020-MSCA-RISE-2015"

References

Bendul C, Rosca E, Pivovarova D (2016) Sustainable supply chain models for base of the pyramid. J Clean Prod. https://doi.org/10.1016/j.jclepro.2016.11.001

Bourlakis M, Maglaras G, Aktas E, Gallear D, Fotopoulos C (2014) Firm size and sustainable performance in food supply chains: insights from Greek SMEs. Int J Prod Econ 152:112–130

Gerbens-Leenes PW, Moll HC, Schoot AJM (2003) Design and development of a measuring method for environmental sustainability in food production systems. Ecol Econ 46(2):231–248

Heller MC, Keoleian GA (2003) Assessing the sustainability of the US food system: a life cycle perspective. Agric Syst 76:1007–1041

Ilbery B, Maye D (2005) Food supply chains and sustainability: evidence from specialist food producers in the Scottish/English borders. Land Use Policy 22:331–344

Jones P, Hillier D, Comfort D (2014) Assurance of the leading UK food retailers' corporate social responsibility/sustainability reports. Corp Gov 14(1):130–138

LEAF (2016) LEAF in 2016: delivering more sustainable food and farming. Internal Report

LEAF (2017) http://www.leafuk.org/leaf/farmers/LEAFmarquecertification/standard.eb

Li D, Wang X, Chan HK, Manzini R (2014) Sustainable food supply chain management. Int J Prod Econ 152:1–8

Mota B, Gomes MI, Carvalho A, Barbosa-Pavoa AP (2015) Towards supply chain sustainability: economic, environmental and social design and planning. J Clean Prod 105:14–27

Prima WA, Xing K, Amer Y (2016) Collaboration and sustainable agri-food supply chain: a literature review. MATEC 5802004:0001. https://doi.org/10.1051/matecconf/20165802004

A Simulink Library for Supply Chain Simulation

Juan Ramón Trapero, Francisco Ramos and Enrique Holgado de Frutos

Abstract Despite the links between forecasting and inventory control, both areas have evolved separately. This work presents a supply chain simulation library developed in Simulink to bridge that gap. To simulate a supply chain is important to define the number of companies/echelons that belong to the same supply chain and the policies that each company employs to demand forecasting and stock control. The potential user can find in this library forecasting and stock control blocks to simulate a supply chain in a modular design. We show how to implement: i) Forecasting policies. For example, the widely used exponential smoothing; ii) Replenishment models. For instance, typical order-up-to-level stock control policies as (*s*, *S*); and finally, how to connect the forecasting and stock control blocks to describe the performance of a company and to extend such relationships to define the whole supply chain.

Keywords Supply chain simulation · Forecasting · Inventory control

1 Introduction

Traditional operations management textbooks (Nahmias and Olsen 2015) widely used in supply chain courses analyse separately the areas of forecasting and stock

J. R. Trapero (✉) · E. H. de Frutos
Department of Business Administration, Facultad de Ciencias y Tecnologías Químicas, Universidad de Castilla-La Mancha, Avda. Camilo José Cela S/N, 13071 Ciudad Real, Spain
e-mail: juanramon.trapero@uclm.es

E. H. de Frutos
e-mail: enrique.holgado.de.frutos@gmail.com

F. Ramos
Department of Electrical, Electronics, Automatica and Communications Engineering, E.T.S. de Ingenieros Industriales, Universidad de Castilla-La Mancha, Avda. Camilo José Cela S/N, 13071 Ciudad Real, Spain
e-mail: Francisco.Ramos@uclm.es

Á. Ortiz et al. (eds.), *Engineering Digital Transformation*, Lecture Notes in Management and Industrial Engineering,
https://doi.org/10.1007/978-3-319-96005-0_41

control. Usually, different forecasting techniques are explained to provide optimal forecasts depending on the demand features as trend and/or seasonality. For instance, the family of exponential smoothing methods, ARIMA models, etc., (Ord and Fildes 2013). On the other hand, stock control techniques are defined assuming different hypothesis about the demand as stationarity, independence, and known statistical distribution, which typically is chosen Gaussian (Nahmias and Olsen 2015). Although, it is acknowledged that forecasting errors are a key variable to determine the safety stocks and the reorder points in well-known stock control policies like order-up-to-level, from the authors' point of view it is not clear how these two areas are connected and it may be a reason behind their isolation from each other.

To bridge that gap this work presents a simulation library developed in SIMULINK that intends to facilitate the task of simulating supply chains to improve decision making. In this sense, this library presents a system dynamics structure with a modular/block design. In other words, the main policies as the forecasting technique and the inventory control method are expressed as a set of inputs/outputs logically connected. Furthermore, in case the user requires trying either different forecasting policies or inventory policies, he/she just ought to change the corresponding block regardless of the rest of the supply chain model.

This supply chain simulation library is useful for both academics and practitioners. Academics can use this library to introduce students to topics as demand forecasting and stock control and, more importantly, to remark the links between both areas. Additionally, practitioners can be benefited from this library because it can be easily utilized to simulate their supply chains and, for example, to analyse what if possibilities to improve their decision making.

The article is structured as follows. Section 2 describes the process to implement the forecasting techniques in Simulink/Matlab, particularly, the exponential smoothing forecasting method. Section 3 explains the implementation of an order-up-to-level stock control policy. Section 4 determines the inputs/outputs of the forecasting and stock control blocks and how they are linked in a system dynamics fashion. Section 5 carries out several simulations to verify the correct performance of the supply chain model, and finally, last section summarizes the main conclusions of the work.

2 Demand Forecasting

Demand forecasting is a key part of supply chain modeling, particularly for those companies where such forecasts drive the manufacturing planning and control. For the sake of simplicity, here we will implement a simple exponential smoothing (SES) technique given its widely use (Ord and Fildes 2013), although more complex forecasting techniques can also be implemented.

The forecasts provided by SES are computed as follows:

$$F_{t+1} = \alpha Y_t + (1 - \alpha) F_t \tag{1}$$

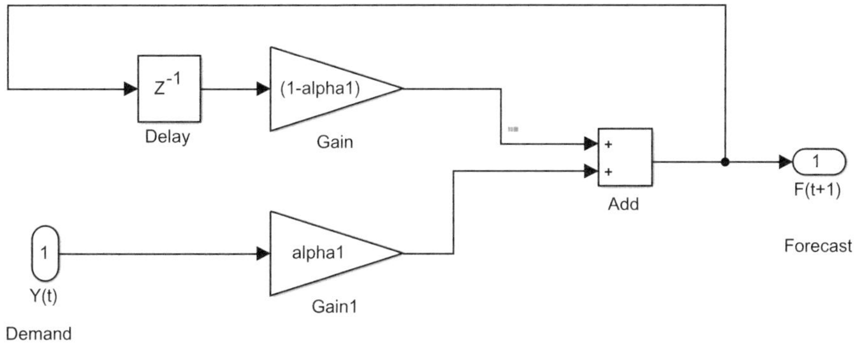

Fig. 1 Simple exponential smoothing implemented in Simulink

Since observations are available up to time t, F_{t+1} is the one step ahead forecast, Y_t is the actual demand observation at time t, and α is the exponential smoothing constant that varies between 0 and 1. In case that the forecasting step is smaller than the lead time, the lead time demand forecast is defined as:

$$F_L = \sum_{i=1}^{L} F_{t+i} = L \cdot F_{t+1} \tag{2}$$

Often, the forecasting step is weekly and the lead time comprises several weeks. Moreover, when using SES it can be shown that $F_{t+i} = F_{t+1}$ for $i = 1, 2, \ldots, L$. (Ord and Fildes 2013). It should be noted that, the lead time (L) is assumed to be known and constant. Figure 1 depicts a possible SES implementation, where *alpha 1* is α in Eq. (1) and Z^{-1} is the delay operator (Z-transform) such as $F_{t+1} \cdot Z^{-1} = F_t$.

Equations (1) and (2) shows the point lead time demand forecast, i.e., the mean estimate of the lead time demand forecast. However, some stock control policies may also require a measure of the forecasting error uncertainty from the forecasting system to compute the safety stock. According to (Silver et al. 1998) if the forecasting errors are Gaussian independent and identically distributed (iid), the safety stock (SS) for a target cycle service level (CSL) can be expressed as:

$$SS = k\sigma_L \tag{3}$$

where $k = \Phi^{-1}(CSL)$ is the safety factor; $\Phi(\cdot)$ is the standard normal cumulative distribution function and σ_L is the lead time forecast of the forecasting error standard deviation. Here, the main problem is how to compute σ_L from the forecasting error. A common approach is firstly, to compute σ_1 (one step ahead forecast of the forecasting error standard deviation) based on the Mean Squared Errors (MSE) and then, to estimate σ_L (Silver et al. 1998). In this sense, MSE at time t is:

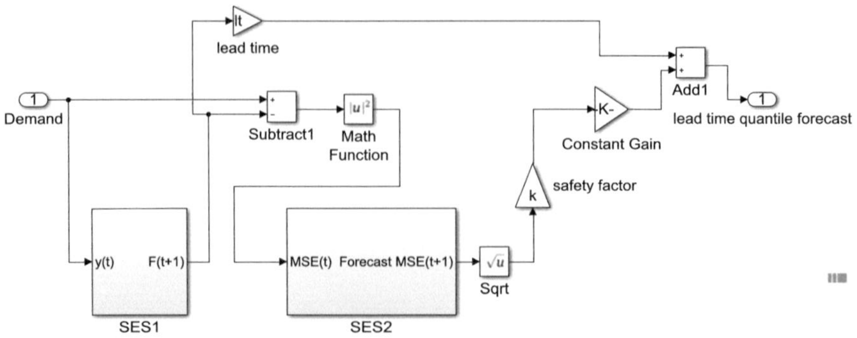

Fig. 2 Forecasting system with two SES. SES1 for the lead time mean forecast and SES2 for the safety stock

$$MSE_t = \frac{1}{n}\sum_{t=1}^{n}(Y_t - F_t)^2 \tag{4}$$

Since the errors variability measured by MSE_t may not be constant, it is possible to apply a second SES to MSE_t such as:

$$MSE_{t+1} = \alpha' Y_t + (1 - \alpha')MSE_t \tag{5}$$

and estimate $\sigma_1 = \sqrt{MSE_{t+1}}$. Secondly, an exact relationship between σ_1y σ_L was found in (Graves 1999), such as:

$$\sigma_L = \sigma_1\sqrt{L}\sqrt{1 + \alpha(L-1) + \frac{1}{6}\alpha^2(L-1)(2L-1)} \tag{6}$$

Figure 2 shows the forecasting system that provides the lead time demand quantile forecast: $F_L + k\sigma_L$, where the last term ($k\sigma_L$) is the safety stock.

3 Stock Control Policy

Once the forecasting policy is defined, the resulting forecasts are fed to the stock control system. Here, we will implement a continuous review, reorder point, order-up-to-level (*s*, *S*) policy (Silver et al. 1998), although the implementation of other stock policies like (S − 1,S), (*s*, *Q*) is straightforward.

Let's assume that we are interested in modelling a retailer with a SES as a forecasting technique and a (*s*, *S*) stock control policy. Figure 3 shows that retailer implemented in Simulink.

We have followed a similar block design developed by (Dejonckheere et al. 2003). Focusing on the stock control system, its inputs are: (i) Demand; (ii) Shipments, i.e.,

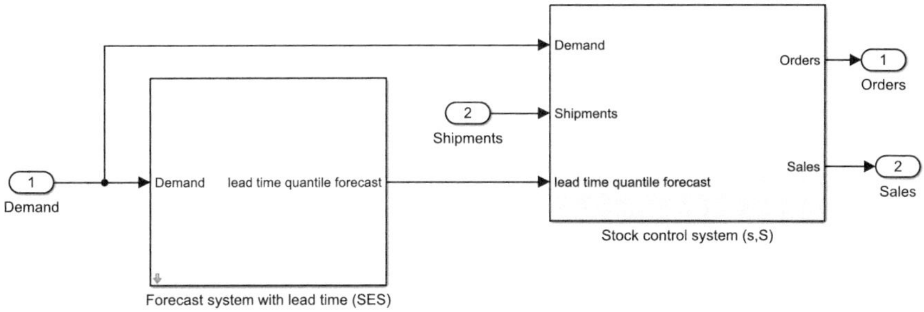

Fig. 3 Retailer that uses SES and (s, S) as forecasting technique and stock control respectively

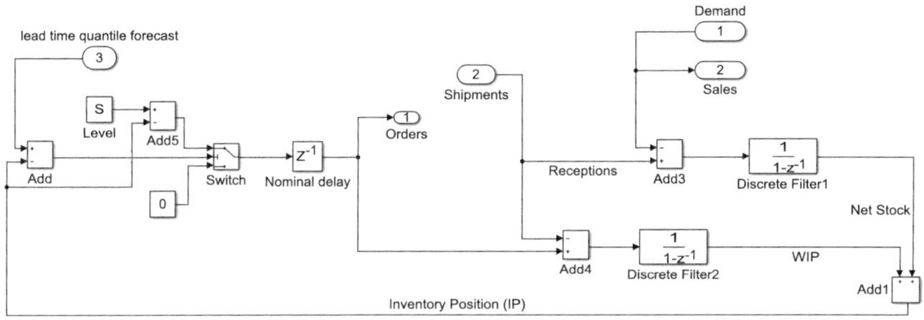

Fig. 4 Block diagram of the stock control system (s, S)

receptions received from an upwards echelon as a manufacturer; (iii) Lead time quantile forecast. This signal is obtained from the forecasting system showed in Fig. 2. Recall that such a lead time demand quantile forecast is the reorder point (s), i.e.:

$$s = F_L + k\sigma_L \tag{7}$$

The outputs of the stock control system are: (i) Orders; and (ii) Sales. Figure 4 depicts how those outputs are calculated. In summary, the retailer satisfies the market demand and with the shipments received updates the Net Stock (NS). Any unfilled demand is backlogged in our model. The Work In Process (WIP) is also updated with the orders and shipments. The Inventory Position (IP) is computed as the sum of NS and WIP. Finally, the IP is compared with the reorder point (s) and if IP is lower than s, then the retailer launches an order, such as:

$$O_t = S - IP_t \tag{8}$$

where S is the order-up-to-level.

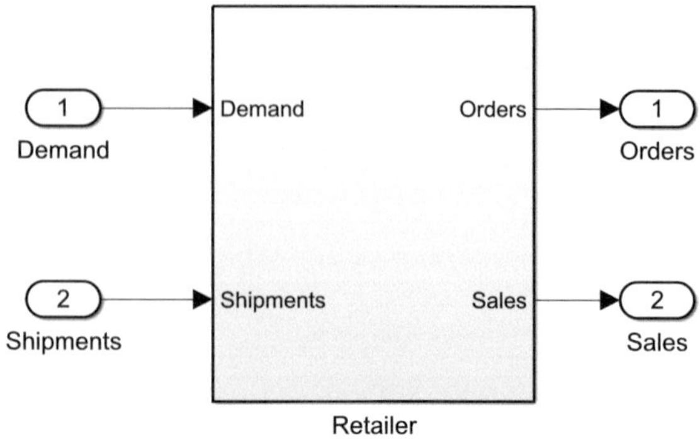

Fig. 5 Example of a retailer system where only inputs and outputs are visible

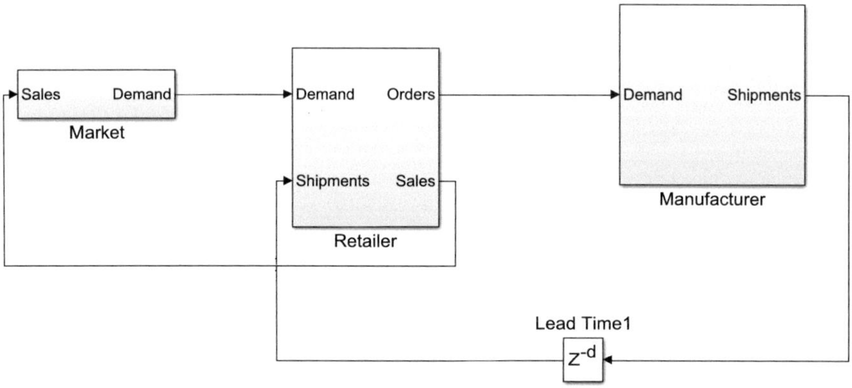

Fig. 6 Example of supply chain with two members serially connected

4 Supply Chain Dynamics

In the previous sections, we have explained the links between the forecasting system and the stock control. In this section, we will use such blocks to simulate a supply chain comprised of a retailer and a manufacturer. Figure 5 shows the retailer system described in Figs. 3 and 4, where only the set of inputs/outputs are visible.

To simulate the supply chain, we should connect those inputs/outputs with the rest of echelons. Figure 6 shows the supply chain. The first block is the market that generates the demand and receives sales from retailer, where the demand is built as a random number generator, although it can be easily changed for a file with actual demand data. Figure 6 also shows the retailer and manufacturer subsystems. The retailer has been previously described in Figs. 3, 4 and 5. The manufacturer follows the same structure than the retailer with two slight differences. Since the manufacturer does not order to upstream supply chain members, that subsystem does not have an orders output. Nonetheless, it requires some time to produce the goods ordered by the retailer, which is modeled by the manufacturing time, see Fig. 7.

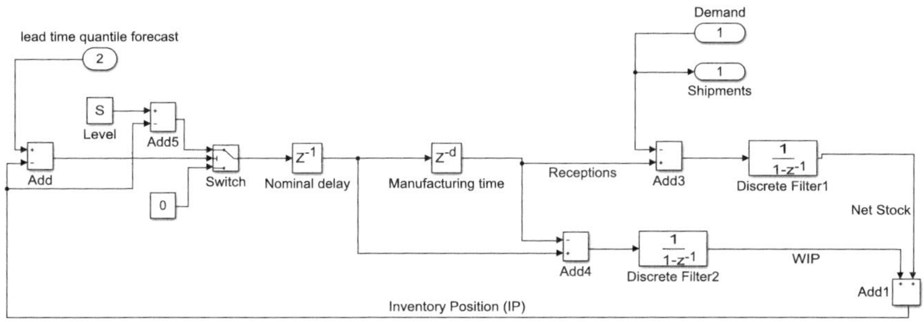

Fig. 7 Block diagram of the stock control system (s, S) for the manufacturer

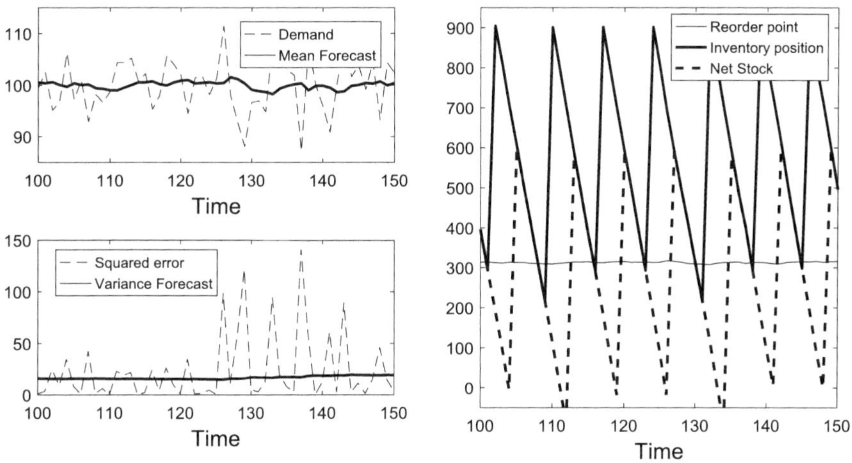

Fig. 8 Simulation results for the retailer

5 Simulation Results

To verify the correct model performance, some simulations were carried out. The market demand was simulated with a normal random number generator with mean 100 and variance 20. The retailer parameters were set to $\alpha = 0.1$, $\alpha' = 0.01$, $CSL = 95\%$, $L = 3$, $S = 1000$. The manufacturer parameters were set to $\alpha = 0.1$, $\alpha' = 0.01$, $CSL = 95\%$, $L = 3$, $S = 2000$. Figure 8 shows some results for the retailer. Left panel depicts the forecasting results for the mean and variance forecasts. Right panel shows stock control results. The figure shows the forecasts (left plots) and inventory variables (right plot).

6 Conclusions

This work aims at developing a simulation library to facilitate the understanding of supply chains. Following the described methodology, different forecasting techniques and stock control policies can be easily implemented. This library can be useful for teaching supply chain concepts and, particularly, improves the links between forecasting and inventory control. Additionally, practitioners can implement their supply chains to improve their decision-making process.

Acknowledgements This work was supported by the European Regional Development Fund and Spanish Government (MINECO/FEDER, UE) under project reference DPI2015-64133-R.

References

Dejonckheere J, Disney SM, Lambrecht MR, Towill DR (2003) Measuring and avoiding the bullwhip effect: a control theoretic approach. Eur J Oper Res 147:567–590

Graves SC (1999) A single-item inventory model for a nonstationary demand process. Manufac Service Operations Manage 1:50–61

Nahmias S, Olsen TL (2015) Production and operations analysis. Waveland Press

Ord K, Fildes R (2013) Principles of business forecasting. Cengage Learning

Silver E, Pyke D, Peterson R (1998) Inventory management and production planning and scheduling. Wiley

MIX
Papier aus verantwortungsvollen Quellen
Paper from responsible sources
FSC® C105338

If you have any concerns about our products,
you can contact us on
ProductSafety@springernature.com

In case Publisher is established outside the EU,
the EU authorized representative is:
Springer Nature Customer Service Center GmbH
Europaplatz 3, 69115 Heidelberg, Germany

Printed by Libri Plureos GmbH
in Hamburg, Germany